Phytoecdysteroids II

Vladimir Nikolayevich Syrov
Ugiloy Yusufovna Yusupova
Durbek Abdikhoshimovich Usmanov
Editors

Phytoecdysteroids II

From Natural Origins to Therapeutic Frontiers

Springer

Editors
Vladimir Nikolayevich Syrov
Department of Pharmacology and Toxicology, Institute of the Chemistry of Plant Substances
Academy of Science of the Republic of Uzbekistan
Tashkent, Uzbekistan

Ugiloy Yusufovna Yusupova
Laboratory of Glycosides Chemistry, Institute of the Chemistry of Plant Substances
Academy of Science of the Republic of Uzbekistan
Tashkent, Uzbekistan

Durbek Abdikhoshimovich Usmanov
Department of Coatings and Polymeric Materials
Materials and Nanotechnology Program
North Dakota State University
Fargo, ND, USA

ISBN 978-981-95-8422-2 ISBN 978-981-95-8423-9 (eBook)
https://doi.org/10.1007/978-981-95-8423-9

This Springer imprint is published by the registered company Springer Nature Singapore Pte Ltd.
The registered company address is: 152 Beach Road, #21-01/04 Gateway East, Singapore 189721, Singapore

Preface

The exploration of natural products has long been one of the most enduring and impactful endeavors in science, yielding transformative advances in medicine, agriculture, and biotechnology. Among the diverse array of bioactive compounds that nature offers, phytoecdysteroids—plant-derived analogs of insect molting hormones—have emerged as particularly intriguing. These compounds, while initially identified for their roles in regulating invertebrate development, are now recognized as multifunctional agents, combining unique ecological significance with broad pharmacological potential. Traditionally, phytoecdysteroids were appreciated for their function in plant defense mechanisms, acting as antifeedants to deter herbivorous insects. However, over the past decades, research has revealed that these molecules exert diverse effects not only within plant systems but also across animal and human biology. Experimental studies have demonstrated that phytoecdysteroids modulate protein, carbohydrate, and lipid metabolism, promote tissue regeneration, enhance stress adaptation, and provide cardioprotective and hepatoprotective effects—all without the androgenic side effects associated with synthetic anabolic agents. Such findings have propelled phytoecdysteroids from a niche botanical interest to a promising subject in pharmacology, biotechnology, and even computational drug discovery.

This book, *Phytoecdysteroids: From Natural Origins to Therapeutic Frontiers*, presents the first comprehensive and interdisciplinary treatment of these fascinating compounds, integrating insights from phytochemistry, pharmacology, toxicology, computational science, and applied biotechnology. Our goal is to bridge the gap between classical natural product research and modern, technology-driven approaches, providing readers with both foundational knowledge and exposure to cutting-edge developments. The volume is organized into nine chapters, each addressing a critical aspect of phytoecdysteroid research. We begin with an exploration of their natural occurrence, biosynthetic pathways, and structural diversity, offering essential context for understanding their ecological and chemical relevance. This is followed by a detailed examination of analytical and spectroscopic methods that have been pivotal in characterizing these complex molecules. We then explore their metabolic, hormonal, stress-protective, cardioprotective, hepatoprotective, regenerative, and toxicological profiles, highlighting their broad therapeutic promise. Importantly, we dedicate focused attention to emerging computational innovations, such as machine learning and artificial intelligence, which are transforming how natural compounds are screened, analyzed, and repurposed for new indications. Finally, we address the translational and industrial potential of phytoecdysteroids, examining applications across medicine, agriculture, cosmetics, and biotechnology.

What distinguishes this book is its interdisciplinary ambition. By bringing together perspectives from international experts—including contributors from Uzbekistan, France, the UK, the USA, and beyond—we offer a unified, forward-looking vision for phytoecdysteroid research and application. Our intended audience includes academic researchers, pharmaceutical and biotechnology professionals, clinicians, and graduate students who are eager to understand both the scientific foundations and the future horizons of this exciting field.

We extend our sincere thanks to the contributing authors and reviewers whose expertise, commitment, and collaborative spirit have shaped this work. We are also grateful to the broader scientific community, whose pioneering efforts continue to expand the boundaries of phytoecdysteroid research. We hope this volume will serve not only as a comprehensive reference but also as an inspiration—a catalyst for future research, innovation, and interdisciplinary collaboration. By integrating the wisdom of nature with the power of modern science and technology, we believe that phytoecdysteroids hold remarkable potential to address some of the most pressing challenges and promising opportunities in twenty-first-century science and medicine.

Tashkent, Uzbekistan — Vladimir Nikolayevich Syrov
Tashkent, Uzbekistan — Ugiloy Yusufovna Yusupova
Fargo, North Dakota, USA — Durbek Abdikhoshimovich Usmanov

Acknowledgment

The editors would like to express their sincere gratitude to all chapter authors for their scientific expertise, dedication, and timely contributions, which collectively shaped the quality and coherence of this volume. We are particularly grateful for their constructive collaboration and responsiveness throughout the editorial process.

We also thank the reviewers for their insightful comments and constructive suggestions, which significantly improved the clarity and scientific rigor of the chapters. Special thanks are extended to the Springer Nature editorial and production teams, in particular Shinichi Koizumi and Reshma Irudayaraj for their professional support, careful coordination, and guidance throughout the publication process.

Contents

Chapter 1
Phytoecdysteroids: Natural Sources, Biosynthesis, and Structural Classification

Ugiloy Yusufovna Yusupova , Nurmurod Ramazonov, and Shamansur Sagdullaev

Abstract This chapter reviews the general chemical properties and ecological roles of phytoecdysteroids. Phytoecdysteroids are steroid compounds found in plants that are structurally similar to insect molting hormones. The chapter highlights their distribution in the plant kingdom, particularly within the families Caryophyllaceae, Amaranthaceae, and Lamiaceae, and their importance in defense against insect predators and in adapting to stress. It also explains the biosynthetic pathway from cholesterol, the aglycone core, and the structural diversity that occurs through various modifications. It shows how environmental conditions, developmental stage, and genetic factors affect their accumulation. Additionally, they are compared with insect ecdysteroids, and the chemical and taxonomic aspects are discussed. Thus, the chapter provides a brief scientific background for the detailed discussions that follow.

Keywords Phytoecdysteroids · Plant-derived steroids · Biosynthesis · Structural classification · Ecological role · Ecdysteroid diversity

1.1 Introduction

Phytoecdysteroids (PEs) are a fascinating group of naturally occurring polyhydroxylated steroids. While they share a close structural similarity with the molting hormones of arthropods (ecdysteroids), their main occurrence and study have been in plants (Arif et al. 2022). In recent years, these compounds have garnered growing interest across phytochemistry, pharmacology, biotechnology, and chemical ecology, primarily due to their diverse range of biological activities and promising

U. Y. Yusupova (✉) · N. Ramazonov · S. Sagdullaev
Institute of the Chemistry of Plant Substances, Academy of Science of the Republic of Uzbekistan, Tashkent, Uzbekistan

V. N. Syrov et al. (eds.), *Phytoecdysteroids II*,
https://doi.org/10.1007/978-981-95-8423-9_1

applications. The term "ecdysteroid" comes from the Greek word *ecdysis* ("shedding" or "molting"), reflecting their central role in controlling growth, metamorphosis, and reproduction in arthropods (Yusupova et al. 2022a).

1.2 Historical Background and Definitions

The study of PEs began in the mid-twentieth century, when Adolf Butenandt and Peter Karlson first isolated α-ecdysone in crystalline form from silkworm pupae (*Bombyx mori*) in 1954 (Areiza 2018). Ecdysteroids, first identified in insects, provided a deeper understanding of their physiological processes. Later, compounds with a similar structure were also found in plants, which began to be called phytoecdysteroids (i.e., ecdysteroids of plant origin) (Dinan 2009). The confirmation of the chemical structure of α-ecdysone using X-ray crystallography in 1965 marked a significant milestone in this field. By the end of the twentieth century, ecdysteroids had been identified not only in insects but also in various plants, as well as in some algae and fungi. The presence of ecdysteroids in plants has raised important evolutionary questions, because, unlike insects, which use them as primary hormones, plants are thought to produce ecdysteroids primarily as defensive metabolites (Fig. 1.1).

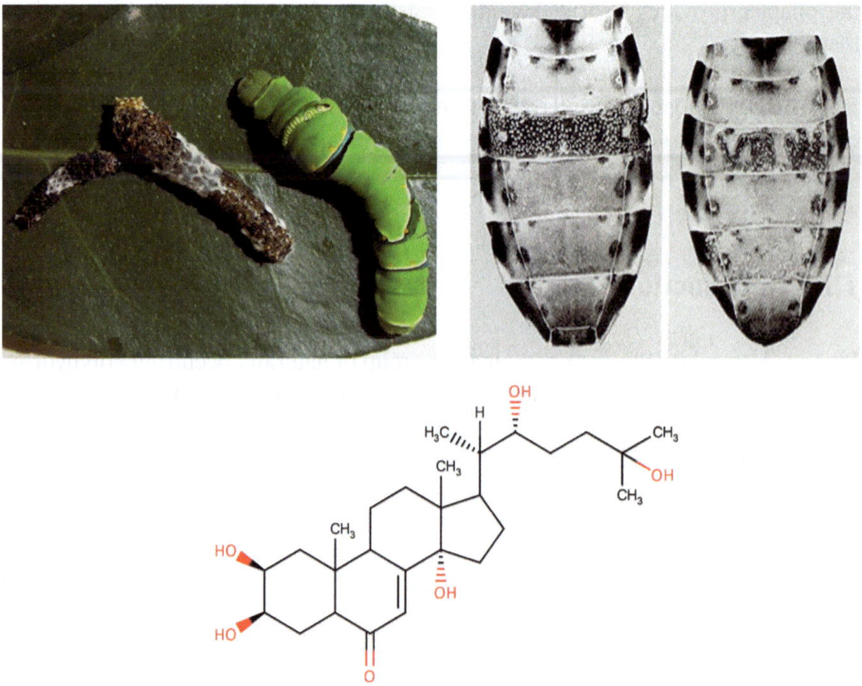

Fig. 1.1 Formation of Bombyx mori silkworm cocoons and the occurrence of α-ecdysone in their composition (Williams 1961, 1976)

By imitating insect hormones, PEs disrupt molting and development in herbivorous insects, making them powerful antifeedants and natural deterrents (Mathur et al. 2025). This places ecdysteroids at the center of plant chemical defense strategies, highlighting their crucial role in plant–insect interactions and coevolution. From a chemical ecology perspective, PEs stand out because, unlike other secondary metabolites such as alkaloids, flavonoids, or terpenoids, they act directly on insect hormonal pathways. This unique mode of action may explain why they are found in more than 100 plant families and over 530 species, with particularly high levels reported in Caryophyllaceae, Amaranthaceae, Asteraceae, and Lamiaceae (Arif et al. 2022). Genera such as *Silene*, *Dianthus*, *Ajuga*, and *Rhaponticum* are vibrant sources, sometimes containing up to 2.5–2.9% of dry mass in PEs (Yusupova et al. 2022b).

The remarkable abundance and structural diversity of PEs have attracted growing attention for both ecological and applied research. Beyond their role in plant defense, they have become the subject of pharmacological studies demonstrating anabolic, adaptogenic, antioxidant, hepatoprotective, and hypoglycemic activities in mammalian systems (Zwetsloot et al. 2014; Todorova et al. 2024). These results have led to the development of dietary supplements. In particular, preparations containing 20-hydroxyecdysone are used to increase physical performance, stimulate muscle growth, accelerate the recovery process, and reduce stress-related metabolic disorders. Importantly, unlike synthetic anabolic steroids (Zádor 2025), phytoecdysteroids do not have androgenic effects and are generally considered safe. Therefore, they are considered promising for the development of therapeutic applications (Graf 2014). PEs are also being studied for their potential applications in biotechnology and agriculture. Their insect growth-regulating properties are important in the development of environmentally friendly pest control strategies (Chowdhury et al. 2024). Furthermore, advances in synthetic biology and metabolic engineering are enabling the reconstruction of PE biosynthetic pathways in microorganisms and plant cell systems, opening up new opportunities for their sustainable and large-scale production (Hidalgo et al. 2018; Patil-Patankar et al. 2022).

The structural diversity of phytoecdysteroids is one of the most important aspects in their study. They are mainly formed on the C27 cholest-7-en-6-one skeleton and are characterized by various modifications such as hydroxylation at positions C-2, C-3, C-20, C-22, and C-25 (Hikino et al. 1975; Almeida et al. 2020), as well as glycosylation, acylation, and lactonization. To date, 600 different natural phytoecdysteroid structures have been identified (ecdybase.org). Studying their complex structures requires modern analytical approaches, including chromatographic methods, mass spectrometry, and nuclear magnetic resonance (NMR) spectroscopy. In recent years, advances in metabolomics and chemoinformatics have expanded the possibilities for identifying new PEs and further elucidating their biological functions (Sarker and Nahar 2012; Yusupova et al. 2022c).

From an evolutionary perspective, the occurrence of ecdysteroids in both plants and insects raises important questions about the convergence and divergence of steroid biosynthesis (Thummel and Chory 2002). Arthropods (insects, spiders, crustaceans, etc.) synthesize ecdysteroids from readily available cholesterol in their diet.

That is, they obtain cholesterol from an external source and use it to produce hormones necessary for development and metamorphosis. Plants, on the other hand, produce ecdysteroids de novo, i.e., entirely through their own internal biosynthetic system. In this process, they utilize a metabolic pathway known as the mevalonate pathway (MVA pathway) to synthesize sterols (e.g., sitosterol or cholesterol) from simple precursor molecules. These sterols then serve as intermediate substrates for ecdysteroid synthesis in subsequent steps (Nes 2011). These differences, specifically the distinct biosynthesis of ecdysteroids in animals and plants, offer a unique opportunity to investigate the evolution of complex metabolic networks and their regulation by environmental and developmental factors. At the same time, PEs clearly demonstrate the advanced chemical defense strategies of plants, as they directly affect the hormonal system of insects, representing defense mechanisms that have evolved in response to strong evolutionary pressures. In addition to their ecological importance, PEs have great potential as a source of promising bioactive molecules for medicine and industry (Sinha and Häder 2020). This chapter provides an overview of the natural occurrence, taxonomic distribution, structural diversity, biosynthetic origin, and practical applications of phytoecdysteroids. Special attention is paid to recent scientific advances in elucidating their biosynthetic pathways, understanding regulatory mechanisms, and exploring novel applications in medicine and biotechnology. In conclusion, this chapter highlights current challenges and promising directions in phytoecdysteroids. In particular, it highlights the potential for accelerating new discoveries and expanding their potential applications in pharmacology and biotechnology by integrating synthetic biology, various omics technologies (including genomics, transcriptomics, and metabolomics), and artificial intelligence approaches (Chamani et al. 2025).

The following sections will review the natural sources and distribution of phytoecdysteroids, drawing on more than six decades of research in phytochemistry. Their biosynthetic pathways in both plants and insects are reviewed, highlighting common principles and unique features in both systems. Finally, future perspectives are outlined—that is, key research opportunities, innovative approaches, and the potential for practical applications of these unique natural compounds in pharmacology and biotechnology.

1.3 Natural Sources and Distribution

PEs are widespread in the plant kingdom and have been identified in more than 100 plant families and over 500 species (Arif et al. 2022). The widespread occurrence of phytoecdysteroids underscores their evolutionary and ecological significance, indicating that they serve diverse functions in plant biology (Fig. 1.2). Although initially identified in insects, they have since been found in ferns, gymnosperms, and angiosperms, as well as some algae and fungi. This suggests that they have deep evolutionary roots and a broad biosynthetic potential across a wide range of organisms (Dinan et al. 2023).

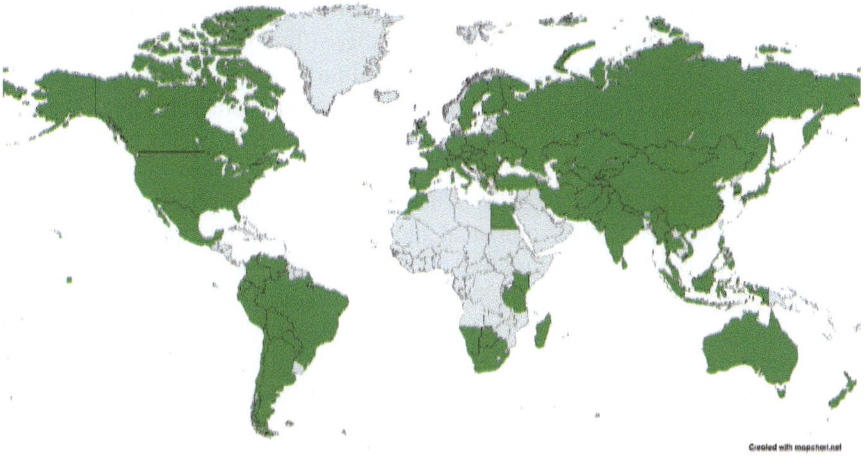

Fig. 1.2 Global distribution of phytoecdysteroids across ferns, gymnosperms, angiosperms, and other organismal groups, highlighting their evolutionary and ecological significance

1.3.1 Plant Sources

Among flowering plants (angiosperms), the largest sources of PEs have been identified in species belonging to the families Caryophyllaceae, Amaranthaceae, Asteraceae, and Lamiaceae (Thiem et al. 2017; Gautam and Khedkar 2024; Acheuk et al. 2022). Within these families, particularly Caryophyllaceae, Amaranthaceae, Asteraceae, and Lamiaceae, the genera *Silene, Dianthus, Ajuga, and Rhaponticum* have been extensively studied due to their richness in PEs, ethnobotanical uses, and pharmacological significance (Yusupova et al. 2019; Park et al. 2023; Głazowska et al. 2018; Yusupova et al. 2020a; Yusupova et al. 2025).

Silene (Caryophyllaceae) is a large genus comprising over 700 species, distributed across Europe, Asia, and North America. Several species, including *Silene praemixta*, *S. tomentella*, *S. popovii*, and *S. claviformis*, have been shown to synthesize various PEs. These compounds include 20-hydroxyecdysone, polypodine B, turkesterone, and their various glycosylated derivatives (Khurramov et al. 2025; Yusupova et al. 2022c). These metabolites have been identified using modern chromatographic and spectroscopic methods and have been shown to have ecological functions. In particular, it has been noted that they may perform functions such as protecting the plant by repelling herbivorous insects (Akhrem et al. 1973).

Dianthus is another large genus in the Caryophyllaceae family, comprising about 2000 species and characterized by the synthesis of important PEs. Viticosterone E, polypodine B, 20-hydroxyecdysone, and ecdysone were identified in the species *Dianthus angrenicus*, *D. hoeltzeri*, and *D. uzbekistanicus*, which occur in Uzbekistan. These results indicate Central Asia as a promising region for the diversity of PEs and their bioprospecting (Yusupova et al. 2020b).

Within the Lamiaceae family, *Ajuga turkestanica* is notable for its long-standing use in traditional medicine as a tonic and adaptogen. Today, it is one of the leading commercial sources of 20-hydroxyecdysone and turkesterone, widely marketed in dietary supplements for athletes and stress resilience (Mamarasulov et al. 2020).

Among the plants of the Asteraceae family, *Rhaponticum carthamoides* (maral root) is distinguished by its accumulation of complex mixtures of ecdysteroids. These compounds have been reported to have anabolic (stimulating muscle growth), adaptogenic (improving adaptation), and immunostimulatory (strengthening immunity) effects. *R. carthamoides* grows naturally in Siberia and has been widely used in Russian and Mongolian folk medicine. Today, it remains one of the primary targets of pharmacological research (Knížková et al. 2025).

In addition to the above examples, PEs are also found in many other plant families, including Ranunculaceae, Chenopodiaceae, and Polypodiaceae. Ponasterones A, B, and C have been isolated from ferns (e.g., *Polypodium vulgare*) and gymnosperms (e.g., *Podocarpus nakaii*). This indicates that the ability to synthesize ecdysteroids is widely distributed among plants and is deeply rooted in their evolutionary history (Lafont et al. 2021; Adler and Grebenok 1999).

1.3.2 Other Sources

PEs have been identified not only in terrestrial plants but also in marine algae. For example, the red alga *Laurencia pinnata* has been found to contain specific PE derivatives, pinnasterol and ecdysteroid-like metabolite (Dziwornu et al. 2017). Fungi also contribute to the diversity: for example, ecdysteroid-like metabolites have been identified in the species *Polyporus umbellatus* (Tomou et al. 2022). These findings suggest that the ability to synthesize or accumulate PE-like substances may have evolved very early in the evolution of eukaryotes. This process may have occurred through horizontal gene transfer, symbiotic relationships, or convergent evolution. Interestingly, ecdysteroids have also been detected in insects that feed on plants rich in PE. Such ecological connections shed light on chemical interactions between links in the food chain and clearly demonstrate the vital role of PEs in plant–insect relationships.

1.3.3 Quantitative Aspects

The amount of PEs in plants varies significantly not only between different species but also in different organs of the same species. In some cases, their concentration can be less than 0.01% of the dry mass, but in highly accumulating species this figure reaches almost 3% (Karatt et al. 2023). In particular, *Ajuga turkestanica* and *Rhaponticum carthamoides* are widely used commercially as the richest sources of PEs, while most other species contain only small or trace amounts. The distribution of PEs across plant organs is also unique, as their levels are typically higher in roots, rhizomes, stems, and leaves than in flowers or seeds. Their synthesis is significantly dependent on external factors, with factors such as light, temperature, soil composition, and insect attack having a significant impact. This indicates that the biosynthetic pathways of PEs are susceptible to environmental conditions (Alami et al. 2024; Holopainen et al. 2025). The growth stage is also an important factor, with higher PE levels observed in young tissues, possibly serving as a protective mechanism. Seasonal changes also have a significant effect on PE levels. These findings suggest that PE biosynthesis is dynamically regulated in response to growth phases and environmental stresses (Chen et al. 2025).

1.3.4 Biogeography and Ethnobotany

PEs are widespread secondary metabolites in plants, occurring in approximately 5–6% of higher terrestrial plant species (Dinan 2001; Lafont and Dinan 2003). The distribution of PEs worldwide follows distinct biogeographical patterns. In particular, Central Asia, Siberia, the Caucasus, and the Mediterranean basin are distinguished as regions rich in PEs. Many plant species native to these regions have long been utilized in folk medicine as stimulants, adaptogens, and tonics. The chemical compounds underlying their biological activity have been identified through scientific research. For example, *Rhaponticum carthamoides* (maral root) was valued by Siberian hunters and athletes as an antifatigue agent, while *Ajuga turkestanica* is known for its anabolic and adaptogenic properties in traditional Central Asian medicine. Such ethnobotanical experiments have stimulated modern pharmacological research and laid the foundation for the development of new nutraceuticals. In terms of biodiversity, the main genera rich in PEs include *Ajuga* (Lamiaceae), *Silene* (Caryophyllaceae), *Rhaponticum* (Leuzea) (Asteraceae), and *Pfaffia* (Amaranthaceae). For example, *Leuzea carthamoides* is endemic to Siberia and the Altai Mountains, and its roots have been found to accumulate up to 1–2% 20-hydroxyecdysone (Slama and Lafont 1995).

Plants of the genus *Silene* are widespread across Eurasia, with their diversity being particularly high in the Balkan Peninsula, the Caucasus, and Central Asia. In these regions, drought conditions and herbivorous insect pressure are considered the primary factors stimulating the biosynthesis of PEs (Zibareva 2000). Native to South America, Pfaffia paniculata ("Brazilian ginseng") grows in the Amazon and Cerrado ecosystems, where it is highly adapted to nutrient-poor soils and seasonal environmental stresses (). Importantly, the growth of the same species under different environmental conditions significantly affects the amount and composition of PEs. In this process, altitude (growth in mountainous or lowland areas), soil composition, and herbivorous insect pressure are the main factors (Baltaev 2000). This biogeographical distribution is directly reflected in folk medicine. For example, in Central Asia and Siberia, the root of *Leuzea carthamoides* has been used as a restorative tonic. In Europe and the Mediterranean region, *Ajuga reptans* and *A.*

Table 1.1 Major PE-containing species

Species	Family	Main PEs	% Dry Weight	Region
Ajuga turkestanica	Lamiaceae	20E, turkesterone, cyasterone	2.5–2.9%	Central Asia
Rhaponticum carthamoides	Asteraceae	20E, inokosterone, ponasterone A	1.5–2.5%	Siberia, Russia, Mongolia
Silene praemixta	Caryophyllaceae	20E, polypodine B, turkesterone	1.5–2%	Central Asia
Dianthus angrenicus	Caryophyllaceae	Viticosterone E, 20E, ecdysone	1–1.5%	Uzbekistan
Podocarpus nakaii	Podocarpaceae	Ponasterone A, B, C	~0.5%	Japan, East Asia
Polypodium vulgare	Polypodiaceae	Polypodine B	~0.3%	Europe, Asia, North America
Laurencia pinnata	Rhodophyta	Pinnasterol, paxyllasterone	Trace	Tropical Marine
Polyporus umbellatus	Polyporaceae	Ecdysteroid-like compounds	Trace	Eurasia

turkestanica have been used for wound-healing, anti-inflammatory, and metabolic regulation. Recent pharmacological studies have confirmed their antioxidant and anti-inflammatory activities. In South America, *Pfaffia paniculata* is valued as a general tonic, aphrodisiac, and immunostimulant. In Asia, *the Ajuga species and Cyanotis arachnoidea have been* used in traditional medicine. They are used in Nepal, Tibet, and China to strengthen bones, protect the liver, and restore hormonal balance.

Therefore, regions with a high diversity of plants rich in PEs are usually closely associated with rich ethnobotanical traditions (Table 1.1). Central Asia and the Altai region are prime examples, where species such as *Leuzea carthamoides* and *Ajuga turkestanica* are of particular importance, not only because of their natural distribution but also because of their widespread use as adaptogens for centuries (Eisenman et al. 2012). However, regions such as Southern Africa and Southeast Asia have not yet been sufficiently studied. Nevertheless, available data indicate that these regions also contain species rich in PEs (Bencheikh et al. 2021). This presents significant opportunities for future research that combines chemical ecology, ethnopharmacology, and biodiversity conservation.

1.3.5 Ecological and Evolutionary Significance and Organ- and Developmental-Specific Synthesis

The ability to synthesize triterpenoids was formed in plants very early in their evolution and has been preserved under intense natural selection pressure (Mathur et al. 2025). In plants, these compounds play essential ecological and physiological roles, including defense against herbivores, regulation of developmental processes, and mediation of responses to abiotic stress, underscoring their importance in plant survival strategies. PEs, a specialized class of triterpenoids, provide a striking example of convergent evolution, as their biosynthesis has evolved independently in plants and arthropods, producing structurally similar molecules with comparable biological functions (Lafont and Dinan 2024). This convergence suggests that steroid signaling plays a crucial role in all eukaryotes, raising new questions about the genetic and enzymatic processes involved in their biosynthesis. Therefore, the natural occurrence and distribution of PEs provide an important basis for research in ecology, evolution, pharmacology, and biotechnology. By identifying PE-producing species and studying the environmental and developmental factors that influence their accumulation, scientists can create new opportunities for the conservation and sustainable use of natural resources, as well as the chemistry of natural products (Kingston 2011).

Owing to their structural similarity to insect molting hormones, PEs disrupt molting, pupation, and metamorphosis in phytophagous insects, reducing survival and reproduction (Dinan 2001; Lafont and Dinan 2003). Laboratory studies have shown that compounds such as 20-hydroxyecdysone have a strong feeding-deterrent effect on insects. Field observations have shown that PE levels in leaves and stems of *Silene* plants attacked by Lepidoptera (butterfly) larvae are increased. This suggests that plants have an inducible chemical defense mechanism (Zibareva 2000). These observations support the "enemy hypothesis," which indicates that the evolution of PEs was shaped by selective pressure from herbivorous insects. However, in addition to this well-studied protective function, preliminary evidence suggests that PEs may also play allelopathic roles (i.e., the chemical inhibition of one plant by another) or influence rhizosphere microbial communities. Nevertheless, ecological evidence on this is currently limited (Suhada et al. 2023). PEs may play a crucial role not only in insect defense but also in enhancing plant resistance to abiotic stresses, such as drought, cold, or intense ultraviolet radiation. For example, populations of *Leuzea carthamoides* growing in high mountain regions have been found to have higher levels of PE; however, the exact mechanisms underlying this are still not fully understood (Kapare et al. 2017).

PEs are not uniformly distributed in plants. Their high concentration in roots serves as a defense against herbivorous organisms in the soil. Increased PE levels in young leaves and stems provide additional protection during sensitive growth stages (Jozdani 2012). In flowers and seeds, transiently high concentrations are observed, which protect them from flower-destroying or seed-eating organisms (Dinan et al. 2023). Quantitative studies have shown that *Leuzea carthamoides* roots contain 0.5–2% PEs by dry weight, while *Cyanotis arachnoidea* roots can produce up to 5% 20-hydroxyecdysone (Stransky et al. 1998; Hunyadi et al. 2016). Young leaves and tissues have high levels of PE, and during flowering, most of it is directed to the reproductive organs, and it decreases as the plant enters the senescence stage. However, the activity of the genes and enzymes that regulate this process has not yet been fully studied (Niwa and Niwa 2014). In general, PEs have evolved as a complex evolutionary adaptation in plants. Their ecological role is reflected in the combination of permanent and conditionally induced defense mechanisms. Additionally, the organ- and developmentally specific distribution of PEs suggests their adaptive role in plant life. Although their synthesis is energy-intensive for the plant, they have been conserved throughout evolution due to their protective effects. Therefore, PEs are not just chemicals, but metabolites that play an important role in plant defense, adaptation, and ecological interactions (Adler and Grebenok 1999). It is essential to continuously combine comparative genomics, environmental field studies, and experimental evolution methods. It is through these approaches that we can fully explain how PE biosynthesis arose, how it diversified, and what role it continues to play in plant adaptation and survival today.

1.4 Biosynthesis Pathways

PEs are produced in plants through complex metabolic networks that reflect the biochemical potential and evolutionary history of plants. Although their structure is similar to that of insect ecdysteroids, in plants, PEs are synthesized de novo from key metabolic precursors via the mevalonate (MVA) pathway. In contrast, insects cannot synthesize PEs independently; they convert dietary sterols, particularly cholesterol, into hormonal ecdysteroids (Sripinyowanich et al. 2023; Behmer and Nes 2003). This section provides a detailed review of the biosynthesis of PEs in plants. It also compares this process with the biosynthetic pathways in insects. Emphasis is placed on the enzymatic steps, metabolic intermediates, and control mechanisms regulating biosynthesis, as well as their evolutionary significance (Fig. 1.3).

1.4.1 Biosynthesis in Plants

In plants, the biosynthesis of PEs begins with acetyl-CoA, a major product of carbohydrate metabolism (Cheng 2017). Acetyl-CoA enters the mevalonate (MVA) pathway, one of the two main pathways of isoprenoid biosynthesis (the other being the methylerythritol phosphate (MEP) pathway). In the MVA pathway, acetyl-CoA is first converted to 3-hydroxy-3-methylglutaryl-CoA (HMG-CoA) and mevalonic acid (MVA) is formed by the enzyme HMG-CoA reductase. Subsequent phosphorylation and decarboxylation reactions produce isopentenyl pyrophosphate (IPP). IPP is the 5-carbon basic unit of all isoprenoids. IPP and its isomer, dimethylallyl pyrophosphate (DMAPP) undergo stepwise condensation to form the following compounds: geranyl pyrophosphate (GPP, C10), farnesyl pyrophosphate (FPP, C15), and finally squalene (C30). Squalene is oxidized to 2,3-oxidosqualene, which is the initial substrate for the synthesis of cycloartenol. Cycloartenol is the main precursor of sterols in higher plants. After cycloartenol, various enzymatic processes such as cyclization, methylation, desaturation, and reduction lead to the formation of sterols such as cholesterol, sitosterol, campesterol, and stigmasterol. These, in turn, serve as the basis for the subsequent biosynthesis of Pes (Talapatra and Talapatra 2015).

Among these, cholesterol (C27 sterol) is the essential precursor for PEs, as it is in insect ecdysteroid biosynthesis. Although cholesterol levels in plants are

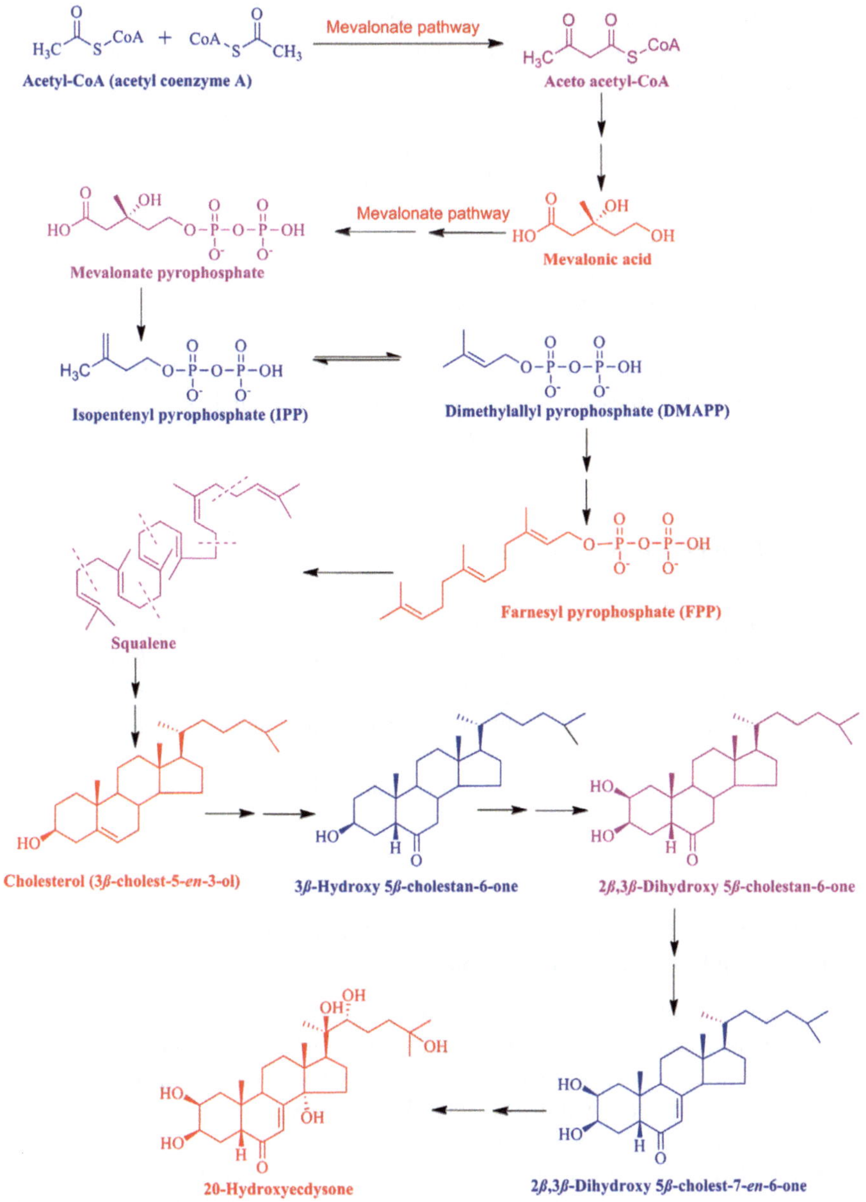

Fig. 1.3 Biosynthesis process of phytoecdysteroids in plants

generally low, it is believed that species rich in PEs have specific pathways that direct metabolic flux toward cholesterol biosynthesis. Several important steps are observed in the conversion of cholesterol to PE. The earliest of these is the formation of a double bond at the C-7 position, followed by an oxidation reaction at the

C-6 position. As a result of these changes, the Δ7-en-6-one structural motif, characteristic of all ecdysteroids, is formed (Dinan et al. 2003). In subsequent steps, the cholesterol skeleton is hydroxylated at various positions, including C-2β, C-3β, C-14α, C-20R, C-22R, and C-25. Glycosylation and acylation reactions further expand the chemical diversity of PEs. As a result, compounds such as 20-hydroxyecdysone, polypodine B, inocosterone, and turkesterone are formed. In addition, plant-specific modifications occur: for example, the formation of a lactone ring between C-20 and C-22 (as in the case of cyasterone) or the addition of acetonide groups (20,22-acetonide derivatives) (Fig. 1.3). These processes further enrich the structural diversity of PEs. The primary enzymes responsible for these modifications include cytochrome P450 monooxygenases, glycosyltransferases, and acyltransferases. Although most of them have not yet been identified, modern studies based on transcriptomics and metabolomics are helping to understand their role and regulatory mechanisms (Arif et al. 2022).

1.4.2 Biosynthesis in Insects

The biosynthesis of ecdysteroids in insects differs from that in plants, but in some respects there is convergence with them (Iga and Kataoka 2012). Insects cannot synthesize sterols de novo, so they rely mainly on dietary sterols, particularly cholesterol (Warren et al. 2009; Blais et al. 1996). Cholesterol is first converted to 7-dehydrocholesterol by the enzyme Δ7-desaturase. This intermediate product then undergoes a series of oxidation, hydroxylation, and dehydrogenation reactions to ultimately form 20-hydroxyecdysone (20E). 20E is the main molting hormone in insects (Ono et al. 2012).

The main pathway involves the following transformations:

- **7-dehydrocholesterol → Δ4-diketol → 5β-ketodiol → 2,22,25-trihydroxy-ecdysteroid → 20-hydroxyecdysone**

The initial intermediates between 7-dehydrocholesterol and 5β-ketodiol have long been described as a "black box" because they are unstable and poorly characterized. However, in recent years, genetic and enzymatic studies, especially in models such as Drosophila melanogaster and Manduca sexta, have gradually shed light on these previously unknown steps (Fig. 1.4).

Several key cytochrome P450 enzymes have been identified that catalyze hydroxylation and oxidation reactions in the formation of active ecdysteroids, including Phantom, Disembodied, Shadow, and Shade. Importantly, the sequence of hydroxylation steps differs between plants and insects: in plants, hydroxylation at the C-2 position usually occurs first, followed by modifications at the C-22 and C-25 positions. In insects, this order is reversed, with C-22 and C-25 modifications occurring first, followed by C-2 hydroxylation. These contrasting processes clearly demonstrate evolutionary differences and reflect the adaptation of each group to its own unique metabolic system.

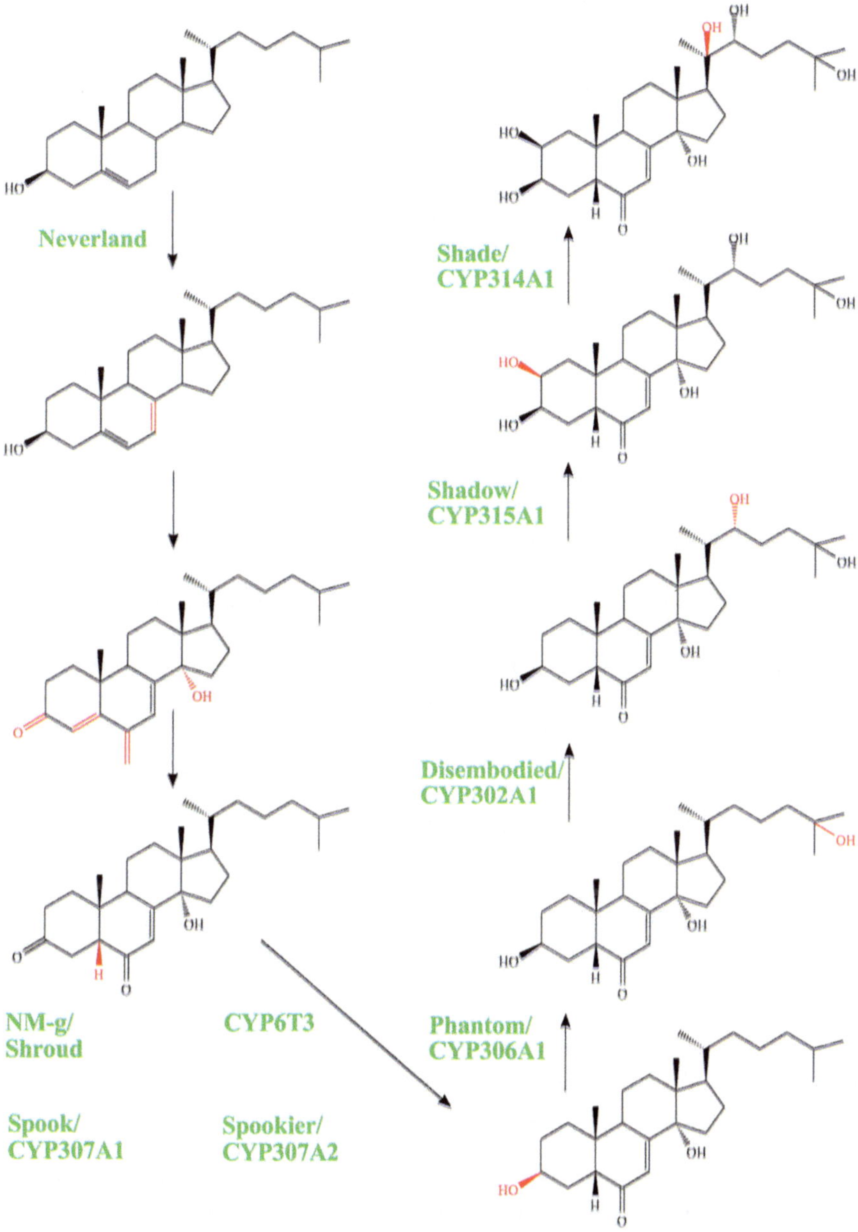

Fig. 1.4 Genes, g enzymes, and biosynthesis of ecdysteroids in insects

1.4.3 Evolutionary Perspectives

The fact that ecdysteroid biosynthesis occurs in both plants and insects raises interesting evolutionary questions. Comparative genomics studies suggest that PEs in plants evolved independently, providing a clear example of convergent evolution that provides a selective advantage through a chemical defense mechanism (Yaktayo-Chang et al. 2020). While in insects, ecdysteroids serve as essential hormones that control molting and reproduction, in plants, they are mainly formed as a defense mechanism against herbivorous insects, disrupting their development. The evolutionary roots of PE biosynthesis likely date back to the earliest land plants, when sterol metabolism diversified in response to various environmental pressures. Their occurrence in ferns, gymnosperms, and angiosperms (flowering plants) supports this view.

In addition, the occurrence of PEs in algae and fungi indicates their more ancient origin or their multiple independent evolution (Lafont 1997). Biochemically, PE biosynthesis evolved from the ancient triterpenoid pathway, in which the mevalonate (MVA) route produces squalene and cycloartenol, precursors of major phytosterols such as sitosterol, stigmasterol, and campesterol. In PE-producing plants, the metabolic flux is directed toward cholesterol biosynthesis rather than toward conventional sterols, as cholesterol is the primary precursor of phytoecdysteroids, such as 20-hydroxyecdysone (Lafont and Dinan 2003). In higher plants, cholesterol is a secondary sterol that is usually found in small amounts. However, the increase in its biosynthesis rate is considered an important evolutionary innovation. Phylogenetic analyses indicate that this ability has evolved independently in plants on multiple occasions. Examples include the identification of phytoecdysteroids in *Polypodium vulgare* (ferns), *Taxus baccata* (gymnosperms—yew), and various angiosperm families (Dinan et al. 2023).

The most powerful evolutionary driver of PE biosynthesis is insect feeding on plants. Structurally similar to insect molting hormones, PEs act as potent antifeedants and endocrine disruptors. They inhibit insect metamorphosis, reducing growth, survival, and reproduction (Dinan 2001; Schmelz et al. 1998). Feeding assays have confirmed the effectiveness of PEs against various insect species. Plants with high levels of PE are usually found in ecological regions with high insect pressure. Examples include species growing in mountain pastures (*Leuzea carthamoides*), arid steppes (*Ajuga turkestanica*), and tropical forests (*Pfaffia paniculata*). Since PE biosynthesis is a metabolically expensive process, its benefits are most evident under conditions of severe biotic stress, particularly insect attack. Recent evidence also suggests that PEs have other ecological functions: they are involved in antimicrobial defense, allelopathic interactions (chemical suppression of the growth of other plants), and the regulation of symbiotic relationships with rhizosphere microbes (Baltaev 2000; Zibareva et al. 2003).

PE biosynthesis is a striking example of convergent evolution. It has evolved independently in diverse plant groups, including ferns, gymnosperms, and unrelated angiosperm families, suggesting that this chemical defense strategy has been selected for repeatedly throughout evolution. Convergence is not limited to plants; insects also synthesize ecdysteroids, but they utilize biosynthetic pathways that are distinct from those of plants, yet functionally similar. Although the overall steroid skeleton and hydroxylation are similar, the underlying genes and enzymes are different, demonstrating the evolutionary plasticity of steroid metabolism. Furthermore, the discovery of ecdysteroid-like compounds in algae, fungi, and some marine invertebrates suggests that they may have evolved independently multiple times or may have evolved through horizontal gene transfer and symbiotic relationships (Lafont and Dinan 2003). However, the adaptive value of PEs is limited by necessary trade-offs. Cholesterol biosynthesis is an energetically demanding process, and diverting resources to secondary metabolism can lead to reduced plant growth or reproduction. In addition, unmodified terpenoids can also be toxic to the plant itself. Therefore, autotoxicity is prevented by glycosylation and compartmentalization (Harrewijn et al. 2012). The high metabolic cost of PE biosynthesis probably explains why only 5–6% of vascular plants produce PE. The remaining plant groups use alternative strategies for defense, including the synthesis of alkaloids, phenolic compounds, or glucosinolates (Karinho-Betancourt 2018).

The evolutionary history of PEs clearly demonstrates how chemical defense mechanisms in plants have evolved in response to the interaction of environmental pressures, metabolic capabilities, and evolutionary constraints. Their repeated and independent emergence in different plant groups confirms the high adaptive value of the hormonal pathways of insects as a defense mechanism. In the future, the integration of genomics, phylogenetics, and ecological studies will be crucial in further understanding the origin and diversification of PE biosynthesis, as well as in revealing its broad applications in ecology, evolutionary biology, medicine, agriculture, and biotechnology.

1.4.4 Regulation and Localization

The biosynthesis and accumulation of PEs in plants are controlled by a complex, multistep regulatory system that operates at the genetic, cellular, tissue- and organ-specific, developmental stage-specific, and environmental levels. A deeper understanding of this process will shed light on plant defense mechanisms and metabolism, as well as open new opportunities in the fields of metabolic engineering and synthetic biology. Recent transcriptome studies have identified genes that may be involved in the accumulation of PEs, including sterol methyltransferases, desaturases, and cytochrome P450 enzymes (Arif et al. 2022). Hormonal signaling pathways, such as those involving jasmonic acid and salicylic acid, further integrate PE biosynthesis into plant-wide defense responses (Mithöfer and Boland 2012). One of the main features of PE metabolism is the spatial and tissue-specific distribution in plants. Typically, the parts most vulnerable to external stresses, such as roots, young leaves, stems, and reproductive organs, have the highest concentrations of PEs. For example, in the roots of *Leuzea carthamoides*, the content of 20-hydroxyecdysone can exceed 1% of the dry weight. In contrast, in *Ajuga turkestanica*, the aboveground parts, in addition to the roots, contribute significantly to the accumulation of PE (Baltaev 2000; Zibareva 2000).

This suggests that plants strategically allocate metabolic resources to maximize their defenses. At the cellular level, PEs are typically stored in glandular trichomes, epidermal cells, or cortical parenchyma. This isolates them from the plant's primary metabolism and protects the plant from autotoxicity (Zibareva et al. 2003). Glycosylation is another protective mechanism that stabilizes PEs and regulates their bioactivity. However, the specific transport proteins that control the intracellular accumulation and transport of PEs have not yet been identified, and they are likely to be ABC or MATE transporters. Their regulation is closely linked to plant hormones and stress signals. For example, jasmonate (*in cell culture of Ajuga reptans*) has been shown to induce PE biosynthesis (Dinan 2001). Salicylic acid, ethylene, ultraviolet (UV) radiation, drought, and nutrient deficiency have also been associated with increased PE levels. Genetic control is only just beginning to be explored, and it is suggested that transcriptional regulation of biosynthetic genes, as well as noncoding RNAs, may play an important role, similar to other secondary metabolic systems (Zhang et al. 2023). The regulation of PEs throughout the plant is not yet well understood. So far, it has been established that their synthesis is increased only in damaged tissues. However, it is still unknown whether PEs, in their glycosylated form, are distributed to other parts of the plant through the phloem and participate in long-range defense processes (Adler and Grebenok 1999). Clarification of this issue is essential for a deeper understanding of the biology of PEs and their effective application in practical contexts.

PE biosynthesis is energy and resource intensive for plants, so they prioritize its production at times and places where it provides the greatest protective benefit. Therefore, understanding the specific organs, developmental stages, and environmental conditions under which PEs are synthesized is essential for elucidating their biological functions and developing biotechnological applications (Table 1.2). Future research should focus on developing predictive models for PE biosynthesis by integrating "omics" approaches, including genomics, transcriptomics, metabolomics, and proteomics. Such models will serve as a basis for designing sustainable production systems in the fields of synthetic biology and metabolic engineering.

Table 1.2 Tissue-specific distribution patterns of phytoecdysteroids across plant species

Tissue type	PE content trend
Roots	Highest concentrations (often >1% dry weight)
Young leaves/stems	Moderate to high, often inducible
Flowers/seeds	Variable; some species show transient peaks
Mature tissues	Lower, consistent with reduced defense needs

1.5 Metabolic Engineering and Synthetic Biology

1.5.1 Heterologous Production (Yeast, Bacteria, and Plant Cells); Precursor Supply (Cholesterol)

Research into the production of PEs outside of natural plant sources uses a variety of microorganisms and plant-based systems. Among the organisms, *Saccharomyces cerevisiae* (yeast) and *Escherichia coli* are widely used, while hairy root cultures and transient gene expression methods in *Nicotiana benthamiana* are being tested as plant systems (Reed and Osbourn 2018; Arif et al. 2022). Microorganism-based platforms enable large-scale production and precisely controlled fermentation, while plant cells provide the natural membrane environment and redox partners necessary for many enzymes involved in PE biosynthesis.

One of the biggest challenges in PE biosynthesis is the supply of precursors. For example, yeast typically produces ergosterol, whereas the main starting compound for PEs is cholesterol (a C27 sterol). Therefore, scientists have created special strains called "cholesterol-yeast," in which the ERG pathway in yeast is reprogrammed and the gene blocks for cholesterol biosynthesis are introduced, similar to those found in animals. However, this reprogramming process requires a delicate balance of metabolic flux: sufficient cholesterol must be produced for PE synthesis, but it must not negatively affect the growth and survival of the yeast. In bacteria, sterol biosynthesis pathways can be introduced in a modular manner; however, there are challenges in completely mimicking eukaryotic features, such as membrane-bound enzymes or redox compatibility (Wei et al. 2016). Plant cell cultures produce PE from natural sterol pools and respond to inducers (elicitors), but their growth rate and yield often limit their potential for widespread industrial application.

1.5.2 Computational/AI-Guided Enzyme Discovery and Pathway Design

Currently, computer technologies support almost all steps of the process of restoring PE biosynthetic pathways. Using machine learning models, various "omics" data (including genomics, transcriptomics, and metabolomics) are analyzed, and potential genes are identified. Among these, P450S, UGTs, and acyltransferases, which are not yet fully characterized, are essential, and they are selected based on coexpression with sterol metabolism, subcellular location, and conserved motifs. Enzyme function prediction programs and structure-based design methods aid in selecting and engineering catalysts that exhibit the desired regio- and stereoselectivity (e.g., differentiation between C-2β and C-22R hydroxylation). At the system level, genome-scale metabolic models simulate precursor availability, energy and redox requirements, and flux balances, allowing for the identification of bottlenecks before conducting experiments. These models are integrated into "design–build–test–learn" cycles, and promoter activity, gene dosage, and enzyme scaffolds are iteratively tuned using biosensors and adaptive laboratory evolution. Overall, such in silico tools reduce the time to identify and optimize new enzymes, reduce research costs, and increase the potential for scaling up PE production to an industrial scale.

1.5.3 Biotechnological and Agricultural Applications: Safety and Ecological Considerations

The laboratory production of PEs has practical applications in several important areas. In the pharmaceutical and nutraceutical fields, compounds such as 20-hydroxyecdysone and turkesterone are valued for their muscle-growth-promoting, adaptogenic, antioxidant, hepatoprotective, and hypoglycemic properties. In agriculture, PEs serve as natural plant protection agents by inhibiting the development of herbivorous insects and reducing their feeding. In addition, the production of PEs under standardized conditions provides a reliable source for scientific research and reduces the need for excessive collection of wild plants. However, caution is required when using them on a large scale: for medical purposes, pharmacokinetic and toxicological evaluation and compliance with production and control standards are required; for agriculture, the effects on beneficial insects and soil microbes, the risk of resistance development, and the shelf life in the environment must be taken into account (Hejran et al. 2024). Also important are environmental safety and ethical principles in the use of biodiversity, including responsible bioprospecting and fair benefit sharing.

1.6 Future Perspectives

Despite numerous years of research, many aspects of PEs remain poorly understood. First, their complete biosynthetic pathway in plants has not yet been fully elucidated. Although key enzymes—such as squalene synthase, cycloartenol synthase, and various hydroxylases—are thought to be involved in this process, many of the specific genes and mechanisms that regulate them have not yet been identified or functionally demonstrated. There are still significant gaps in the scientific literature on PEs. First, multi-omics approaches (combining genomics, transcriptomics, proteomics, and metabolomics data) are needed to fully map their biosynthesis. Second, the ecological functions of PEs are not yet fully understood. In addition to protection against herbivorous insects, their roles in plant–microbe interactions, allelopathy (chemical inhibition of the growth of other plants), and adaptation to various stresses require further study. Third, the pharmacokinetics and pharmacodynamics of PEs in mammals (i.e., how they are absorbed, distributed, metabolized, excreted, and what biological effects they exert) are still poorly understood. PEs are generally considered safe and do not have hormonal (androgenic) effects. However, further research is needed for their use in clinical medicine and veterinary medicine. In particular, the bioavailability, metabolism, specific intracellular molecular targets, and long-term safety of PEs require further investigation. Additionally, the geographical and taxonomic expansion of research will facilitate the identification of new PE species. Plants from regions that have not yet been studied sufficiently—for example, Central Asia, South America, Africa, and Oceania—may produce PEs with unique structures and biological activities (Cheng 2017).

Systematic bioprospecting and ethnobotanical research are essential strategies for identifying new bioactive compounds. Synthetic biology presents significant opportunities for developing sustainable sources of PEs. In particular, their biosynthetic pathways can be reconstructed in microorganisms such as yeast or bacteria, as well as by optimizing plant cell cultures for industrial production. The primary priorities in this regard are to identify enzymes that slow down biosynthesis, increase the availability of precursors, balance metabolic flux, and reduce unwanted byproducts. If these problems are solved, a reliable and sustainable supply of rare and valuable PEs for pharmaceuticals and agriculture will be created. At the same time, a systematic and in-depth study of their pharmacological applications remains an urgent task. High-throughput screening, target identification, molecular docking, and structure-activity relationship (SAR) studies play an essential role in unlocking the pharmaceutical potential of PEs. These approaches enable us to identify new therapeutic areas of PEs, enhance lead compounds, and inform the drug development process. Currently, promising areas include metabolic diseases, neurodegenerative syndromes, inflammatory processes, muscle wasting, and immune system dysfunction. Thus, PEs are emerging as potential therapeutic agents for various diseases, thanks to the advancements in modern bioinformatics and pharmacological studies (Shahbaz et al. 2023).

To expand the practical potential of PEs, their combination with synthetic modifications or their use in novel delivery systems such as nanoparticles may further enhance their efficacy. In agriculture, PEs are of great importance as an environmentally friendly and sustainable alternative. The development of PE-based biopesticides, the selection of PE-rich crops, and the development of transgenic plants with high PE levels are promising directions in pest control. At the same time, it is necessary to study the mechanisms of insect resistance in depth and their effects on nontarget organisms, as this is crucial for ensuring long-term efficacy. Artificial intelligence and machine learning (ML) are expected to elevate PE research to a new level. By utilizing predictive models, it will be possible to identify new PE-producing plants, locate biosynthetic gene clusters, and design optimized synthetic pathways (Usmanov et al. 2025). ML technologies are helping to further unlock the pharmacological potential of PEs. They can analyze existing data to identify promising biological activities, predict side effects, and rationally design new compounds. However, the evolutionary and ecological origins of PEs are still poorly understood. Comparative phylogenetic studies can help elucidate the origins and diversification of biosynthesis. Understanding how environmental factors—particularly insect predators, drought, and temperature stress—affect PE production could pave the way for new strategies to increase their natural and agricultural production.

In summary, phytoecdysteroids were first discovered in the mid-twentieth century, and their ecological, biological, and pharmacological significance has been widely studied. PEs are polyhydroxylated steroids isolated from plants, structurally similar to insect molting hormones. In insects, they act as essential hormones that regulate metamorphosis, molting, and reproduction. In plants, PEs have evolved primarily as chemical defense mechanisms, providing necessary adaptations against herbivorous insects, various pathogens, and environmental stresses. Due to these properties, PEs are of particular interest in science as multifunctional natural compounds. In addition to their ecological functions, PEs have great potential in the fields of pharmacology and biotechnology. They have shown anabolic, adaptogenic, antioxidant, hepatoprotective, hypoglycemic, immunomodulatory, and neuroprotective properties. Therefore, PEs are considered promising candidates for dietary supplements, pharmaceuticals, and cosmeceuticals. Additionally, their natural insect-repellent properties suggest their potential use as sustainable biopesticides in agriculture. Overall, PEs are a striking example of chemical innovation in plants—they simultaneously serve as ecological protectors and have practical value. With the increasing scientific evidence of their benefits and the emergence of new technological tools, the study and production of PEs are entering a new phase. They are expected to have a significant impact on science, medicine, and industry in the future as natural pesticides, adaptogenic nutraceuticals, new therapeutics, or as models to explain evolutionary processes.

References

Acheuk F, Basiouni S, Shehata A, Dick K, Hajri H, Lasram S, Yilmaz M, Emekci M, Tsiamis G, Spona-Friedl M, May-Simera H, Eisenreich W, Ntougias S (2022) Status and Prospects of Botanical Biopesticides in Europe and Mediterranean Countries. Biomolecules *12*:311. https://doi.org/10.3390/biom12020311

Adler JH & Grebenok RJ (1999) Occurrence, Biosynthesis, and Putative Role of Ecdysteroids in Plants. Critical Reviews in Biochemistry and Molecular Biology 34:253–264. https://doi.org/10.1080/10409239991209282

Akhrem AA, Levina IS, Titov YuA (1973) Ecdysones are insect steroid hormones. p 232.

Alami MM, Guo S, Mei Z, Yang G, Wang X (2024) Environmental factors on secondary metabolism in medicinal plants: exploring accelerating factors. Medicinal Plant Biology 3:16. https://doi.org/10.48130/mpb-0024-0016

Almeida A, Dong L, Appendino G, Bak S (2020) Plant triterpenoids with bond-missing skeletons: biogenesis, distribution and bioactivity. Natural Product Reports 37:1207–1228. https://doi.org/10.1039/c9np00030e

Areiza M (2018) Regulation of juvenile hormone synthesis by 20-hydroxyecdysone in the yellow-fever mosquito, Aedes aegypti. FIU Electronic Theses and Dissertations 3804. https://doi.org/10.25148/etd.FIDC006829.

Arif Y, Singh P, Bajguz A, Hayat S (2022) Phytoecdysteroids: Distribution, Structural Diversity, Biosynthesis, Activity, and Crosstalk with Phytohormones. Int J Mol Sci *23*:8664. https://doi.org/10.3390/ijms23158664

Baltaev UA (2000) Phytoecdysteroids: Structure, Sources, and Biosynthesis in Plants. Russian Journal of Bioorganic Chemistry 26:799–831. https://doi.org/10.1023/A:1026662505403

Behmer ST, Nes WD (2003) Insect sterol nutrition and physiology: a global overview. Adv Insect Physiol 31:1–65.

Bencheikh N, Elbouzidi A, Kharchoufa L, Ouassou H, Alami Merrouni I, Mechchate H, Es-safi I, Hano C, Addi M, Bouhrim M, Eto B, Elachouri M (2021) Inventory of Medicinal Plants Used Traditionally to Manage Kidney Diseases in North-Eastern Morocco: Ethnobotanical Fieldwork and Pharmacological Evidence. Plants *10:*1966. https://doi.org/10.3390/plants10091966

Blais C, Dauphin-Villemant C, Kovganko N, Girault JP, Descoins C, Lafont R (1996) Evidence for the involvement of 3-oxo-Δ^4 intermediates in ecdysteroid biosynthesis. Biochem J 2:413–419. https://doi.org/10.1042/bj3200413.

Chamani M, Dadpour M, Dehghanian Z, Panahirad S, Chenari Bouket A, Oszako T, Kumar S (2025) From Digestion to Detoxification: Exploring Plant Metabolite Impacts on Insect Enzyme Systems for Enhanced Pest Control. *Insects 16:*392. https://doi.org/10.3390/insects16040392

Chang Y, Tang JP, Mendoza HV, Christensen J, Block AK (2020) Plant Defense Chemicals against Insect Pests. Agronomy *10*:1156. https://doi.org/10.3390/agronomy10081156

Chen D, Xiao Y, Zheng X, Sun H, Zhang C, Zhu J, Xue T (2025) Seasonal dynamics and molecular regulation of flavonoid biosynthesis in *Cyclocarya paliurus* (Batal.) Iljinsk. Frontiers in Plant Science 16:1525226. https://doi.org/10.3389/fpls.2025.1525226

Cheng DK (2017) Primary and secondary metabolites. A Phytochemical Perspective 3:111–114. https://doi.org/10.1002/9781118961933.

Chowdhury SK, Banerjee M, Basnett D, Mazumdar T (2024) Natural pesticides for pest control in agricultural crops: an alternative and eco-friendly method. Plant Science Today 11:433–450. https://doi.org/10.14719/pst.2547

Dinan L (2001) Phytoecdysteroids: biological aspects. Phytochemistry 57:325–339. https://doi.org/10.1016/S0031-9422(01)00078-4

Dinan L (2009) Phytoecdysteroids: what use are they? Archives of Insect Biochemistry and Physiology: Published in Collaboration with the Entomological Society of America 72:126–141. https://doi.org/10.1002/arch.20334

Dinan L, Bourne P, Whiting P, Tsitsekli A, Saatov Z, Dhadialla TS, Hormann RE, Lafont R, Coll J (2003) Synthesis and biological activities of turkesterone 11α-acyl derivatives. Journal of Insect Science 3:11pp. https://doi.org/10.1093/jis/3.1.6

Dinan L, Lafont F, Lafont R (2023) The Distribution of Phytoecdysteroids among Terrestrial Vascular Plants: A Comparison of Two Databases and Discussion of the Implications for Plant. Insect Interactions and Plant Protection. Plants *12*:776. https://doi.org/10.3390/plants12040776

Dziwornu GA, Caira MR, Mare JAL, Edkins AL, Bolton JJ, Beukes DR, Sunassee SN (2017) Isolation, Characterization, and Antiproliferative Activity of New Metabolites from the South African Endemic Red Algal Species *Laurencia alfredensis*. Molecules *22:*513. https://doi.org/10.3390/molecules22040513

Eisenman SW, Zaurov DE, Struwe L. (Eds.) (2012) Medicinal plants of central Asia: Uzbekistan and Kyrgyzstan. Springer Science & Business Media pp 46.

Gautam K, Khedkar R (2024) Amaranthus: Biology, Functional Potential and Sustainable Utilization. In: Thakur M. (eds) Millets: The Multi-Cereal Paradigm for Food Sustainability. World Sustainability Series. https://doi.org/10.1007/978-3-031-64237-1_12

Głazowska J, Kamiński MM, Kamiński M (2018) Chromatographic separation, determination and identification of ecdysteroids: Focus on Maral root (*Rhaponticum carthamoides*). Journal of separation science 41:4304–4314. https://doi.org/10.1002/jssc.201800506

Graf BL (2014) Phytochemistry, Pharmacology, and Socioeconomic Implications of Quinoa *(Chenopodium quinoa Willd.)*. Rutgers The State University of New Jersey. Diss Doct Phylos pp 24

Harrewijn P, Oosten VAM, Piron PG (2012) Natural terpenoids as messengers: A multidisciplinary study of their production, biological functions, and practical applications. Springer Science & Business Media. p 141.

Hejran AB, Azizi A, Saken K, Ardak K, Sawicka B, Niazi P (2024) Plant-derived bio-pesticides for eco-friendly control of the *Prodenia litura* (Fabricius) insect pest. J Entomol Zool Stud 12:192–205. https://doi.org/10.22271/j.ento.2024.v12.i5c.9397

Hidalgo D, Sanchez R, Lalaleo L, Bonfill M, Corchete P, Palazon J (2018) Biotechnological production of pharmaceuticals and biopharmaceuticals in plant cell and organ cultures. Current medicinal chemistry 25:3577–3596. https://doi.org/10.2174/0929867325666180309124317

Hikino H, Okuyama T, Konno C, Takemoto T (1975) Carbon-13 nuclear magnetic resonance spectra of phytoecdysones. Chemical and Pharmaceutical Bulletin 23:125–132. https://doi.org/10.1016/S0015-3796(17)30100-2

Holopainen JK, Himanen SJ, Yuan JS, Chen F, Stewart CN (2025) Ecological Functions of Terpenoids in Changing Climates. In: Ramawat KG, Mérillon JM (eds) Natural Products. Springer, Berlin, Heidelberg. https://doi.org/10.1007/978-3-642-36202-6_129-1

Hunyadi A, Herke I, Lengyel K, Báthori M, Kele Z, Simon A, Tóth G, Szendrei K (2016) Ecdysteroid-containing food supplements from *Cyanotis arachnoidea* on the European market: evidence for spinach product counterfeiting. Sci Rep 6:37322. https://doi.org/10.1038/srep37322

Iga M, Kataoka H (2012) Recent studies on insect hormone metabolic pathways mediated by cytochrome P450 enzymes. Biol Pharm Bull 35:838–843. https://doi.org/10.1248/bpb.35.838

Jozdani EP (2012) Partition, sorption & biodegradation of dialkoxybenzenes that modulate insect behavior. Thesis M Sci p 96.

Kapare V, Satdive R, Fulzele DP, Malpathak N (2017) Impact of Gamma Irradiation Induced Variation in Cell Growth and Phytoecdysteroid Production in *Sesuvium portulacastrum*. J Plant Growth Regul 36:919–930. https://doi.org/10.1007/s00344-017-9697-3.

Karatt TK, Sathiq MA, Laya S, Ajeebsanu MM, Karakka Kal AK, Subhahar MB, Perwad Z (2023) An extensive screening method for the identification and quantitation of ecdysteroids in equine urine and plasma using liquid chromatography coupled with mass spectrometry. Rapid Communications in Mass Spectrometry 37:9611. https://doi.org/10.1002/rcm.9611

Karinho-Betancourt E (2018) Plant-herbivore interactions and secondary metabolites of plants: Ecological and evolutionary perspectives. Bot. sci 96:35–51. https://doi.org/10.17129/botsci.1860.

Khurramov A, Yusupova U, Bobakulov K, Syrov V, Egamova F, Usmanov D, Quadir M, Rasulev B (2025) Secondary metabolites of *Silene claviformis* Litv. and the anti-inflammatory activity of its butanol extract. Nat Prod Res 1–6. https://doi.org/10.1080/14786419.2025.2495850

Kingston DG (2011) Modern natural products drug discovery and its relevance to biodiversity conservation. Journal of natural products 74:496–511. https://doi.org/10.1021/np100550t

Knížková K, Cihlářová P, Kuchař M (2025) Determination of Selected Biologically Active Substances in Dietary Supplements and the Adaptogenic Herb *Rhaponticum carthamoides* Using HPLC–MS/MS. Food Anal Methods 18:1514–1531. https://doi.org/10.1007/s12161-025-02805-y

Lafont R (1997) Ecdysteroids and related molecules in animals and plants. Archives of Insect Biochemistry and Physiology: Published in Collaboration with the Entomological Society of America 35:3–20. https://doi.org/10.1002/(SICI)1520-6327(1997)35:1/2<3::AID-ARCH2>3.0.CO;2-X

Lafont R, Balducci C, Dinan L (2021) Ecdysteroids. Encyclopedia *1:*1267–1302. https://doi.org/10.3390/encyclopedia1040096

Lafont R, Dinan L (2003) Practical uses for ecdysteroids in mammals including humans: An update. Journal of Insect Science 3:1–30. https://doi.org/10.1093/jis/3.1.7

Lafont R, Dinan L (2024) Insect Sterols and Steroids. In: Advances in Experimental Medicine and Biology. Springer, Cham. https://doi.org/10.1007/5584_2024_823

Mamarasulov B, Davranov K, Jabborova D (2020) Phytochemical, pharmacological and biological properties of *Ajuga turkestanica* (Rgl.) Brig (Lamiaceae). Ann Phytomed 9:44–57. https://doi.org/10.21276/ap.2020.9.1.6

Mathur V, Dokka N, Raghunathan G, Rathinam M, Parashar M, Srivastava Sh, Sreevathsa R (2025) Beyond bitter: plant triterpenoids in the battle against herbivorous insects. Journal of Experimental Botany 76:4441–4457. https://doi.org/10.1093/jxb/eraf238

Mithöfer A, Boland W (2012) Plant defense against herbivores: chemical aspects. Annual review of plant biology 63:431–450. https://doi.org/10.1146/annurev-arplant-042110-103854

Nes WD (2011) Biosynthesis of cholesterol and other sterols. Chemical reviews 111:6423–6451. https://doi.org/10.1021/cr200021m

Niwa R, Niwa Y (2014) Enzymes for ecdysteroid biosynthesis: their biological functions in insects and beyond. Bioscience, Biotechnology, and Biochemistry 78:1283–1292. https://doi.org/10.1080/09168451.2014.942250

Ono H, Morita S, Asakura I, Nishida R (2012) Conversion of 3-oxo steroids into ecdysteroids triggers molting and expression of 20E-inducible genes in *Drosophila melanogaster*. Biochem Biophys Res Commun 421:561–566. https://doi.org/10.1016/j.bbrc.2012.04.045.

Park I, Park K, Lee HS, Hong SM, Sriramulu DK, Hwang H, Kwon HC (2023) Ecdysteroids from the Korean endemic species *Ajuga spectabilis* with activities against glucocorticoid receptors and 11β-hydroxysteroid dehydrogenase type 1. ACS omega 8:26191–26200. https://doi.org/10.1021/acsomega.3c02421

Patil-Patankar V, Yadav-Bhagwat P, Kedari P (2022) Valorization of secondary metabolites in plants. Handbook of biomass valorization for industrial applications p 287–313. https://doi.org/10.1002/9781119818816.ch13

Reed J, Osbourn A (2018) Engineering terpenoid production through transient expression in *Nicotiana benthamiana*. Plant Cell Rep 37:1431–1441. https://doi.org/10.1007/s00299-018-2296-3

Sarker SD, Nahar L (2012) An Introduction to Natural Products Isolation. In: Sarker S, Nahar L (eds) Natural Products Isolation. Methods in Molecular Biology 864. https://doi.org/10.1007/978-1-61779-624-1_1

Schmelz EA, Grebenok RJ, Galbraith DW, Bowers SW (1998) Damage-Induced Accumulation of Phytoecdysteroids in Spinach: A Rapid Root Response Involving the Octadecanoic Acid Pathway. J Chem Ecol 24:339–360. https://doi.org/10.1023/A:1022588610232

Shahbaz M, Imran M, Momal U, Naeem H, Alsagaby SA, Al Abdulmonem W, Jbawi EA (2023) Potential effect of kaempferol against various malignancies: recent advances and perspectives. Food and Agricultural Immunology 34:2265690. https://doi.org/10.1080/09540105.2023.2265690

Sinha RP, Häder DP (2020) Natural bioactive compounds: Technological advancements. Academic Press. p 462.

Slama K, Lafont R (1995) Insect hormones-ecdysteroids: their presence and actions in vertebrates. Eur J Entomol 92:355–377. https://www.eje.cz/pdfs/eje/1995/01/43.pdf

Sripinyowanich S, Petchsri S, Tongyoo P, Lee TK, Lee S, Cho WK (2023) Comparative Transcriptomic Analysis of Genes in the 20-Hydroxyecdysone Biosynthesis in the Fern *Microsorum scolopendria* towards Challenges with Foliar Application of Chitosan. International J Mol Sci *24:*2397. https://doi.org/10.3390/ijms24032397

Stransky K, Nemec V, Sláma K (1998) Lipid composition of the seeds of an ecdysteroid-containing plant, *Leuzea carthamoides* (Willd.) DC (Asteraceae). Russ J Plant Phys 45:333–338.

Suhada KM, Nidha Fathima TP, Rabin JM, Fathima Rinsha VP, Hiba Fathima P, Rashida CN, Mubashira CP (2023) Effect of allelochemicals on digestive and biomolecules of rice weevil: sitophilus oryzae. Fundamental Applications of Biochemistry in the Environment. p 9.

Talapatra SK, Talapatra B (2015) Steroids: Cholesterol and Other Phytosterols. In: Chemistry of Plant Natural Products. Springer, Berlin, Heidelberg. https://doi.org/10.1007/978-3-642-45410-3_11

Thiem B, Kikowska M, Maliński MP, Kruszka D, Napierała M, Florek E (2017) Ecdysteroids: production in plant in vitro cultures. Phytochem Rev. 16:603–622. https://doi.org/10.1007/s11101-016-9483-z.

Thummel CS, Chory J (2002) Steroid signaling in plants and insects—common themes, different pathways. Genes & development 16:3113–3129. https://doi.org/10.1101/gad.1042102.

Todorova V, Ivanova S, Chakarov D, Kraev K, Ivanov K (2024) Ecdysterone and Turkesterone—Compounds with Prominent Potential in Sport and Healthy Nutrition. Nutrients *16:*1382. https://doi.org/10.3390/nu16091382

Tomou C, Barda P, Lymperis, Lytra K, Skaltsa H (2022) In Edible FungiChemical Composition, Nutrition and Health Effects ed. D Stojković and L Barros. The Royal Society of Chemistry 7:273–303. https://doi.org/10.1039/9781839167522-00273

Usmanov D, Yusupova U, Syrov V, Casanola-Martin GM, Rasulev B (2025) A Quantitative Structure–Activity Relationship Study of the Anabolic Activity of Ecdysteroids. Computation 13*:*195. https://doi.org/10.3390/computation13080195

Warren JT, O'Connor MB, Gilbert LI (2009) Studies on the black box: incorporation of 3-oxo-7-dehydrocholesterol into ecdysteroids by *Drosophila melanogaster* and *Manduca sexta.* Insect Biochem Mol Biol 39:677–687. https://doi.org/10.1016/j.ibmb.2009.08.004.

Wei JH, Yin X, Welander PV (2016) Sterol synthesis in diverse bacteria. Frontiers in microbiology 7:990. https://doi.org/10.3389/fmicb.2016.00990.

Williams CM (1961) The juvenile hormone. II. Its role in the endocrine control of molting, pupation, and adult development in the cecropia silkworm. Biological Bulletin 121:572–585. https://doi.org/10.2307/1539456.

Williams CM (1976) Juvenile Hormone in Retrospect and in Prospect. In: Gilbert, L.I. (ed.) The Juvenile Hormones. Springer. https://doi.org/10.1007/978-1-4684-7947-8_1

Yusupova U, Usmanov D, Azamatov A, Ramazonov N, Rejepov J (2020b) Phytochemical constituents and biological activities of *Dianthus helenae* Vved., growing in Uzbekistan. Nat Prod Res 36:3480–3484. https://doi.org/10.1080/14786419.2020.1862834.

Yusupova UY, Bobakulov KM, Girault JP, Usmanov DA, Abdullayev ND, Ramazonov NSh (2022c) New Phytoecdysteroid from *Silene popovii.* Chem Nat Compd 58:1082–1084. https://doi.org/10.1007/s10600-022-03874-9

Yusupova UY, Bobakulov KM, Khurramov AR, Usmanov DA, Quadir M, Karuth A, Rasulev B (2025) Phytochemical constituents isolated from *Silene popovii* Schischk. Med Chem Res 34:423–431. https://doi.org/10.1007/s00044-024-03345-6

Yusupova UY, Ramazonov NS, Syrov VN, Sagdullaev SS (2022a) Phytoecdysteroids. In: Phytoecdysteroids. Springer, Singapore. https://doi.org/10.1007/978-981-16-6711-4_1

Yusupova UY, Ramazonov NS, Syrov VN, Sagdullaev SS (2022b) Technologies for Producing Drugs and Biologically Active Additives Based on Ecdysteroids from Plants *Rhaponticum Carthamoides*, *Silene Praemixta* and *Ajuga.* In: Phytoecdysteroids. Springer, Singapore. https://doi.org/10.1007/978-981-16-6711-4_7

Yusupova UY, Usmanov DA, Ramazonov NS (2019) Phytoecdysteroids from the Plant *Dianthus helenae.* Chem Nat Compd 55:393–394. https://doi.org/10.1007/s10600-019-02701-y

Yusupova UY, Usmanov DA, Ramazonov NS (2020a) Phytoecdysteroids from the Aerial Part of *Silene popovii.* Chem Nat Compd 56:562–563. https://doi.org/10.1007/s10600-020-03092-1

Zádor E (2025) Molecular Targets of 20-Hydroxyecdysone in Mammals, Mechanism of Action: Is It a Calorie Restriction Mimetic and Anti-Aging Compound? Cells 6:431. doi:https://doi.org/10.3390/cells14060431.

Zhang L, Lin T, Zhu G, Wu B, Zhang Ch, Zhu H (2023) LncRNAs exert indispensable roles in orchestrating the interaction among diverse noncoding RNAs and enrich the regulatory network of plant growth and its adaptive environmental stress response. Horticulture Research 10:234. https://doi.org/10.1093/hr/uhad234

Zibareva L (2000) Distribution and levels of phytoecdysteroids in plants of the genus *Silene* during development. Archives of Insect Biochemistry and Physiology: Published in Collaboration with the Entomological Society of America 43:1–8. https://doi.org/10.1002/(SICI)1520-6327(200001)43:1<1::AID-ARCH1>3.0.CO;2-D

Zibareva L, Volodin V, Saatov Z, Savchenko T, Whiting P, Lafont R, Dinan L (2003) Distribution of phytoecdysteroids in the Caryophyllaceae. Phytochemistry 64:499–517. https://doi.org/10.1016/S0031-9422(03)00376-5

Zwetsloot KA, Shanely AR, Merritt EK, McBride JM (2014) Phytoecdysteroids: a novel, nonandrogenic alternative for muscle health and performance. J Steroids Hormon Sci 12:2. https://doi.org/10.4172/2157-7536.S12-e001

Chapter 2
Spectroscopic Techniques for the Characterization of Ecdysteroids

Pierre Le Pogam and Mehdi A. Beniddir

Abstract This chapter critically evaluates modern analytical methods—including NMR, mass spectrometry, and chiroptical methods—used to characterize ecdysteroids. Combining dissertation experience with literature advances, it showcases how optimized spectroscopic workflows enable precise structure elucidation and stereochemical determination, setting standards for future research.

2.1 Introduction

Ecdysteroids are a structurally diverse class of polyhydroxylated steroids, defined by a cyclopentano[a]perhydrophenanthrene core and a characteristic 7-en-6-one chromophore in ring B (Báthori et al. 2008). Originally recognized as arthropod molting hormones, ecdysteroids are now known to be widely distributed across both animal and plant kingdoms. In arthropods, they regulate critical developmental processes such as molting and metamorphosis, while in plants, their biosynthesis has evolved independently, resulting in a rich array of analogues termed phytoecdysteroids (Dinan et al. 2021). Over 600 ecdysteroid structures have been identified to date (Lafont et al. 2002), with more than 400 occurring in plants alone.

The primary ecological role of phytoecdysteroids is plant defense against herbivorous arthropods and other phytophagous invertebrates (Lafont et al. 1991). By mimicking endogenous arthropod molting hormones, phytoecdysteroids act as antifeedants or endocrine disruptors in non-adapted insect species, inducing premature molting, metabolic dysfunction, reduced fecundity, and increased mortality. This defensive strategy is a product of sophisticated co-evolutionary dynamics between plants and herbivorous insects, with monophagous and oligophagous predators exerting stronger selective pressure for chemical diversification than polyphagous species.

P. Le Pogam · M. A. Beniddir (✉)
Université Paris-Saclay, Orsay, France
e-mail: mehdi.beniddir@universite-paris-saclay.fr

V. N. Syrov et al. (eds.), *Phytoecdysteroids II*,
https://doi.org/10.1007/978-981-95-8423-9_2

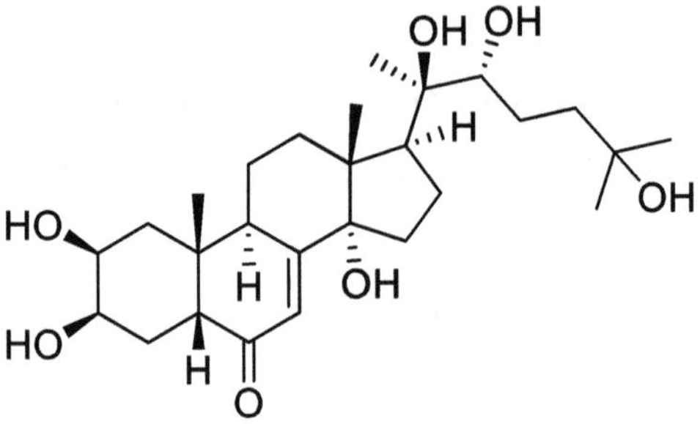

Fig. 2.1 Structure of 20-hydroxyecdysone

Beyond their ecological significance, phytoecdysteroids exhibit a diverse array of pharmacological activities in mammals. In skeletal muscle, these compounds promote protein synthesis through calcium-dependent activation of Akt and related signaling cascades, resulting in anabolic effects comparable to some synthetic anabolic–androgenic steroids (Gorelick-Feldman et al. 2010). Additional beneficial properties include anti-diabetic effects, modulation of adipose tissue metabolism, and enhancements in physical performance and muscle mass. Notably, phytoecdysteroids are characterized by exceptionally low toxicity in mammals, with LD_{50} values exceeding 9 g/kg for ingested 20-hydroxyecdysone (Fig. 2.1) and 6.4 g/kg for intraperitoneal administration, making them attractive candidates for pharmaceutical and nutritional applications (Dinan et al. 2021).

The structural diversity, multiple sources, and diverse bioactivities of ecdysteroids necessitate robust and comprehensive analytical methods for their isolation, characterization, and quantification. Four years after the publication of the previous edition of the book of this same series (i.e., Phytoecdysteroids: Properties, Biological Activity, and Applications), this chapter reviews, through selected examples, modern approaches to dereplicating and elucidating ecdysteroid molecules. In the first section of this report, the ecdysteroid frame is being decomposed into its subsequent components to (i) present the main structural variations encountered for each of them and (ii) the spectroscopic pieces (nuclear magnetic resonance (NMR), ultraviolet (UV), infrared (IR), and electronic circular dichroism (ECD) spectroscopy features) of evidence used to address them (configurations included). In the second section, discussions on mass spectrometry (MS)-based targeting and dereplication methods will be developed.

2.2 Ecdysteroids Structure Elucidation Logic and Strategies

The process of elucidating the structure of ecdysteroids *sensu lato* is similar to that of steroids. As such, their gross structure can be determined using the standard set of 1D and 2D nuclear magnetic resonance (NMR) sequences required for the *de novo* elucidation of the structure of any new natural product. The steroidal frame of ecdysteroids results in extensive spin systems, enabling most cycles and side chains

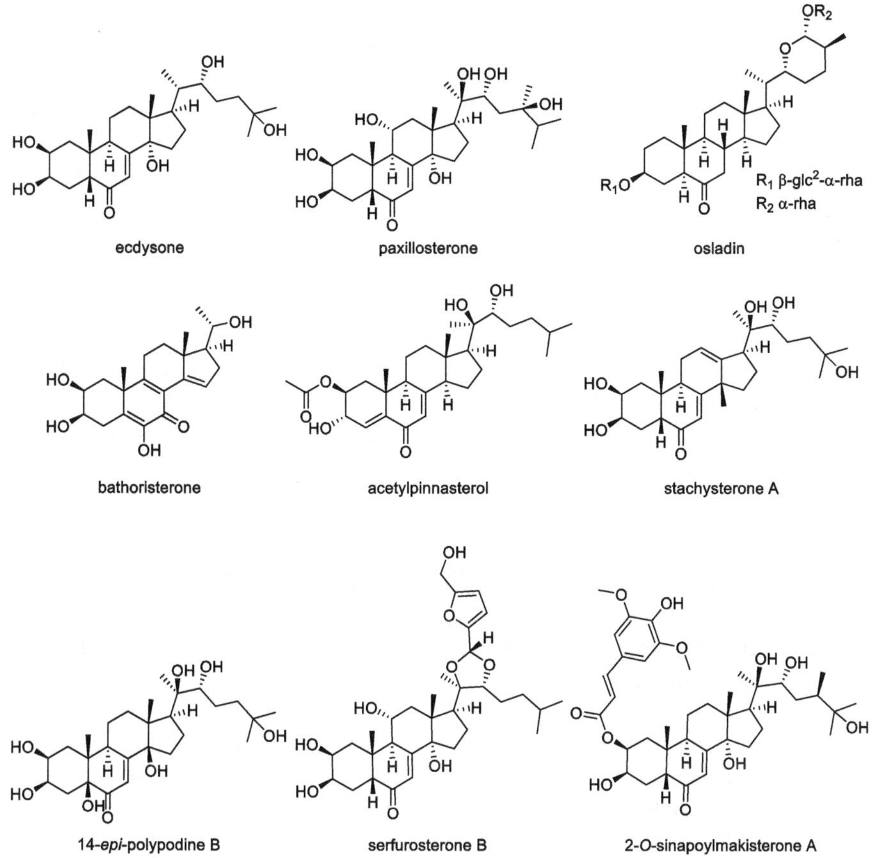

Fig. 2.2 Examples of some ecdysteroids illustrating the structural diversity within this structural family (see Fig. 2, 13, 14, and 15 for examples of cyclizations)

to be easily assigned using COSY/TOCSY analyses. HMBC crosspeaks are then usually considered to connect these spin systems and the few hydrogen-deprived positions of these structures. HMBC correlations originating from angular methyl groups are especially useful for establishing connections between the aforementioned spin systems.

Unfortunately, the structural diversity of ecdysteroids is quite difficult to circumscribe (Fig. 2.2), and a distinction is often made between "true" ecdysteroids and ecdysteroid-like compounds. True ecdysteroids include a *cis*-junction between A and B rings, a 7-en-6-one group and a 14 α-hydroxy functionality. Ecdysteroid-like compounds can lack some of these features (e.g., a *trans*-junction between rings A and B, no 14 α-OH group, no $\Delta^{6,7}$ functionality) and may contain additional unsaturations (e.g., $\Delta^{4,5}$, $\Delta^{14,15}$, a $\Delta^{8,9}$, $\Delta^{5,6}$-7-one) and hydroxylation sites (e.g., 1, 2, 3, 4, 9, 11, 12, 15, 16, and/or 19). Structural modifications to the side chain are also an important way to diversify ecdysteroid structures. The side chain may also contain a variable number of additional carbon atoms, mostly appearing at C-24. Analogues with shorter or no side chains have also been reported. Hydroxylations may occur at almost every position (C-20, C-21, C-22, C-23, C-24, C-25, C-26, and/or C-27), possibly generating polyhydroxylated variants. Unsaturations may appear in the side chain as well at different positions such as $\Delta^{24,25}$, $\Delta^{25,26}$, or $\Delta^{26,28}$. Finally, the side chain may yield a 5-, 6-, or 7-membered oxygen-containing ring. Overall, these structural modifications result in a potentially elevated number of asymmetric centers, which undoubtedly present the greatest challenge when determining the structure of a new ecdysteroid. As a way to describe the logic underlying the structure elucidation and stereochemical assignment of ecdysteroids, a thorough literature analysis was conducted to mine for spectroscopic landmarks (i.e., ^{13}C and ^{1}H NMR chemical shifts, *J* coupling constant magnitudes, 2D NOE contacts, UV, IR, and ECD absorptions) as well as computational and derivatizing strategies that were commonly applied. This knowledge will be summarized below following an ecdysteroid skeleton-centered analysis.

2.2.1 Spectroscopic Features Related to A and B Rings

2.2.1.1 Spectroscopic Strategies to Assign the Configuration of the A/B Ring Junction

While most ecdysteroids exhibit an A/B *cis*-junction, some analogues possess a *trans*-junction due to a 5α-hydrogen atom. Both series occasionally co-occur within a single plant (Báthori et al. 2000; Wu et al. 2010). It is unclear whether 5α-ecdysteroids are genuine natural products, particularly given that 5β-ecdysteroids slowly equilibrate with the 5α form. It has been observed that steric hindrance between 2β-OH and 19-CH_3 favors the 5β-form; however, the 5α-form predominates in 2-deoxyecdysteroids (Horn 1971; Mamadalieva et al. 2002). X-ray crystallography has shown that H-5 adopts an axial orientation in both cases (Fábián et al. 2002; Kalász et al. 2002), so the α- or β- orientation must be determined before reliable stereochemical conclusions can be drawn from the magnitude of the coupling constant values relating to the protons appearing on ring A. The C-5 stereochemistry is readily determined from the ^{13}C NMR chemical shift of the C-19 methyl group. The 5α-H configuration typically shields C-19 by 10–11 ppm (≈ 13 ppm) compared with 5β-H-containing analogues (≈ 24 ppm) (Russell et al. 1981; Kim et al. 1988; Wang et al. 2009). While this general guidance is undeniably useful for defining the nature of the junction from tabulated ^{13}C NMR data in many cases (see, for example, Wu et al. 2010; Tóth et al. 2021; Laczkó et al. 2025), the chemical shift of this methyl group appears to be affected by other structural elements. For instance, a $\Delta^{9,11}$ functionality deshields the Me-19 group up to more than 30 ppm (Tóth et al. 2021), while a 5β-OH configuration causes an upfield shift of 5–6 ppm via the *γ*-gauche effect, resulting in ~17 ppm resonance (Krepinsky et al. 1977; Píš et al. 1994; Odinokov et al. 2002; Jayasinghe et al. 2003b; Tóth et al.

2008). Thus, it is safer to double-check the configuration of the junction using relevant NOE correlations. An A/B *cis*-junction shows diagnostic NOE contacts between H-5 and CH_3-19, and between H-2α (if applicable) and H-9 (Simon et al. 2007; Tóth et al. 2021), whereas a 5α-ecdysteroid displays NOE contacts between H-5 and H-1α, H-3α, H-4α, and H-9 (Wu et al. 2010; Csábi et al. 2015) (Fig. 2.3). Finally, subtle shifts in the ^{1}H NMR spectrum were noted when 5-α and 5-β ecdysteroids were compared. In the 5α-H series, the methylenic protons at H-1, H_{eq}-2, H_{eq}-4, and H-5 are expected to be downfield shifted, while H_{ax}-3, H_{ax}-9, and to a lesser extent, H_{ax}-4 are expected to shift upfield (Girault et al. 1990). A marked downfield shift at H_3-19 is also observed in the case of *trans*-linkage between rings A and B (Saatov et al. 1987) (δ_H 1.34 vs 1.10 for *cis*-A/B junction). A more modest downfield shift (0.05 to 0.1 ppm) of the singlet corresponding to the CH_3-19 protons was sometimes noted (Russell and Fraser 1973; Dzhukharova et al. 1993) relative to 5β-H analogues, a finding that is also reported in later literature (Tóth et al. 2008).

Some rare ecdysteroids, such as 4-dehydroecdysterone (Searle and Molinski 1995) and diaulusterols A–B (Williams et al. 1986), also reveal a $\Delta^{4,5}$ functionality. These compounds are more complex since ring A can exist in two different half-chair conformations (Černý et al. 1973). In this context, the configuration of these substituents cannot be derived from ^{1}H NMR data alone, and the final proof of the stereostructure of such compounds should be based on NOE effects (Kovganko and Sokolov 2001).

In addition to these characteristic NMR landmarks, ECD-based chiroptical diagnostic values have been established by Nakanishi's pioneering work in 1970, which observed that ecdysteroids with an A/B *cis* ring junction show a pair of relatively

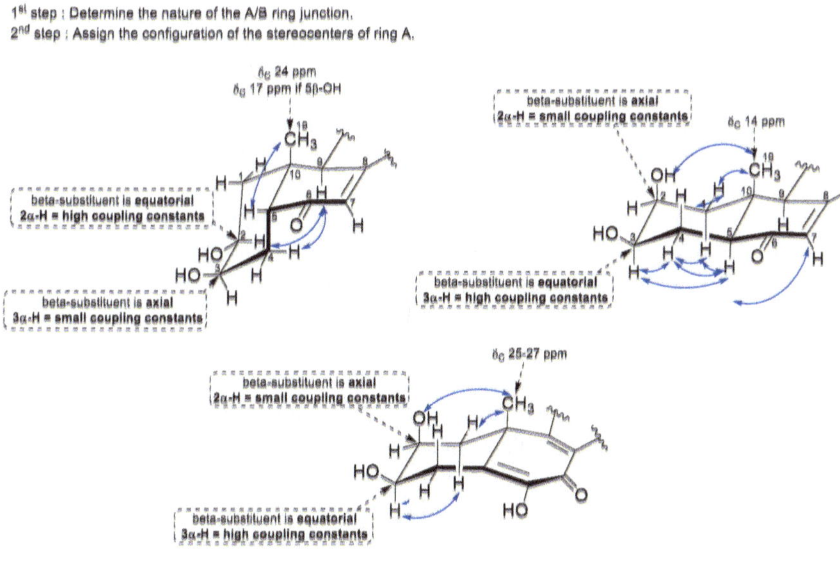

Fig. 2.3 ^{1}H and ^{13}C NMR chemical shifts, *J* coupling constants values, and key 2D NOE contacts landmarks observed depending on the A/B ring junction

low amplitude Cotton effects for the enone chromophore, with a negative $\pi \rightarrow \pi*$ band around 242 nm ($\Delta\epsilon \approx -3.5$) and a positive $n \rightarrow \pi*$ band around 340 nm ($\Delta\epsilon \approx +1.6$), clearly weaker than the corresponding bands in A/B *trans* systems (ca. 240 nm $\Delta\epsilon \approx -7.1$; 340 nm $\Delta\epsilon \approx +3.3$) (Koreeda and Nakanishi 1970).

2.2.1.2 *Cis*-A/B Rings and Their Geometrical and Electronic Effects on A and B Rings Substituents

A characteristic structural feature of many ecdysteroids is their 2,3-diol system, and the vicinal coupling constant values of the oxymethine protons convey valuable stereochemical information for defining substituent orientations in compounds with ***cis*-A/B ring junctions** (Fuchino et al. 1997; Vokáč et al. 1998; Csábi et al. 2015).

In *cis*-A/B ecdysteroids with 3β-OH, the hydroxy group assumes an axial orientation, with the proton at C-3 (equatorially positioned) exhibiting small coupling constants typically less than 3 Hz, detected as a quadruplet or multiplet with $W_{1/2} \approx 8$ Hz (Fuchino et al. 1997; Vokáč et al. 1998; Csábi et al. 2015). In the case where an ecdysteroid contains a 2β-OH group and C-1 is unsubstituted, the proton at C-2 is axially oriented and exhibits elevated *trans*-diaxial coupling constants of approximately 12 Hz, or is detected as a multiplet with $W_{1/2} \approx 22$–24 Hz, reflecting coupling with the diastereotopic methylenic protons at C-1 (Vokáč et al. 1998; Hang et al. 2015; Csábi et al. 2015). In contrast, **paristerone** represents an antifacial arrangement of hydroxy groups (2α,3β-OH), where both oxymethine protons are equatorially oriented and detected as broad singlets with $W_{1/2} \approx 7$ Hz (Singh and Thakur 1982). An alternative antifacial configuration occurs in **3-*epi*-ecdysone**, which possesses 2β,3α-dihydroxy groups. In this instance, both hydrogen atoms are axially oriented, resulting in elevated coupling constant values ($W_{1/2} \approx 20$ Hz) (Dinan and Rees 1978; Rudel et al. 1992). The ECD spectrum of paristerone in dioxane exhibits a positive $n \rightarrow \pi^*$ Cotton effect at 343 nm $[\theta] = +5070$ and a negative $\pi \rightarrow \pi^*$ Cotton effect at 246 nm $[\theta] = -16{,}490$, preserving the *cis*-type sign pattern and wavelength region but with slightly bathochromically shifted maxima and enhanced negative ellipticity attributable to the antifacial A-ring substitution. In methanol, the same bisignate pattern is maintained, with the positive band shifted to 332 nm $[\theta] = +6820$ (hypsochromic) and the negative band to 252 nm $[\theta] = -17{,}060$ (bathochromic), with both bands increasing in magnitude, consistent with solvent stabilization (Singh and Thakur 1982).

Other ecdysteroids display synfacial 2α,3α-dihydroxy substitution such as **ponasterone B** and **lesterone**. In these compounds, the oxymethine proton H-3 exhibits elevated *trans*-diaxial coupling constants, while the equatorial proton H-2 exhibits low coupling constant values (Nakanishi et al. 1968; Borovikova and Baltaev 1999). For 1,2,3-trihydroxylated compounds with β-oriented hydroxy positions at each site, such as **zoanthusterone**, the equatorial proton H-1 is detected as a broad singlet and the H-2 signal appears as a broad singlet or triplet with low coupling constants, reflecting its involvement in two axial-equatorial couplings (Jayasinghe et al. 2002; Suksamrarn et al. 2002a). H-5 shifts significantly downfield in such compounds due to *trans*-diaxial interactions between the axial 1β-hydroxy function and H-5β (+ 0.26 ppm) (Girault et al. 1990). Similarly, compounds with a 2α,3α,5β-trihydroxy substitution pattern, such as **rapisterone D**, exhibit coupling patterns characteristic of synfacial 2,3-diol systems (Baltaev 1995).

An unusual example is **palythone A**, which features a single hydroxy group at C-4. Although the H-4 signal was recorded as a multiplet with undefined coupling constants, its NOE correlations with H-5 supported their synfacial orientation. Notably, this configuration was also supported by TDDFT-ECD simulated spectra for both epimers at C-4, which provided the best fit with the experimental spectrum (Lee et al. 2016).

The substitution of a 5β-OH group produces characteristic spectroscopic changes. A 5β-OH causes an upfield shift of approximately 2 ppm for the C-6 carbonyl carbon (Krepinsky et al. 1977). More significantly, a characteristic 12 nm hypsochromic shift of the positive n → π* band to approximately 328 nm (compared to the Nakanishi reference of 340 nm) has been observed for **polypodine B**, **dacrysterone** (Russell and Fraser 1973), **sengosterone**, and **ponasterone C**. X-ray structural analysis of **brahuisterone** (Dzhukharova et al. 1993), which bears a 3β-hydroxy group and 5β-OH, revealed that the A ring adopts a chair conformation similar to that of ecdysteroids with a *cis*-A/B ring junction, consistent with the chemical coupling patterns observed for similarly substituted ecdysteroids (Shigemori et al. 1999; Hunyadi et al. 2004; Tóth et al. 2008).

The 2,3-diol functionality of ecdysteroids has sometimes been chemically transformed to enable their stereochemical assignment when ^{1}H NMR signals overlap or coupling constants are unreadable. Derivatization through acetylation or conversion into 2,3-acetonides enables clearer spectroscopic interpretation; the same coupling constant interpretation applies after chemical modification, as in the case of the **2,3,20,22-diacetonide of makisterone C** (Nakanishi et al. 1968; Simon et al. 2009). A specialized class comprises **ecdysonelactones A–D**, which feature either an H-5β or an OH-5β group and a distinctive γ-lactone ring involving C-2 and the oxygen atom at C-3 (Guillen et al. 2018). The configuration at C-3 is assigned from the H-3 splitting pattern, while the configuration at C-2 is determined using diagnostic NOE correlations between the protons of the bicyclic structure (Guillen et al. 2018) (Fig. 2.4).

2.2.1.3 *Trans*-Arranged A/B Rings and Their Geometrical and Electronic Effects on A and B Rings Substituents

The orientation of the substituents on the A ring is completely different for 5α-ecdysteroids compared with 5β-counterparts. X-ray crystallographic analysis of **5α-2-deoxyintegristerone A** has revealed these structural distinctions (Kalász et al. 2002). In compounds with a 3β-OH group and a *trans*-A/B junction, the hydroxyl group assumes an equatorial orientation, and the proton at C-3 is expected to exhibit *trans*-diaxial coupling constant values, a pattern characteristic of

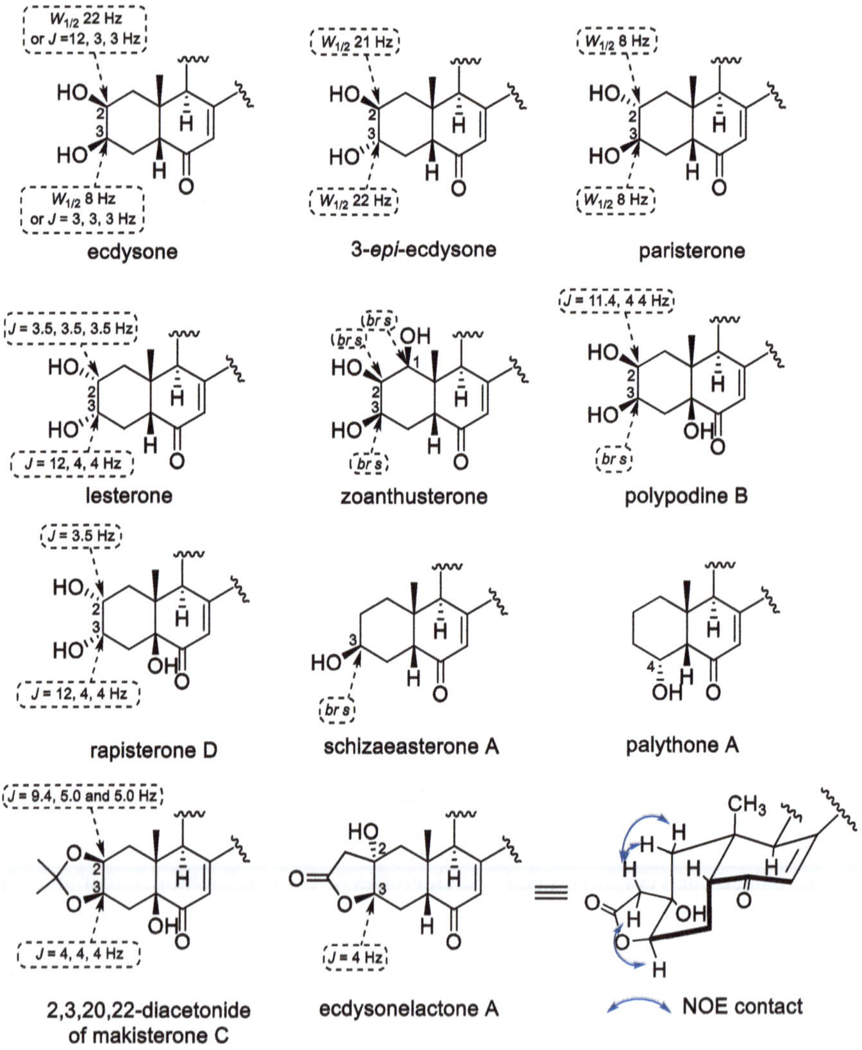

Fig. 2.4 Expected coupling patterns for the various mono-, di-, or trihydroxylated A rings of *cis*--A/B-con ecdysteroids

terpenic compounds featuring *trans* ring junctions (Csábi et al. 2015; Mbeng Obame et al. 2026). **Tenuifolioside A**, which has a β-oriented *O*-glycosyl residue at C-3, has a similar coupling constant value (Mimaki et al. 1989). Conversely, when a 2β-OH group is present, this oxymethine proton assumes an equatorial orientation and reveals small coupling constant values, frequently detected as a quadruplet with $J = 3$ Hz (Wu et al. 2010, Yokota et al. 1982; Csábi et al. 2015). In the case of **bombycosterol 3-phosphate**, a similar coupling constant appears, although this structure contains a 5α-OH group (Hiramoto et al. 1988). The

5,6-diol functionality in **bombycosterol** is an unusual structural trait. Based on the width of the ^{1}H NMR signal for H-6, it is unsafe to assign the configuration at C-6. It was noted that a homoallylic coupling appears between H-6 and H-9 ($J \approx 1.5$ Hz). Additionally, coupling involving 6-OH may occur in certain solvents. For example, differences in the H-6 signal width in NMR spectra recorded in pyridine-d_6 were presumed to result from traces of water in the solvent. Therefore, although there is a general trend for the 6β-isomer to have a larger $J_{6,7}$ value than the corresponding 6α-isomer, C-6 assignment should be approached with caution. A safer strategy based on pyridine-induced deshielding of 4-methylenic protons has been proposed. Accordingly, the 4α-hydrogen resonance of the 6α-isomer appears at approximately 3 ppm, while the 4β-proton resonance of the 6β-isomer resonates at approximately 3 ppm (the 4α and 4β hydrogens can be easily identified thanks to their characteristic coupling patterns) (Fujimoto et al. 1985). It was also noted that the chemical shift of CH_3-19 is influenced by the configuration at C-6, being detected at 1.5–1.6 ppm for 6β-isomers and at ca. 1.1 ppm for 6α-isomers (Fujimoto et al. 1985). In the case of sphenocentrosides A and B, the ROE correlation between H-6 and CH_3-19 could be used to determine the configuration at this stereocenter (Ajayi et al. 2019).

Castasterone is an interesting example of an ecdysteroid with a *trans*-A/B ring junction and synfacial, α-oriented hydroxy groups at positions 2 and 3. The splitting patterns of H-2 and H-3 denote their axial and equatorial orientations, respectively, consistent with the previously discussed geometric features. The downfield chemical shift of H-5 is also related to a 1,3-diaxial relationship between H-5 and the α-hydroxy group at C-3 (Yokota et al. 1982) (Fig. 2.5).

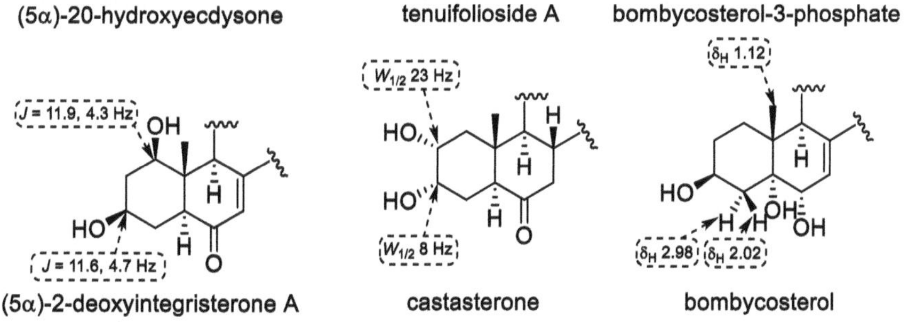

Fig. 2.5 Expected coupling patterns for the various hydroxylation pattern of A ring of *trans*-A/B-con ecdysteroids

2.2.1.4 $\Delta^{4,5}$ and $\Delta^{5,6}$ Functionalities and Their Geometrical and Electronic Effects on A and B Rings Substituents

Compounds featuring a $\Delta^{5,6}$-7-one-$\Delta^{8,9}$ functionality, such as **calonysterone** and **brainesteroside C**, appear to adopt a similar conformation of their A ring as 5α-H-containing ecdysteroids, such that the usual 2β,3β-dihydroxy groups result in a pseudoequatorial H-2 and a pseudoaxial H-3 (Wu et al. 2010; Tóth et al. 2023; Yuan et al. 2025). It is straightforward to define the existence of this motif based on HMBC correlations from CH_3-19 that are expected to be established with several sp^2-hybridized carbons (whose expected chemical shift ranges are summarized in Fig. 2.6). For these compounds, the chemical shift value of C-19 appears to be around 26.5–27 ppm (Tóth et al. 2021; Tóth et al. 2023).

A $\Delta^{4,5}$ functionality has been reported to occur for a few compounds (Kawahara et al. 1995), and these variants present distinct stereochemical challenges. In some instances, these compounds do not present any particular difficulty due to the absence of a hydroxy group at position 3 and thus of any stereocenter on the A ring, as exemplified by **calvasterol A**. A closely related compound with a keto group at C-3 and a *trans*-A/B ring junction, namely **cyathisterone**, was reported from the same fungal source (Kawahara et al. 1994). Notably, calvasterol A displays a negative Cotton effect at 275 nm and a positive Cotton effect at 234 nm, characteristic of a $\Delta^{4,5}$-3-one-$\Delta^{8,9}$-containing ecdysteroid (Kawahara et al. 1995).

Conversely, 4-dehydroecdysteroids containing hydroxy groups at positions 2 and 3 present more challenging configurational issues, as the orientation of these substituents cannot be determined solely from the splitting pattern of the oxymethine

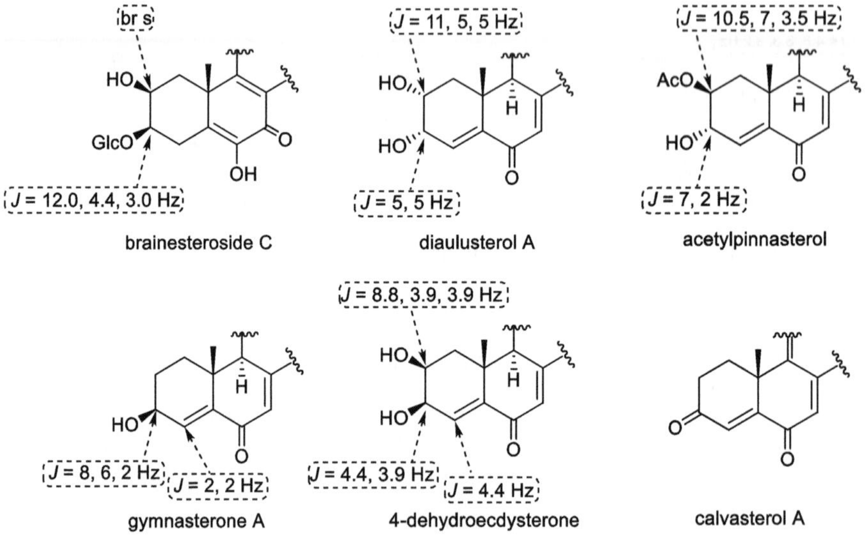

Fig. 2.6 Expected coupling patterns for the various hydroxylation patterns of $\Delta^{4,5}$ and $\Delta^{5,6}$ functionalities-containing ecdysteroids

protons H-2 and H-3 (Kovganko and Sokolov 2001). To date, few such examples appear to have been reported. Searle and Molinski reported **4-dehydroecdysterone**, with relative stereochemistry determined through a combination of coupling constant analysis and NOE measurements (Searle and Molinski 1995). The small coupling constant between H-2 and H-3 ($J = 3.9$ Hz) indicated their *cis* orientation, and a molecular model demonstrated that a half-chair conformation with a pseudoaxial orientation of H-2 and, thus, 2β,3β-dihydroxyl substituents was the only configuration consistent with an NOE effect between H-2 and H-9 (Searle and Molinski 1995).

Acetylpinnasterol represents another $\Delta^{4,5}$ compound with antifacially oriented groups, the structure of which was determined using X-ray crystallography (Fukuzawa et al. 1981). The specific case of **diaulusterols A and B** and **hyousterones A-D** deserves particular attention, as these compounds possess both a $\Delta^{4,5}$ functionality and two α-oriented hydroxy groups at C-2 and C-3. In diaulusterol A, the coupling constant between H-2 and H-3 ($J = 5$ Hz) is quite small, consistent with *cis*-oriented protons, a finding especially notable given that the *trans*-arranged OH groups of analogues reveal higher coupling constant values (Fukuzawa et al. 1981; Williams et al. 1986). NOE effects between H-2 and H-3, and between H-2 and H_3-19 in a triacetylated derivative, confirmed both their synfacial arrangement and their β-orientation (Williams et al. 1986). The coupling patterns of the H-2 and H-3 signals of **hyousterones** are less well-defined than in the former case (reported as multiplets), but the authors indicate that they are in close agreement with those of **diaulusterols**, suggesting a synfacial orientation (Miyata et al. 2007). The ROESY spectrum indicated the α-configuration of the 2,3-dihydroxy system via a correlation between H-2 and H_3-19 (Miyata et al. 2007).

Gymnasterone A exemplifies the interpretive challenges posed by 4-dehydroecdysteroids bearing a single hydroxy group at C-3. Initially, the coupling constants between H-3 and H_2-2 led the authors to assign a 3α-OH configuration. However, a reconsideration of this structure by the same authors 10 years later led them to revise the configuration at C-3, based on NOE correlations between H-3 and both H-1α and H-2α, and between H-2β and H-1β with CH_3-19, indicating that 3-OH and the CH_3-19 groups were synfacially oriented (Amagata et al. 1998; Amagata et al. 2007) (Fig. 2.6).

2.2.2 Spectroscopic Features Related to Rings B and C

Most ecdysteroids exhibit a hydrogen atom H-9 that adopts an axial, α-orientation (Fábián et al. 2002). In this context, a β oriented H-11 atom results in a characteristic *trans*-diaxial orientation ($J \approx 9$ Hz), which can be further backed up by NOE/ROE contacts with the synfacial Me-18 and Me-19 (Buděšínský et al. 2008). The α-axial orientation of H-9 is supported by its NOE correlations: with H-2α in ecdysteroids with a *cis*-A/B junction (Takács et al. 2010), and with H-1α and H-5α in *trans*-configured ecdysteroids between rings A and B (Wu et al. 2010). A derivative named **9α,20-dihydroxyecdysone**, containing a 9α-hydroxy group has been reported from *Silene italica* ssp. *nemoralis*. The orientation of this group was determined based on NOE correlations between the CH_3-19 and H-11β groups (Pongrácz et al. 2003). One year later, **9β,20-dihydroxyecdysone** was reported from the same plant. The stereo orientation of OH-9 was deduced from ROE responses between 9-OH/Me-19 and 9-OH/Me-18, observed in DMSO-d_6 (Simon et al. 2004). In this specific context, it was noted that H-5 is equatorially oriented ($J = 4.5$ and 2.3 Hz). Thus, the upfield shifts of signals C-4 (δ_C 26.2) and C-19 (δ_C 21.6) compared to the epimeric 9α,20-dihydroxyecdysone chemical shifts (δ_C 36.3 and 29.6, respectively) were related to the γ-gauche effect of the axial methyl group (Simon et al. 2004).

2.2.2.1 Ring B: Primary Structure and Chromophoric Features

Ring B offers a more limited degree of structural variation compared with many other structural elements of the ecdysteroid scaffold, as it contains a recurrent α,β-enone type 7-en-6-one chromophore. This functionality produces a maximum absorption wavelength of 242 nm (log ε = 4.093) (Butenandt and Karlson 1954; Liu et al. 2025).

However, sometimes a $\Delta^{5,6}$-7-one-$\Delta^{8,9}$ chromophore appears instead, and $\Delta^{7,8}$-6-one-$\Delta^{9,11}$-containing ecdysteroids have been described as well (Fig. 2.7). The ECD spectrum of ecdysteroids that contain this latter system (i.e., $\Delta^{7,8}$-6-one-$\Delta^{9,11}$), such as **dacryhainansterone** (Songsheng et al. 1982), features a pronounced 44 nm bathochromic shift of the negative Cotton effect observed around 300 nm, compared with the corresponding non-extended $\Delta^{7,8}$-6-one system. This shift is directly attributed to the introduction of a second double bond at C-9, which extends the conjugation, lowering the energy of the $\pi \rightarrow \pi*$ transition and moving both the ECD band and the UV absorption by a similar magnitude (Hunyadi et al. 2007). Likewise, a bathochromic shift had been reported for compounds displaying a $\Delta^{5,6}$-7-one-$\Delta^{8,9}$ feature, with UV maximal absorption occurring at approximately 290 nm. A significant shift in IR absorption for these compounds has also been noted (1616 cm^{-1} for a $\Delta^{5,6}$–7-one-$\Delta^{8,9}$-containing ecdysteroid vs. ca. 1650 cm^{-1} for a compound with a regular $\Delta^{7,8}$-6-one function) (Hou et al. 2023).

A unified reactivity scheme has been suggested as a way of accounting for the diversity of these structural variants (Csábi et al. 2015). It has already been shown that *ortho*-quinol B-ring-containing ecdysteroids spontaneously oxidize to form a compound with 9,15-diol and a $\Delta^{8,14}$ olefin (Tóth et al. 2023).

The occurrence of additional double bonds involving ring B can be inferred based on the double bond equivalent count, and additional unsaturations can be easily located based on HMBC correlations from Me-19, which usually provide clear structural insights (Hang et al. 2015). Although both structural types have sp^2-hybridized carbons, the ^{13}C NMR chemical shift value of C-9 differs significantly between the two subtypes (δ_C 136 for $\Delta^{7,8}$-$\Delta^{9,11}$-containing ecdysteroids against δ_C 165-169 for *ortho*-quinol type compounds) (Girault and Lafont 1988; Pongrácz et al. 2003; Csábi et al. 2015). The resonance of the carbonyl-type carbon in $\Delta^{7,8}$-6-one-containing ecdysteroids (δ_C 206) differs from that in compounds with a

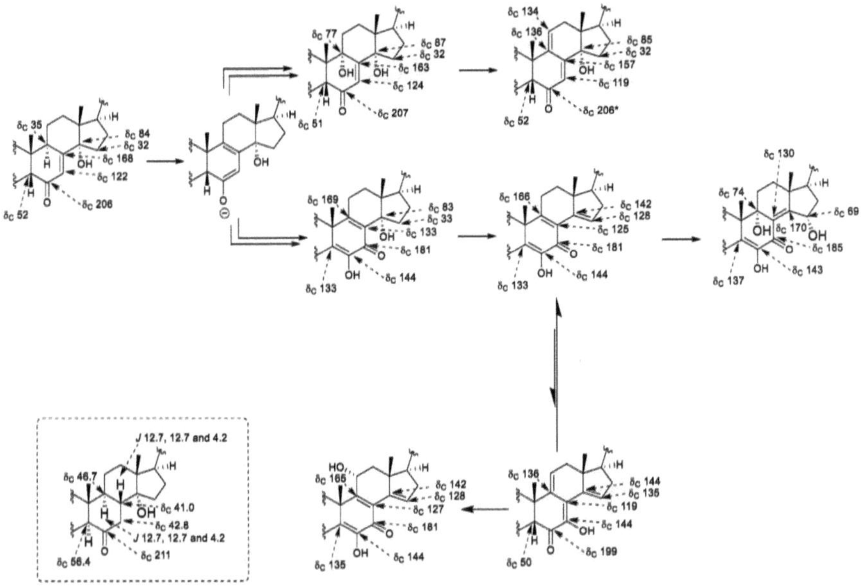

Fig. 2.7 Main structural variants affecting cycle B. Chemical shift values have been rounded for readability and are extracted from (Csábi et al. 2015). Ecdysone chemical shift values are extracted from (Girault and Lafont 1988), chemical shifts for 9α,20-dihydroxy ecdysone are taken from (Pongrácz et al. 2003), and chemical shifts for 25α-hydroxydacryhainansterone (Hunyadi et al. 2007) are taken from here, except for C-6 which remained undetected in this compound and was thus taken from the spectroscopic data of dacryhainansterone (Bourne et al. 2002). The 9α,15α-diol chemical shifts are taken from (Tóth et al. 2023)

$\Delta^{5,6}$-$\Delta^{8,19}$-motif (δ_C 181) (Girault and Lafont 1988; Pongrácz et al. 2003; Hunyadi et al. 2007; Csábi et al. 2015). Quite notably, **calonysterone** was shown to occur as a desmotrope (type of tautomerism where both forms may be isolated) with a $\Delta^{7,8}$-$\Delta^{9,11}$-containing ecdysteroid (Csábi et al. 2015). The carbonyl functionality of this desmotrope discloses a diagnostic value of its carbonyl-type carbon, owing to the shielding effect of the *ortho*-phenolic group (δ_C 199). Canonica et al. reported a λ_{max} value of 285 nm (ε = 9435), which shifts to 340 nm (ε = 7570) in 5% NaOH in MeOH due to the presence of an α-diketone (Canonica et al. 1973; Hou et al. 2023).

2.2.2.2 Saturated B Ring-Containing Ecdysteroids

A few ecdysteroids lack the typical ring B unsaturations, exemplified by **tenuifoliosides A** and B, **osladin,** and **castasterone** (Jizba et al. 1971; Yokota et al. 1982; Mimaki et al. 1989; Yamada and Nishizawa 1995). The NMR data for these compounds differ significantly from those for canonical ecdysteroids, the olefinic proton H-7 and sp^2 quaternary resonance at C-8 disappear, while an additional diastereotopic methylene group (CH_2–7) and a further methine group (CH–8) appear (Mimaki et al. 1989). The UV absorption of such compounds appears to be very limited in relation to other possible structural variants (Mimaki et al. 1989). Some derivatives reveal a secondary alcohol at C-6 (e.g., **sphenocentrosides A** and B). The ROE correlation between H-6 and CH_3-19 could be used to determine the configuration at this stereocenter (Ajayi et al. 2019).

2.2.2.3 Ring C Structural Variations

A hydroxy group at C-11 is a very common feature of ecdysteroids. Most of the time, the hydroxy group is α-oriented so that H-11 reveals *trans*-diaxial couplings with both H-9 and H-12_{ax} (Vokáč et al. 1998). It was noted that an 11α-hydroxy group is responsible for the low-field shift of H-1β (δ_H 2.5 vs. 1.8) (Minli et al. 1992; Vokáč et al. 1998). The presence of an 11β-hydroxy group has been reported in a derivative of **shidasterone**, namely **11β-20-deoxyshidasterone**. In this instance, the H-11 signal is a sharp multiplet ($W_{1/2}$ = 5 Hz) and shows an NOE correlation with H-9 (Ochieng et al. 2012).

The presence of a hydroxy group at C-12 in 12α,20-dihydroxyecdysone is also unusual. Its orientation was determined by its small coupling constant values, which indicated an equatorial (and hence β) orientation, and was further validated by the ROE correlation between H-12 and CH_3-18 (Guo et al. 2021). 12-Oxoecdysteroids have also been reported (e.g., spectasterone A) (Park et al. 2023) or proposed to act in the biosynthesis of rearranged ecdysteroids (see below for the example of diploclidine) (Jayasinghe et al. 2003a; Cai et al. 2017).

2.2.3 *Spectroscopic Features Related to Rings C and D*

2.2.3.1 Spectroscopic Features Related to Ring D

The 15-hydroxy group in **15-hydroxyponasterone A** is identifiable by the low-field shift of C-16 (δ_C 35.6 vs. 21.5 in **ponasterone A**). The 15β-OH group exerts a strong influence on the CH_3-18 group (δ_H 1.14 in 15-hydroxyponasterone A vs. 0.89 in **ponasterone A**), resulting in a low-field shift of the olefinic proton H-7 (δ_H 6.47 vs. 5.81 in ponasterone A). This configuration is supported by NOE correlations between H-15 and H-7, and between H-15 and H-16α (Buděšínský et al. 2008). It was deduced that an oxidized derivative of 14β,15-dihydrocalonysterone contained a 15α-OH group. The H-15β configuration was confirmed by its ROE correlation with CH_3-18; however, the multiplicity and coupling constant of H-15 were not reported (Tóth et al. 2023).

Hydroxylation may also occur at C-16, as seen in the 16β-OH containing **malacosterone**. In this case, H-16α is expected to reveal two coupling constants of similar magnitude (7.8–8.0 Hz) and a smaller one (5.5 Hz), typical for five-membered rings with two $^3J_{cis}$ and one $^3J_{trans}$ coupling constants (Garcia et al. 1986; Yokosuka et al. 2021). The proximity of H-22 to the 16β-OH group results in a large downfield shift of the former (δ_H 3.86 vs. 3.43), and a notable downfield shift of 18-CH_3 (δ_H 1.11 vs 0.87) also accompanies this substituent (Garcia et al. 1986). In the case of 16α-hydroxytaxisterone, an NOE correlation between H-16 and H_3-18 could be used to determine the configuration at this stereocenter. The oxymethine H-16 then emerges as a doublet of doublets (J = 8.3 and 6.1 Hz) (Yokosuka et al. 2021).

The vast majority of ecdysteroids exhibit a 17α-oriented hydrogen, frequently observed as a triplet with a coupling constant of approximately 9 Hz (or a doublet of doublets with similar coupling constants) (Suksamrarn et al. 2002b; Buděšínský et al. 2008; Hunyadi et al. 2016; Kakumu et al. 2023). Consequently, the configuration of this center is often presumed to be α-disposed unless stated otherwise. NOE crosspeaks between H-17/H-15α and/or between H_3–18/H_3–21 can confirm that H-17 is α-oriented (Park et al. 2023). However, a few ecdysteroids contain a 17β-H, such as **14β,17α-14-deoxypoststerone**. Steric proximities between H-14/H_3-18 and H-14/H-17 inferred from NOE contacts, along with 17-Hβ selective ROE (selROE) responses distinguishing nearby α- and β-methylenic hydrogens, support this assignment (Tóth et al. 2021). Some compounds without a side chain have been reported to exist, with an OH group or ketone functionality present at the C-17 position. An NOE correlation between H-17 and H-12α could then be used to define the β-orientation of the OH group, with H-17 being detected as a doublet of doublets (J = 9, 6 Hz) (Báthori et al. 2000; Hang et al. 2015).

2.2.3.2 *Trans*-Arranged C/D Rings and Their Geometrical and Electronic Effects on Rings C and B Substituents

Most ecdysteroids, like **ecdysone** itself, possess a *trans*-junction between rings C and D and a 14α-OH functional group. In this context, a NOE correlation is expected between axial H-12α and H-17 (Liktor-Busa et al. 2008; Kakumu et al. 2023). In solvents like DMSO-d_6, NOE correlations from OH-14 provide direct spectroscopic evidence of its α-orientation (Simon et al. 2004). Some compounds, such as **Diaulusterols A and B,** feature a 14 α-hydrogen. Their stereochemical assignment was based primarily on comparison with cholesterol chemical shifts and NOESY considerations (Williams et al. 1986).

2.2.3.3 *Cis*-Arranged C/D Rings and Their Geometrical and Electronic Effects on Rings C and B Substituents

A few derivatives reveal a *cis*-junction between rings C and D due to a 14 β-OH group, exemplified by **14-*epi*-polypodine B** and **14-*epi*-ponasterone A 22-glucoside**. Such derivatives often result from photo-transformation reactions of *trans*-annulated C/D compounds (Harmatha et al. 2002). The presence of a 14β-OH group is inferred from the significant downfield shift of Me-18 (δ_H 1.26 vs. δ_H 0.89)

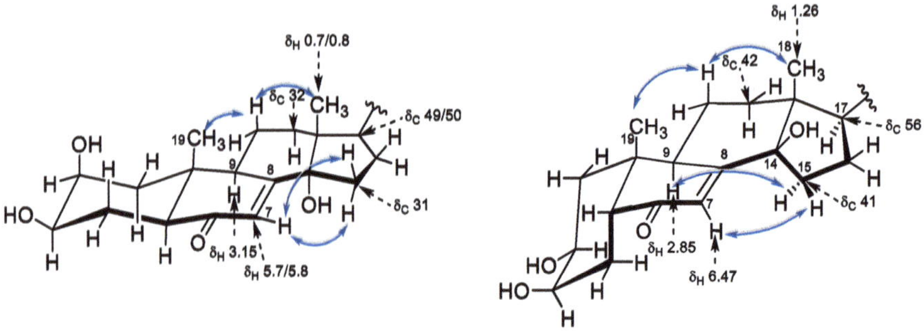

Fig. 2.8 Characteristic NMR features (chemical shifts and NOE correlations for *trans* vs. *cis* C/D ecdysteroids)

and the shielding of H-9 (δ_H 2.85 vs. δ_H 3.15) (Girault et al. 1990; Buděšínský et al. 2008; Jiang et al. 2021). While the ^{13}C NMR chemical shift of Me-18 tends to be downfield shifted (δ_c 19.1/19.2 vs. δ_c 18.0), this value is influenced by other structural features; for example, an 11α-hydroxy group in a **11α-hydroxyecdysteroid** can deshield this position to values expected for a **14β-OH**-containing derivative(Buděšínský et al. 2008). More reliable variations appear at C-15 ($\delta_c \approx 41$ ppm) and C-17 (δ_c 56–57 ppm), which are deshielded by regard to their usual values (31–32 ppm and 50–52 ppm, respectively) (Buděšínský et al. 2008; Jiang et al. 2021). It is also suggested that C-12 is deshielded in the 14β-OH configuration (δ_C 41.5–42.0 vs. δ_C 32.5–33.0 for 14α-OH, respectively) (Harmatha et al. 2002; Simon et al. 2009; Jiang et al. 2021).

NOESY data supports the 14β-OH assignment: H-7 shows NOE crosspeaks with H-15β (whereas with 14α-OH, H-7 correlates with both H-15 diastereotopic methylene protons), and a NOE contact between H-9 and H-15α also implies a 14β-OH group (Harmatha et al. 2002; Buděšínský et al. 2008; Jiang et al. 2021) (Fig. 2.8). For H-14 containing ecdysteroids, a strong NOE crosspeak between H-14 and H_3-18 indicates a *cis* C/D ring junction (Tóth et al. 2021). Regarding ECD, to the best of our knowledge, no robust *cis* vs. *trans* C/D ring junction rule analogous to Nakanishi's A/B *cis* rule has emerged in the literature. However, **14-*epi*-20-hydroxyecdysone** (14β-OH) exhibited two Cotton effects of decreased intensities at 247.4 [θ] = −9834 and 328.4 nm [θ] = +1169.

2.2.3.4 14,15 Epoxy-Containing Ecdysteroids

The occurrence of a hydroxy group at C-14 allows for structural diversification, including 14,15-epoxides. Distinguishing a dihydroxy group from the corresponding epoxide is a common source of structure elucidation error (Bulcock et al. 2024). The $^1J_{C\text{-}15,H\text{-}15}$ coupling constant (expected value ≈ 190 Hz) can reliably prove the existence of an epoxide ring (Simon et al. 2007). For **14α,15α-epoxy-14,15-dihydrostachysterone B**, the *trans*-type C/D junction was assigned using NOE correlations: H-12β/H_3-18,H-12β/H_3-21, and H-12α/H-17α (Simon et al. 2007). The H-15β orientation was confirmed via NOE correlation between H-15 and Me-18 (Wu et al. 2010). Notably, **gymnasterone B** features a 14,15β-epoxy group, deduced from NOE crosspeaks between H-9 and both H-12α and H-15, between H_3-18 and H-12β, and between H-17 and H-12α. These spectroscopic features determined the synfacial orientation of H-9 and H-15, and established that H_3-18 is antifacially oriented with H-9, H-15, and H-17 (Amagata et al. 1998). This structure was later ascertained by total synthesis (Li et al. 2006).

2.2.3.5 $\Delta^{12,13}$, $\Delta^{14,15}$, and 14β Methyl-Containing Ecdysteroids

Structural diversification of rings C and D includes dehydration reactions yielding a $\Delta^{14,15}$ moiety. It has long been known that treatment with HCl-MeOH (for example) results in facile loss of a 14α-hydroxy functionality (Karlson et al. 1965). This results in compounds disclosing a $\Delta^{14,15}$ functionality such as **stachysterone B** or **sphenocentrosides A** and **B** (Imai et al. 1970; Ajayi et al. 2019). The olefinic proton H-15 is detected either as a triplet with low coupling constants or as a narrow multiplet (Imai et al. 1970; Ajayi et al. 2019). A few compounds reveal an angular methyl group at position 14 rather than position 13 (Imai et al. 1970; Wang et al. 2011). **Stachysterone A** biosynthesis is proposed to proceed from a precursor with a 13β-methyl and a 14α-hydroxyl group, involving simultaneous methyl migration and hydroxyl elimination (Nakanishi et al. 1970a). **Niuxixinsterone B** exhibits a similar $\Delta^{12,13}$ functionality and 14β-methyl group (Wang et al. 2011). Its 14β-CH_3 configuration was confirmed by ROE crosspeaks between H-7 and H-15β, and between H-9 and H-15α (Wang et al. 2011). In the pyridine-containing ecdysteroid **diploclidine** (Jayasinghe et al. 2003a), the reaction sequence may be driven by a nucleophilic attack by the nitrogen atom. In both cases, the β-orientation of the methyl group is supported by biogenetic considerations of the original C-13 methyl orientation. From a spectroscopic standpoint, the 14-Me orientation in **diploclidine** was supported by an NOE correlation with axial H-11β (Jayasinghe et al. 2003a).

2.2.4 Spectroscopic Features and Strategies Related to Linear Side Chains

The side chain of ecdysteroids, extending from C-17 of the steroid core, represents a major axis of structural diversification and poses major challenges for stereochemical assignment (Fig. 2.9). A detailed literature survey reveals that the linear side chain of ecdysteroids can undergo various diversifications, including hydroxylations at C-20, C-22, C-23, C-24, C-25, and C-26, alkylation (e.g., methyl, ethyl, or ethylidene/vinyl groups) at C-24, oxidation at C-22, and length reduction. These transformations introduce multiple new asymmetric centers whose configurations

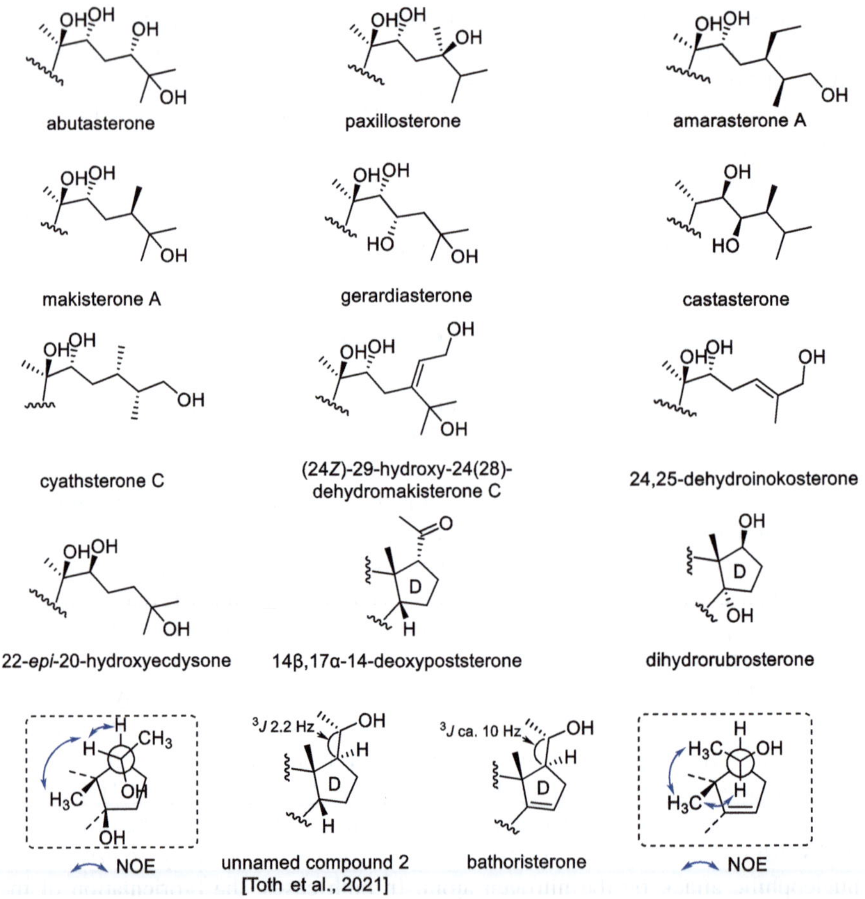

Fig. 2.9 Overview of ecdysteroids' linear side chain structural diversity. The model structures shown in the boxes were taken from (Marat et al. 1993)

must be determined through a combination of classical NMR analysis, comparative spectroscopy against reference compounds, modern computational methods, and chiroptical techniques. The following sections provide a comprehensive landscape of these structurally diverse natural products, categorized by established structure elucidation strategy.

2.2.4.1 Derivatization Approaches Coupled to Spectroscopic Methods

A significant challenge in elucidating ecdysteroid structures is correctly assigning the configuration of the chiral centers in the side chain. It is not uncommon for ecdysteroids to have asymmetric centers that are initially left unassigned (Pinheiro et al. 1983). Effective strategies have been developed to overcome this limitation,

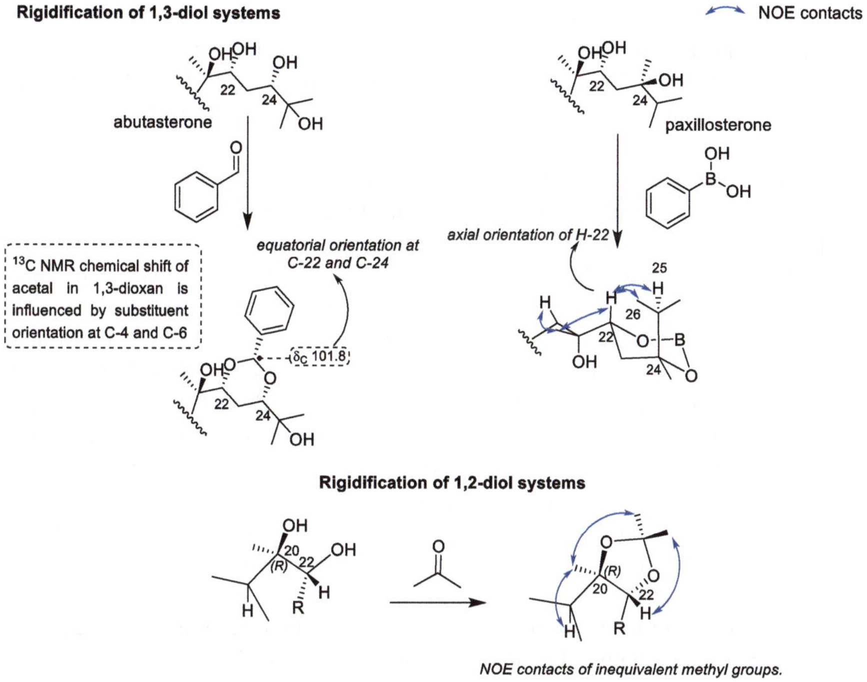

Fig. 2.10 Example of absolute configuration assignment via transformation of diols into rigid systems

typically involving the production of derivatives in which the stereocenters of interest are incorportaed into a conformationally rigid, often cyclic, system. A variety of derivatizing reagents have been employed, including acetone, benzaldehyde (Blunt et al. 1979) or phenylboronic acid (Vokáč et al. 1998) (Fig. 2.10).

For example, the configuration of the hydroxy group at C-24 in **abutasterone** cannot be established based on ^{1}H and ^{13}C NMR data alone (Coll et al. 1994). Blunt et al. recommended interpreting the ^{13}C NMR data of its 22,24-benzylidene derivative (Blunt et al. 1979). The ^{13}C NMR chemical shift value of the phenyl-bearing carbon, which favors an equatorial orientation (Eliel and Knoeber 1968), can be used to determine substituent orientations at positions 4 and 6 of the 1,3-dioxane ring. If all the substituents are equatorially oriented, the acetal carbon appears at approximately 101 ppm. If an axial substituent is introduced at C-4 or C-6, an upfield shift of 7 ppm is expected (Bock and Wiebe 1973). Since then, it has been common practice to assign C-24 configuration by comparison with the NMR data of compounds with a defined stereochemistry at this center (Sarker et al. 1996; Suksamrarn et al. 2000). Specific NMR chemical shifts have been reported for compounds with an epimeric side chain at C-24 (Nishimoto et al. 1987; Ohta et al. 1996; Suksamrarn et al. 1997) (Fig. 2.10).

Paxillosterone, containing both a hydroxy group and a methyl group at C-24, forms a rigid system upon reaction with phenylboronic acid, resulting in a six-membered ring adopting a chair conformation. The magnitude of the vicinal coupling constants of H-22 (J = 12.0, 3.0 Hz) determined its axial orientation (Vokáč et al. 1998). ROE crosspeaks between H-22 and both H-25 and Me-26 determined its 24*R* configuration, assuming the C-22 configuration to be *R*, as in the vast majority of natural ecdysteroids reported to date.

Most ecdysteroids contain two OH groups at C-20 and C-22. The configurations at these stereocenters are consistent across most reported ecdysteroids, though chirality labels may be inverted depending on substituent priorities around C-20 or C-22 (Buděšínský et al. 2009; Lafont et al. 2021). Nevertheless, it is interesting to check whether these expected configurations can be supported by spectroscopy. The conversion of 20,22-diol moieties to isopropylidene (acetonide) derivatives upon acetone treatment elegantly rigidifies the side chain, enabling NOE/ROE correlation analysis (Vokáč et al. 1998). Such **20,22-acetonide** derivatives, frequently reported as natural products, (Tóth et al. 2008; Simon et al. 2009), will be detailed in the section dedicated to the cyclic side chain. The induced circular dichroism (ICD) method using $Mo_2(OAc)_4$ has been employed to determine C-20 and C-22 configurations. A positive Cotton effect at 310 nm indicates a clockwise dihedral twist angle of the *vicinal* diols, supporting the 20*R*, 22*R* configuration (Liu et al. 2010; Guo et al. 2021).

Ecdysteroids with an additional asymmetric center arising from alkylation pose greater stereochemical challenges. Initial strategies involved chemical conversion to compounds whose absolute configuration could be easily assigned via chiroptical considerations. For example, Danieli et al. assigned the C-24 configuration of **makisterone A** as *R* based on its oxidation to (+)-*R*-methylsuccinic acid (Danieli et al. 1974) (Fig. 2.11). A closely related compound with a similar spin system but slightly different chemical shifts around C-24 (i.e., **24-*epi*-makisterone A**) provided reference ^{1}H and ^{13}C NMR datasets for both epimers (Miller et al. 1985; Maurer et al. 1993). This approach was subsequently applied to other C-24 methyl-containing compounds (e.g., **24-*epi*-atrotosterone A**) (Crouzet et al. 2009).

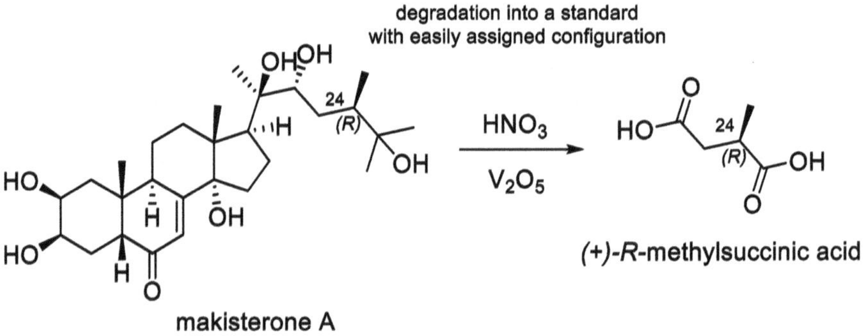

Fig. 2.11 Example of absolute configuration assignment via degradation of the side chain into a component whose configuration can be easily determined

2.2.4.2 Reference Compounds Synthesis and Spectroscopic Features Comparisons

Hydroxylation of a terminal methyl group creates an additional asymmetric center. **Amaresterone A**, bearing both this structural feature and a C-24 methyl group, has been reported from two distinct plant sources. Analogues of all four possible diastereoisomers (**a–d**, Fig. 2.12) were synthesized for subsequent comparison with the natural product's NMR data (Hirayama et al. 2014). **Achyranthesterones A and B** similarly possess a hydroxylated terminal methyl residue at C-25. Stereochemical assignment was achieved by comparing NMR data, particularly chemical shifts at C-24, C-25, and C-26, with 25*R*- and 25*S*-**inokosterone** (Ying-Ying et al. 2022).

Gerardiasterone exemplifies an ecdysteroid with a tetrahydroxylated side chain and three **contiguous** asymmetric centers at C-20, C-22, and C-23 (Guerriero et al. 1986). Assuming the 20*R* configuration on the basis of biosynthetic considerations, the configurations at C-22 and C-23 were assigned after four diastereoisomeric candidates **1–4** (Fig. 2.12) were synthesized and compared with the natural compound's ^{1}H NMR data (Tsubuki et al. 1996).

Although in the minority, a few ecdysteroids with a 22*S* configuration have been reported. Synthetic studies provided spectroscopic landmarks to distinguish 22*S* from 22*R* compounds. The ^{1}H NMR signal of CH_3–21 is deshielded by approximately 0.1 ppm in the 22*S* configuration (Hedtmann et al. 1989). Lafont et al.

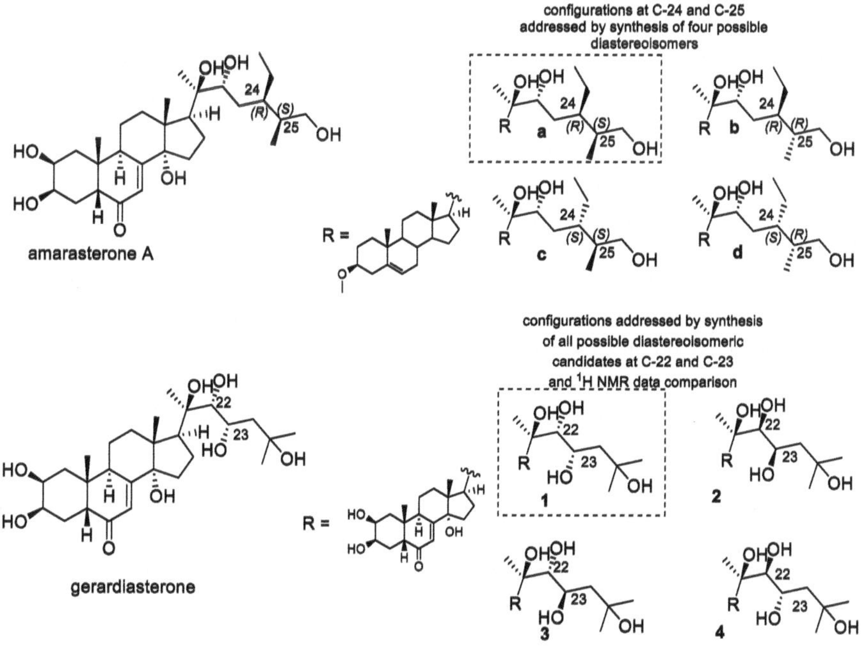

Fig. 2.12 Examples of compounds where the configuration of the asymmetric sites had been determined through the synthesis of possible diastereoisomers

reported on **22-*epi*-20-hydroxyecdysone**, noting that the only difference is a 0.2 ppm downfield shift of the H-17 signal compared to 20-hydroxyecdysone, subsequently confirmed against a synthetic standard (Bathori et al. 1998). Another such ecdysteroid, namely **22-*epi*-14α-hydroxyacetylpinnasterol**, was reported several years later (Fukuzawa et al. 1986). Both compounds were co-isolated with 22-oxo-derivatives, suggesting that an oxidation-reduction sequence may be responsible for their formation. Notably, the 22*S* configuration is nominal for brassinosteroids because brassinosteroid biosynthesis involves stereoselective hydroxylation at the 22*S* position (Clouse 1996). **Castasterone** (Yokota et al. 1982) underscores the possible biosynthetic connection between ecdysteroids and brassinosteroids, emphasized by its side-chain composition. Its asymmetric center configurations were determined by comparison with brassinolide (assigned via X-ray analysis (Grove et al. 1979) and stereoselective synthesis (Fung and Siddall 1980)). The 22*R* configuration arises from a change in priority labels due to C-23 hydroxylation. Several epimers differing in side-chain asymmetric center configurations have been reported, with expected coupling constants previously discussed (Stoldt et al. 1997; Bajguz and Tretyn 2003). Identifying these epimers remains complex and is often done chromatographically against a standard (Yokota et al. 1987).

2.2.4.3 Direct Spectroscopic Analysis and Computational Workflows

Some analogues reveal a shorter side chain that may contain asymmetric centers, such as C_{21}-ecdysteroids with an alcohol function at C-20 (see **14β,20(s)-14-deoxy-20-hydroxy-poststerone** and **14β,17α-14-deoxypoststerone**). The configuration of this center is quite challenging to address and is more commonly encountered in pregnane-type steroids (Tóth et al. 2021). Two such ecdysteroids have recently been reported: one with a *cis*-fused C/D ring junction (unnamed compound 2) and the other featuring a $\Delta^{14,15}$ functionality (bathoristerone) (Tóth et al. 2021). In both cases, relevant coupling constants related to H-20 were resolved after selective decoupling the CH_3-21 protons. For the *cis* C/D compound, ROE contacts between H-20 and both H-17α and CH_3-18, as well as the magnitude of $J_{H_{17}\text{-}H_{20}}$ (2.2 Hz), supported *S* configuration at C-20. These data align with NMR studies of C-20 substituted pregnanes, which determined the conformational preferences of both epimers at C-20 (Marat et al. 1993). Similarly, the 20*S* configuration of the unnamed compound was assigned based on ROE contacts between CH_3-18 and both H-20 and CH_3-21, as well as between CH_3-21 and H-12β. The elevated magnitude of $J_{H_{17}\text{-}H_{20}}$ ($\approx$ 10 Hz) denoted their antiperiplanar arrangement. Once again, these NMR data are consistent with those reported for analogous $\Delta^{14,15}$-containing 20-hydroxypregnanes (Marat et al. 1993). A few ecdysteroids feature an OH group at C-17 instead (e.g., **callecdysterols A-C**). The β-orientation of the hydroxyl group at C-17 was assigned based on NOE correlations between H-17 and H-12α and H-16α (Hang et al. 2015).

Cyathsterones C and D have highly branched side chains with a hydroxy group at C-22, a methyl group at C-24 for cyathsterone C and at C-23 for cyathsterone D, and an additional stereocenter at C-25 due to C-26 hydroxylation. Despite the freely rotatable nature of the bonds in the side chain, the configurations were determined using NOE correlations between the various asymmetric centers (Yang et al. 2022). Some branched analogues exhibit side-chain unsaturation, resulting in trisubstituted ethylidene-type double bonds (e.g., (24Z)-29-hydroxy-24(28)-dehydromakisterone C) (Vokáč et al. 2002). In such cases, the configuration is easily determined via NOE correlations. For the aforementioned example, the *Z* configuration was assigned based on NOE correlations between H-28 and H-22/H-23, and between H_2–29 and H_3–26 and H_3–27. A trisubstituted double bond also appears in **$\Delta^{24,25}$ dehydroinokosterone**. Its configuration was determined using NOE correlations between olefinic protons H-24 and H_2-26, and between H_2-23 and H_3-27 (Kumpun et al. 2011).

Chiroptical considerations have sometimes been used to assign C-20 and C-22 configurations. An instance of the octant rule applied to C-20 (when C-22 appears as a carbonyl) was reported by Tóth et al. for **22-dehydro-20-deoxy-ajugasterone C** (Takács et al. 2010). More recent studies on **spectasterone A**, **14-*epi*-polypodine B** (Jiang et al. 2021), and **22-oxo-hydroxyecdysterone** combined DP4+ probability analysis of DFT-NMR calculations with Mosher's method for the assignment of C-20 and C-22 absolute configurations (Park et al. 2023; Franco et al. 2023).

2.2.5 Spectroscopic Features and Strategies Related to Cyclic Side Chains

Ecdysteroid side chains frequently undergo oxidative cyclization, lactonization, and etherification reactions that generate complex polycyclic and heterocyclic appendages. These diversifications include various motifs such as dioxolanes, epoxides, lactones, lactols, tetrahydrofurans, as well as rare cycles. The following sections summarize established derivatization strategies, spectroscopic diagnostic criteria, and structural exemplars organized by cyclization motif, providing a comprehensive framework for structure elucidation of these structurally diverse natural products.

2.2.5.1 Dioxolane-Containing Side Chains

Various aldehyde-containing compounds react with vicinal diols (Fig. 2.13). Close derivatives of isopropylidene have been reported using alternative alkyl building blocks, including ethylidene (**ajugasterone c 2,3;20,22-diacetonide**) (Odinokov et al. 2005), butylidene (**20-hydroxyecdysone-20,22-butylidene acetal**) (Ling et al. 2003), *para*-hydroxybenzylidene (**paxillosterone 20,22-*para*-hydroxybenzylidene acetal** and **ponasterone A 20,22-*para*-hydroxy-benzylidene acetal**), 5-hydroxymethylfurfural (**niuxixinsterone A** and **serfurosterones A and B**), and 4-hydroxybutyraldehyde (**ajugasterone-20,22-dioxolanes A and B**)

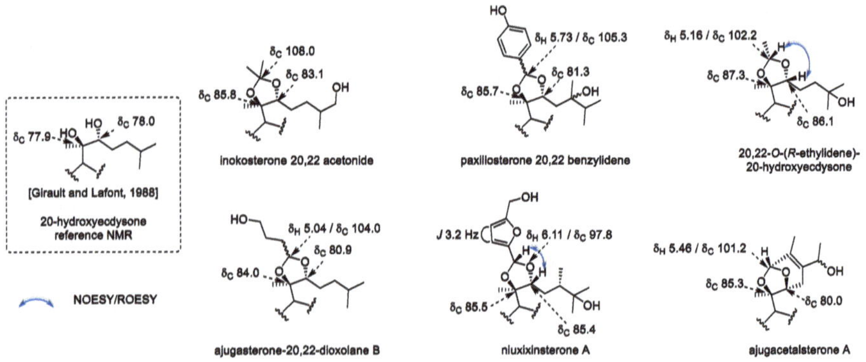

Fig. 2.13 Structurally diverse dioxolane-containing ecdysones, along with the key spectroscopic features needed to determine their (stereo)structure

(Olatunde et al. 2022). The nature and location of these aliphatic chains are easily determined through COSY and HMBC correlations, the latter occurring between the ketalic proton and both C-20/C-22.

Para-hydroxybenzylidene derivatives are rarely reported. However, the 1,4-disubstituted ring is readily identified via characteristic chemical shift values and coupling patterns (Vokáč et al. 1998). The hemiketal carbon configuration could be determined using the NOE correlations between H-1′ and H-22, as well as between aromatic protons and α-oriented CH_3-21 (Fang et al. 2010).

The incorporation of 5-hydroxymethylfurfural (Ranoux et al. 2013) represents a common building block obtained by the elimination of three water molecules from hexoses (see: **serfurosterones A and B**) (Liktor-Busa et al. 2008; Wang et al. 2011). The furanic ring is strongly suggested by aromatic resonances disclosing small coupling constants ($J \approx 3$ Hz). This extraneous unit introduces a new asymmetric center, with its configuration determined based on an NOE correlation between H-1′ and H-22 (Liktor-Busa et al. 2008; Wang et al. 2011).

The side chain of **ajugacetalsterone C** features a unique bicyclic appendage, a dioxabicyclo[3.2.1]octene acetal, where the aldehyde group stems from a C_{26} aldehyde-containing precursor rather than an external unit. Diagnostic chemical shift values at C-26 (δ_H 5.26, δ_C 101.2) and HMBC correlations to C-20 and C-22 demonstrated ether connectivity, with characteristic downfield-shifted signals at C-20 (δ_C 85.3) and C-22 (δ_C 80.0) (Castro et al. 2008). The spectroscopic features underlying the configuration of the asymmetric centers are unclear, as is the reason why the oxygenated functionality at C-22 differs from that of most other ecdysteroids.

2.2.5.2 Epoxide-Containing Side Chains

Various compounds reveal additional side-chain cycles from the etherification of hydroxy groups, forming heterocycles of different sizes (Fig. 2.14). A few epoxide-containing side chains have been reported, with an ether group typically connecting

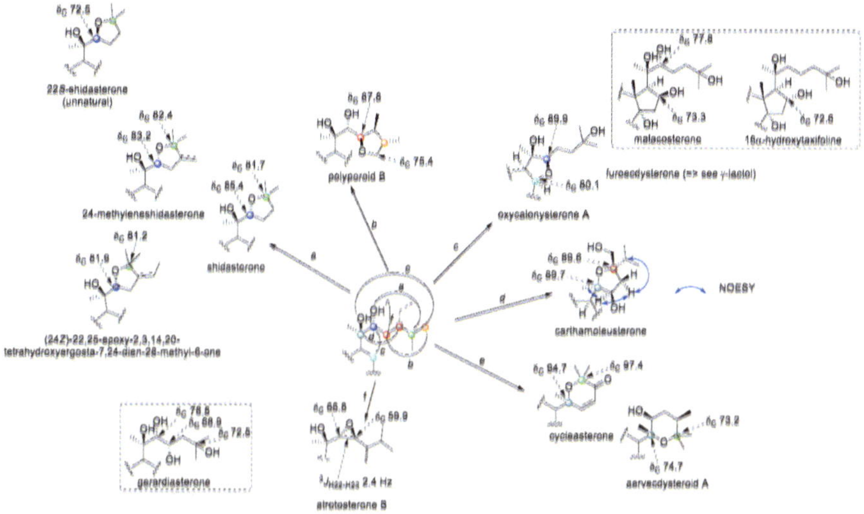

Fig. 2.14 Overview of furan-, pyran-, oxacycloheptane-, and epoxide-containing rings encountered in ecdysteroid side chains

C-22 and C-23 (Ohsawa et al. 1992; Vokáč et al. 1998). The $^{1}J_{C\text{-}H}$ coupling constant identifies such features (see **gymnasterone B**) (Amagata et al. 1998). Characteristic infrared (IR) bands (907 and 880 cm^{-1}) identify epoxide functionality (Ohsawa et al. 1992). Shielded oxymethine protons ($\delta_{H\text{-}22}$ 2.6–3.0 and $\delta_{H\text{-}23}$ 2.5–2.9) and the associated ^{13}C NMR chemical shifts ($\delta_{C\text{-}22}$ 64–66 and $\delta_{C\text{-}23}$ 58–60) confirm the epoxide (Ohsawa et al. 1992; Vokáč et al. 1998). The $^{3}J_{H22\text{-}H23}$ value of 2.4 Hz in **atrotosterone B** indicates *trans* proton orientation (Vokáč et al. 1998).

2.2.5.3 Tetrahydrofuran (THF) C-22-to C-25-Containing Side Chains

The most common side-chain heterocycle is the tetrahydrofuran ring. The ability of the side chain to undergo hydroxylation at various positions, coupled with the variable aliphatic-chain length, which may incorporate C-24 alkyl groups, results in diverse structures with furanic rings. C-22 and C-25 cyclization is the most common strategy for this scaffold. A comparison between **ecdysterone** (two hydroxy groups at C-22 and C-25) and **shidasterone** (ether connection between these sites) revealed downfield shifts of approximately 8 and 6 ppm, respectively, upon ring closure (Hikino et al. 1975). Synthesis confirmed that C-22 exhibits the typical *R* configuration (Roussel et al. 1995). The unnatural 22*S*-containing shidasterone exhibited significantly different chemical shifts at C-22 (δ_C 77.9 for (22*S*)-shidasterone vs. 83.9 for shidasterone) and different signals for *gem*-dimethyl groups (δ_H 1.56 and 1.58 vs. 1.24 and 1.25 for shidasterone) (Baltayev et al. 1999). Structural variants closely related to these compounds include **24-methyleneshidasterone,** with an exomethylenic functionality at C-24, identifiable by characteristic chemical shift values

($\delta_C \approx$ 103–104 ppm, $\delta_H \approx$ 4.8–4.9). HMBC and NOE/ROE correlations establish this group's location (Simon et al. 2007; Jia et al. 2010). A close analogue, **24Z,22,25-epoxy-2,3,14,20-tetrahydroxy-28-methyl-ergosta-7,24(28)-dien-6-one**, possesses a similar ring structure and ethylidene substituent at C-24. Although the configuration of this newly introduced double bond could be determined from NOE correlations of H-28 and H_3-29, it is unclear how the chosen configuration was established (Jin et al. 2025).

2.2.5.4 Alternative THF Couplings-Containing Side Chains

Alternative THF ring couplings have been reported. **Polyporusterone I/polyporoid B** features a THF ring from etherification between C-23 and one of the terminal *gem*-dimethyl groups. The branched nature of the side chain (a methyl group at C-24) yields two contiguous methyl groups whose orientations were tentatively determined via NOE contacts (Sun and Yasukawa 2008). In **(23*R*, 24*R*, 25*R*) 23, 26-epoxy-3β, 14α, 21α, 22α-tetrahydroxyergost-7-en-6-one**, the THF configuration was determined from NOE contacts between H-24 and H_3-27, and between H-25 and H_3-28. However, the 23*R* configuration relied solely on NOE contact between H-17 and H-23, which may prove unreliable. Similarly, the reported 20*S* configuration is unexpected, as it was assigned based on an NOE contact between H_3–18 and H_3–21, a feature also anticipated for the 20*R* epimer. Thus, this assignment merits caution (Zhou et al. 2007).

An alternative THF formation involves C-16 and C-22 ether connectivity, creating a ring junction between two five-membered rings (Sun and Yasukawa 2008; Tóth et al. 2023). The chemical shift of C-22 upon formation of the C-16-*O*-C-22 connection is deshielded by approximately 13.5 ppm, as in **oxycalonysterone A** (Tóth et al. 2023). Although these modifications create additional stereocenters, the largely conserved C-17 and C-22 configurations across ecdysteroids and the near universal *cis* junction between two five-membered rings in natural product chemistry (which admits an extremely limited number of exceptions (Köck et al. 2007)) make configurational assignment straightforward. Biosynthetic hypotheses are supported by NOE/ROE contacts (Tóth et al. 2023). **Furoecdysterone** exhibits a similar structure, but produces a γ-lactol-type product, which will be discussed in the relevant section (Wang et al. 2024). **(20*S*, 22*R*, 24*R*)-16, 22-epoxy-3β, 14α, 23β, 25-tetrahydroxyergost-7-en-6-one** is a related compound with two additional asymmetric centers: a C-23 hydroxy group and a C-24 methyl group (Zhou et al. 2007). The *cis* D/E ring junction was determined from H_3-21/H-16 and H_3-21/H-17 NOE crosspeaks. The C-23 configuration was determined from the $^3J_{H22\text{-}H23}$ value and molecular dynamics and mechanics calculations in CHARMM force field. The C-24 assignment appears fragile, relying on an NOE contact between H-24 in the flexible side chain and distant H_3-18 (Zhou et al. 2007).

Carthamoleusterone reveals a peculiar THF ring with the ether bridge connecting C-20 and C-24 (Buděšínský et al. 2008). This connection, involving two hydrogen-deprived positions, precluded direct 2D NMR evidence. The authors used in situ TAI (trichloroacetyl isocyanate)-acylation methodology to remedy this (Samek and Buděšínský 1979). Of all the candidate structures, only the THF ring between C-20 and C-24 with one primary and one secondary OH group was consistent with observed acetylation shifts (Buděšínský et al. 2008).

2.2.5.5 Tetrahydropyran (THP)-Containing Side Chains

Cycleasterone A appears to result from etherification between C-20 and C-25, yielding a rare six-membered heterocycle. The hydrogen-deprived status of these positions precluded direct HMBC evidence, but hydrogen deficiency and *O*-bonded status supported the aforementioned connection (Wang et al. 2019). **Aervecdysteroids A** and **B** are further examples; however, the similar NMR chemical shift values in the open diol form and presumed pyran ring form are surprising (Saleem et al. 2013).

2.2.5.6 *δ*-Lactone-Containing Side Chains

Lactonization results in additional side-chain rings (Fig. 2.15). The size of these rings and the nature of their substituents depend on available side-chain carbons (often involving C-24-derived substituents). Many compounds reveal *δ*-lactones, formed by lactonization between OH-22 and a terminal side-chain carbon that has been oxidized to a carboxylic acid, typically derived from oxidation of one of the *gem*-dimethyl groups, as exemplified by **ajugalactone** and **capitasterone**.

Capitasterone shows a characteristic IR band at 1730 cm^{-1} for a δ-lactone (Takemoto et al. 1968). That C-22 forms the lactone was demonstrated by H-22's insensitivity to acetylation (Takemoto et al. 1968). The C-22 configuration was assigned as *R* based on ECD Cotton effects comparable to formerly assigned analogues (Nakanishi et al. 1970b). This configuration results in axially oriented H-22, indicated by *trans*-diaxial coupling constants (Calcagno et al. 1996; Zhu et al. 2009). Capitasterone-type side chains typically feature an ethyl residue at C-24 and an asymmetric center at C-25. The relative stereochemistry of δ-lactone moieties has not been assigned consistently. Assuming that *Ajuga reptans*-derived ecdysteroids are derived from the 24*S*-configured **clerosterol**, Calcagno et al. surmised a 24*S* configuration (Calcagno et al. 1996). Conversely, the authors of **lygodiumsteroside A** used a NOE contact between H-22 and H-24 to establish its structure, while the lack of an NOE contact between H-24 and H-25 supported their antifacial orientation (Zhu et al. 2009).

Zenkersterone differs from **capitasterone** only by a methyl group instead of ethyl side chain at C-24 (Tabekoueng et al. 2020). Its ring relative configuration was determined from NOESY correlations between H-22/H-24, H-24/H-25, and H_3–27/H_3–28 (Tabekoueng et al. 2020). **Ajugacumlactone** has a longer side chain with a propyl group at C-24 (Olatunde et al. 2022). The presence of a $\Delta^{24,25}$ functionality

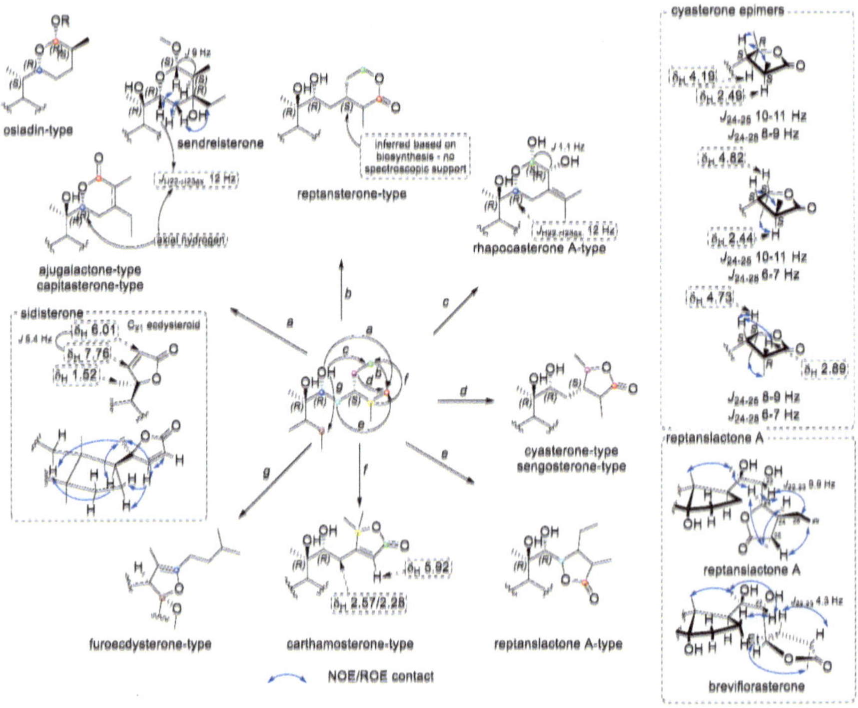

Fig. 2.15 Chemical logic underlying the structural diversity of ecdysteroids with a cyclized side chain. Where available, spectroscopic features that underpin some interesting stereochemical assignments are presented

simplifies the process of elucidating the structure. However, it is notable that C-22 was left stereochemically unassigned, and moreover that the coupling constant of H-22 (J = 3.8 Hz) is lower than expected for an axially oriented H-22 (Olatunde et al. 2022), as is the case with other δ-lactones containing a $\Delta^{24,25}$ function (Calcagno et al. 1996).

Leuzeasterone exhibits a hydroxypropyl group at C-24. C-22 is the sole δ-lactone-ring stereocenter due to $\Delta^{24,25}$ functionality (Vokáč et al. 2002). In line with its β-axial orientation, H-22 exhibits an elevated *trans*-diaxial coupling constant relative to one of the C-23 diastereotopic methylene protons.

An alternative ***δ*-lactone** forms between C-26 or C-27 (one *gem*-dimethyl converted to carboxylic acid) and an alcohol group at C-29, as in **reptansterone** (Calcagno et al. 1996). The relative configuration of the lactonic cycle has not been determined spectroscopically (only the *S* configuration at C-24 was proposed based on biogenetic considerations) (Calcagno et al. 1996). Additionally, ^{1}H and ^{13}C NMR signals at C-24 and C-25 were not reported by Calcagno et al.

2.2.5.7 *δ*-Lactol-Containing Side Chains

Sendreisterone and **ajugacetalsterone A** are δ-lactol equivalents of **capitasterone/ajugalactone**-type side chains (Coll et al. 2007; Ványolós et al. 2009). The hemiketal position C-26 is easily identified by NMR chemical shifts ($\delta_C \approx$ 95–105 ppm, δ_H 4.5–5.5 ppm). The elevated $^3J_{H22\text{-}H23ax}$ assigned the *R* configuration at C-22 in both cases. For **sendreisterone**, other lactone asymmetric center configurations were inferred from NOE crosspeaks between H-22/H-26, H-26/H_3-27, H-23$_{ax}$/H-25, and H-23$_{eq}$/H-29, with $^3J_{H25\text{-}H26}$ consistent with *trans*-diaxial disposition (Ványolós et al. 2009). Similar NMR landmarks were noted for **ajugacetalsterone A**, with coupling constant values similar to those of **sendreisterone**. However, no spectroscopic evidence supported the C-24 configuration; the authors favored an equatorial ethyl due to steric constraints, opposite to **sendreisterone** (Coll et al. 2007). Several further δ-lactols have been reported since, differing in configurational assignments. Hu et al. reported two ecdysteroids with different C-25 methyl orientations based on NOE considerations. Some NMR features are difficult to reconcile with the retained configuration, particularly the $^3J_{H25\text{-}H26}$ value showing *trans*-diaxial magnitude despite protons being determined synfacially oriented (Hu et al. 2014). Subsequent work reached similar conclusions through strict analogy of new analogues (notably **cyathsterone B**) without NOE/ROE consideration. Notably, although the authors intended to define similarly configured δ-lactol rings, the ethyl group orientation differed (Yang et al. 2022).

Some **δ-lactols** originate from similar reactions with non-branched C-24 side chains. **Osladin** exemplifies this; its structure was revised through X-ray analysis and total synthesis, leading to the reassignment of the absolute configuration of its ring component (Jizba et al. 1971; Yamada et al. 1992; Yamada and Nishizawa 1995). The NOE correlation between H-22 and H-26 revealed synfacial orientation, while the *trans*-diaxial coupling constant (J = 8.8 Hz) between H-25 and H-26 supported antifacial disposition. Synthesis of initial and revised **osladin** structures revealed significantly different ^{13}C NMR chemical shifts at C-26 and the rhamnoside C-1, consistent with previously reported glycosylation shifts (Nishizawa et al. 1994a). Closely related heterosides (**polyposides A-C**) described to date show excellent agreement with these key NMR values (Kim et al. 1988; Kim and Kinghorn 1989). However, these compounds also required structure revision (Nishizawa et al. 1994b). Under P450 monooxygenase action, **20,26-dihydroxyecdysone** oxidizes to form a 26-aldehyde, which then converts to a *δ*-lactol with C-25 hydroxy group. The approximately 4 ppm downfield shift of C-22 indicates ring closure, while H-22/H-26 NOE establishes synfacial orientation. The methyl group orientation at C-25 determined the entire cycle's configuration, addressed by calculating expected ^{13}C NMR shifts based on published data of related compounds. This product converts to a minor form (3:1 equilibrium), differing in C-26 configuration, revealing notable H-22 downfield shift (3.86 ppm vs. 3.54 ppm) due to its synaxial orientation with the 26-OH group (Kayser et al. 2002).

Rhapocasterones A and B exhibit a δ-lactol ring formed by nucleophilic attack from a C-22 hydroxy group and an oxidized C-29. The additional ring in these compounds reveals two adjacent oxygenated groups at C-28 and C-29. The small $^3J_{28-29}$ value (1.2 Hz) and intense NOE correlations between H-22/H-29 and H-29/H-28 suggest synfacial orientation for these groups in **rhapocasterone A**. This was further ascertained by X-ray analysis (Zughdani et al. 2020).

2.2.5.8 γ-Lactone-Containing Side Chains

The most straightforward approach to identifying lactone elements is to observe HMBC crosspeaks between the oxygenated methine and the carbonyl carbon. Even without these, chemical shifts of the oxygenated methine are typically downfield shifted compared with open analogues (Ványolós et al. 2009).

Several ecdysteroids contain γ-lactones, apparently via similar biosynthetic precursors and reaction sequences similar to those proposed for **ajugalactone** and **capitasterone**, but with C-28 rather than C-29 involved in lactonization for the first such derivatives (Calcagno et al. 1996); alternative reactivities have been described since then (Fig. 2.15). IR bands indicative of γ-lactones (1778–1752 cm^{-1}) supported this assignment in early reported derivatives (Hikino et al. 1968).

Different **cyasterone** derivatives contain a **γ-lactone** resulting from ring closure between 28-OH and 26-COOH groups. **Cyasterone** reveals quasi-axial hydrogens at C-24, C-25, and C-28, yielding $^3J_{24-25}$ and $^3J_{24-28}$ values of approximately 11 and 9 Hz, respectively (Okuzumi et al. 2005). 28-***epi*-cyasterone** differs in H-28 resonance (downfield shifted) and lower $^3J_{24-28}$ (approximately 7 Hz) (Calcagno et al. 1996; Okuzumi et al. 2005). NOE contacts between H_3-29/H-25, and H_3-27/H-24 validated its relative configuration, supported by X-ray analysis (Okuzumi et al. 2005). **25-*epi*-28-*epi*-cyasterone** shows lower H-25 coupling constants than cyasterone ($^3J_{24-25}$ = 9.0 Hz; $^3J_{24-28}$ = 6.7 Hz), indicating quasi-axial-quasi-equatorial interactions, and thus a synfacial orientation (Okuzumi et al. 2005). Based solely on these coupling constants, assigning a *syn*/*anti* relationship in γ-lactones remains difficult. H-25 and H-28 chemical shifts provide clearer structural information, particularly in CD_3OD, where the values differ markedly depending on C-25/C-28 configuration, and in $CDCl_3$ for acetylated derivatives (Okuzumi et al. 2005). Nevertheless, lactonic ring proton coupling constants in **sengosterone** and **28-*epi*-sengosterone** align with established guidelines (Yi et al. 2004). Related C_{28} ecdysteroids (**29-norsengosterone** and **29-norcyasterone**) contain branched C-24 methyl groups, reducing **γ-lactone** ring asymmetric centers by one. The $^3J_{24-25}$ value (10.5 Hz) is consistent with the 24*S*, 25*S* configuration of **norcyasterone**, as determined by X-ray analysis of its 2,3,20,22-diacetone derivative (Camps et al. 1982).

Other derivatives, such as the C_{29}-type **reptanslactones A and B**, contain an alternative five-membered **γ-lactone** that involves both a 23-OH group and, as usual, a 26-carboxy group. The five-membered lactone ring is evidenced by an HMBC correlation from H-23 to C-26. The H-22/H-23 coupling constant defines their antiperiplanar arrangement (J = 9.9 Hz), with NOE crosspeaks (Fig. 2.15) defining the *trans* arrangement of Me-27 and Et-24 (Ványolós et al. 2009). PM3 semi-empirical calculations compared candidate structures of **reptanslactone A** differing at C-23/C-24 configurations against the experimental NOESY spectrum and NMR data. Of all candidates, only the 22*R*, 23*R*, 24*S*, 25*S* structure was consistent with NOE correlations between H-23/H-17, and the antiperiplanar interaction requirement between H-22/H-23. The configuration of the structurally similar **breviflorasterone** was determined from key NOE correlations (Fig. 2.15), with the H-22/H-23 coupling constant (J = 4.3 Hz) denoting their gauche arrangement (Ványolós et al. 2009).

Carthamosterone and related analogues represent unusual ecdysteroids where the side chain lactone involves a C-25 hydroxy group and C-29 carboxylic acid. Besides an IR band at 1745 cm^{-1} indicating α,β-unsaturated **γ-lactone**, its location was initially proposed from downfield ^{1}H NMR chemical shifts of diastereotopic methylenic protons H_2–22 (δ_H 2.57/2.28). The ethylenic proton H-28 chemical shift (δ_H 5.92) is consistent only with an α-position in an α,β-unsaturated carbonyl system (Girault et al. 1988). **Acaulesterone** has a similar lactone ring with two contiguous hydroxy groups at C-24 and C-28 (Zughdani et al. 2020). Although acetylation shifts demonstrated hydroxy group location, the C-24/C-25 configurations of the peracetate derivative remained unassigned.

Furoecdysterone is a unique example of γ-lactol that exhibits a $\Delta^{20,22}$ functionality and a 16α-methoxy group. The methoxy orientation is expected given the likely *cis* D/E ring junction and is supported by NOE crosspeak with H-17α (Wang et al. 2024). The formation of close analogues via ether connections between C-16 and C-22 had already been discussed in the tetrahydrofuran section of this review. In that case, it can be presumed that C-16 disclosed a ketone moiety before undergoing a nucleophilic attack by OH-22.

Sidisterone, the first reported C_{24} ecdysteroid, showed IR evidence of a γ-lactone and a side chain with a single low-field methyl signal (δ_H 1.51/ δ_C 25.2), and two ethylenic signals with a large chemical shift difference (δ_H 7.76 (d, J = 5.4 Hz) and δ_H 6.01 (d, J = 5.4 Hz)) (Girault et al. 1996), consistent only with a γ-lactone-type side chain. The side chain methyl orientation was determined using displayed NOE correlations and molecular mechanics-derived models. This structure was later confirmed by total synthesis (Zhylitskaya et al. 2011). A closely related derivative, **11-hydroxysidisterone**, was subsequently reported, but the C-18 configuration remained unassigned (Guibout et al. 2015).

2.2.5.9 Rare Cyclization Motifs

The structurally intriguing **diploclidine** contains a pyridine ring between its side chain and the tetracyclic ecdysteroid frame (Jayasinghe et al. 2003a). The reactivity that led to this compound remains unclear, but it may proceed from a 12,22-diketone derivative. This would be consistent with the known reactivity of 1,5-dicarbonyl compounds with amines (Cai et al. 2017). The presence of a pyridine ring is evident

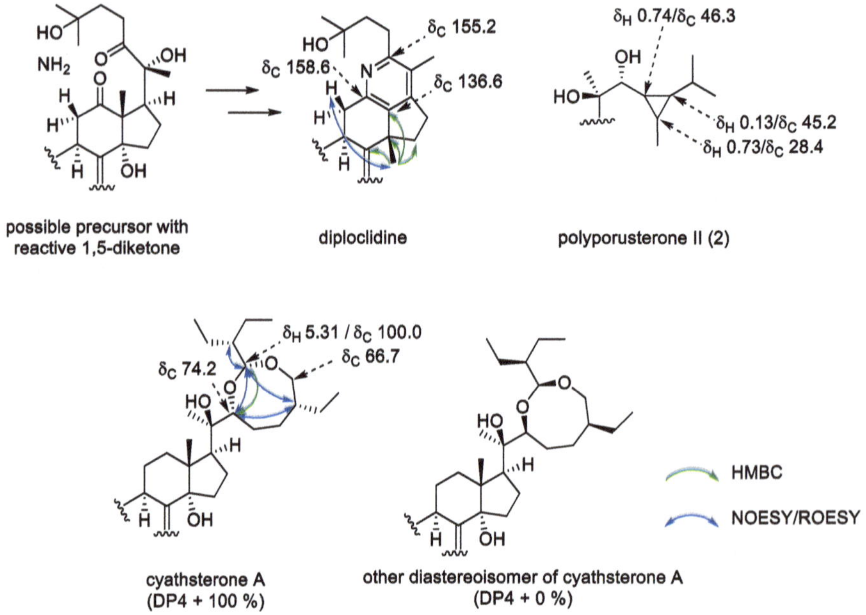

Fig. 2.16 Overview of rare cyclization motif encountered in ecdysteroid side chains

from the numerous sp^2-hybridized carbons in the structure, particularly the *N*-bonded ones (Fig. 2.16). UV absorption at 270 nm has also been suggested to be related to the pyridine ring.

Polyporusterone II discloses a cyclopropane-containing side chain (Zheng et al. 2004). This moiety was first inferred from the degree of unsaturation and the unusually upfield ^{1}H NMR signals at δ_H 0.74, δ_H 0.73, and even δ_H 0.13 (Zheng et al. 2004). Sterols with similar side chains reported from marine sources reveal consistent NMR landmarks (Giner and Djerassi 1992).

Complex polycyclic side chains require advanced computational approaches for configurational assignment, as in the case of **cyathsterone A**. Although this compound has an alternative counterpart containing an aldehyde, the biosynthetic pathway leading to it is probably related to that leading to the dioxolane compounds discussed earlier. The gross structure of this intriguing side chain can be easily identified through the key COSY and HMBC correlations outlined in Fig. 2.16, while the hemiketal proton lies within the typical range reported for dioxolanes. However, it appears that the positions involved in ether connections are less deshielded than those in dioxolanes. **Cyathsterone A**'s eight-membered ring relative configuration was determined using NOE correlations between H-22 and H-25, and between H-22 and H-28. Although the asymmetric center at C-32 is distant from the aforementioned centers via a freely rotatable bond, NOE correlations between H-28/H-32 and H-22/H-32 determined the synfacial orientation of H-32 (Yang et al. 2022). A DP4+ workflow subsequently assigned the absolute configuration of these blocks in relation to the rest of the ecdysteroid appendage. This process successfully retained the diastereoisomeric candidate with a 22*R* carbon, which is presumed to be constant across ecdysteroids (Yang et al. 2022).

2.2.6 Spectroscopic Features Related to Various Substituents of the Ecdysteroid Skeleton

A variety of different substituents has been reported on ecdysteroids (Fig. 2.17). Most of them are common building blocks in natural product chemistry and do not create major challenges for structure elucidation. They are generally attached through hydroxy groups located either on the tetracyclic nucleus or the side chain, most commonly forming *O*-heterosides or ester functionalities. In the latter case, these connect ecdysteroids to biosynthetically diverse acyl units.

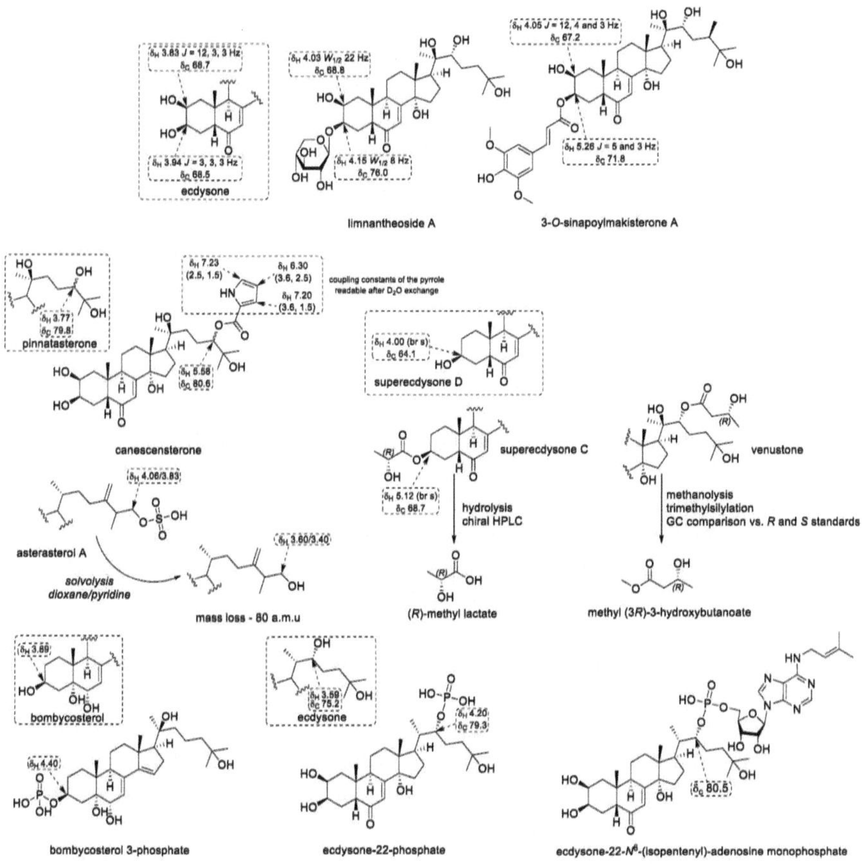

Fig. 2.17 Some miscellaneous substituents along with key spectroscopic features and strategies to assign their structures

2.2.6.1 *O*-Heterosides: Detection and Assignment

The identification of heterosides is often initiated by mass spectrometry, where the molecular mass suggests attachment of a pentose or a hexose. In ^1H NMR spectra, sugars are usually recognized by additional resonances in the region typical of protons on oxygenated carbons (δ_H 3.3–3.8) and, most diagnostically, an anomeric signal at lower field. In ^{13}C NMR spectra, additional signals are expected in the 50–110 ppm region, with the anomeric carbon typically appearing near ~100 ppm.

The glycosylation site is classically inferred from glycosylation shifts: a modest change in the ^1H NMR chemical shift of the involved oxymethine proton (often ~0.1–0.5 ppm) and more pronounced changes in the corresponding ^{13}C chemical shift (often ~5–12 ppm) (Maria et al. 2005). Sugar identity is then usually established from coupling constants within the sugar spin system. When needed, sugar absolute configuration may be determined after acid hydrolysis by measuring its specific rotation after isolation (Nakamura et al. 2010; Wu et al. 2010). Alternative approaches can also be used, such as ^1H NMR analysis after hydrolysis in D_2SO_4 (Giner et al. 2016).

Septanoecdysone contains an unusual oxepan ring attached at C-22 position of 20-hydroxyecdysterone (Ben Nejma et al. 2015). Despite not being a sugar unit, it does bear some resemblance to one. This was deduced from an HMBC correlation between the oxymethine H-22 and the hemiketal carbon C-1 of the oxepan (δ_C 105.2). The ^1H and ^{13}C NMR chemical shift values at C-22 are reported to be identical in both free and conjugated forms. The oxepan's relative configuration (four contiguous asymmetric sites) was determined solely from NOE correlations; legible coupling constants were not discussed, though they may not support the claimed synfacial proton orientation (DeMatteo et al. 2005).

2.2.6.2 *O*-Acyl Esters: Acetates, Aromatic, and Aliphatic Acyl Groups

Acetyl groups are frequently encountered on ecdysteroid hydroxyls and can be localized through characteristic acetylation shifts (Hikino et al. 1975) and diagnostic HMBC correlations between an oxygenated methine proton and an ester-type carbonyl carbon. Beyond acetates, benzoates and cinnamate, coumarat, or sinapate esters are clearly recognized by the appearance of additional aromatic and/or olefinic resonances (Russell et al. 1971; Russell et al. 1972; Baltaev et al. 1987; Saatov et al. 1987). A glycolate-type ester has also been reported (Bückmann et al. 1986). With modern 2D NMR methods, the attachment site of these substituents is typically confirmed by HMBC correlations (Zeng et al. 2022; Kakumu et al. 2023).

A broad set of long-chain fatty acyl esters of ecdysones has been described in insects. Early characterization approaches relied heavily on HPLC separation of esterase-derived hydrolysis products and comparison with fatty acid references (Hoffmann et al. 1985). In NMR spectra, such compounds reveal a wealth of additional aliphatic methylene signals and an additional ester-type carbonyl resonance, and the acylation site (often C-22) can be located using characteristic acetylation shifts at that position (e.g., $\delta_{C\text{-}22}$ 78.0 and $\delta_{H\text{-}22}$ 5.4–5.5 vs. $\delta_{C\text{-}22}$ 74.4 and $\delta_{H\text{-}22} \approx 4.0$) (Dinan 1988).

Ecdysteroids bearing an (*R*)-methyllactic acid group have been reported from *Dianthus superbus* (Hou et al. 2023). In that work, the methyllactic acid absolute configuration was determined after basic hydrolysis and chiral HPLC analysis by comparison with *R*- and *S*-methyllactic acid standards. Historically, before routine access to 2D NMR, the location of side chain substituents was sometimes inferred from characteristic mass spectrometric fragments. Similarly, **venustone** contains an ester of hydroxybutanoic acid and therefore has a chiral substituent. The absolute configuration was determined following methanolysis, after which a GC comparison was performed using the trimethylsilylated derivatives of methyl (3*R*)- and methyl (3*S*)-3-hydroxybutanoate (Roth et al. 1995).

A pyrrole-carboxylate substituent at C-24 is present in **canescensterone** and was initially detected using acetylation shift reasoning (Suksamrarn et al. 1995). The substituent was supported by three additional aromatic resonances displaying coupling constants consistent with a pyrrolic system, with clearer splitting after D_2O exchange. The presence of this moiety was further confirmed by an EI-MS signal at *m/z* 94 (COC_4H_4N).

2.2.6.3 Sulfates

Sulfate conjugates are rare among ecdysteroids (Saatov et al. 1984; De Marino et al. 1997). They are commonly inferred from mass spectrometry by an *S*-containing molecular formula and a characteristic sulfate neutral loss of 80 Da, observed as $[M+H-80]^+$ or $[M-H-80]^-$ (Perrone et al. 2007; Otogo N'Nang et al. 2023). Sulfation generally causes downfield shifts of the ^1H and ^{13}C NMR resonances at the substituted carbon compared with the corresponding alcohol (Govindan et al. 1994). For example, the H-22 proton of alpha-ecdysone-22-sulfate experiences a significant downfield shift ($\Delta\delta = 1.11$ ppm) when transitioning from ecdysone ($\delta_H = 4.10$ ppm) to ecdysone-22-sulfate ($\delta_H = 5.21$ ppm). The authors validated the position of the sulfate group by identifying its peracetylated derivative as the 2,3,25-triacetate (Saatov et al. 1984). Compared with the NMR data of α-ecdysone, the C-22 signal of alpha-ecdysone-22-sulfate shifted downfield by +8.5 ppm (δ_C 82.9). C-20, C-23, and C-24 were found to shift upfield by 3.5, 1.9, and 1.4 ppm, respectively, in line with previous reports (Kitagawa et al. 1978; Kuramoto et al. 1981). For **asterasterols**, which feature a sulfate group at C-26, solvolysis (dioxane/pyridine) led to a high-field shift of the diastereotopic methylenic protons H_2–26, supporting sulfate loss (De Marino et al. 1997). The occurrence of a sulfate group had been related to IR absorption at 1235 cm^{-1} (Saatov et al. 1984).

2.2.6.4 Phosphates and Nucleotide Conjugates

Phosphorylated ecdysteroids can show subtle, reproducible changes in ^1H NMR chemical shifts. For example, **ecdysone 3-phosphate** was reported to differ from **ecdysone** by an upfield shift of approximately 0.2 ppm for H_3–19 and a downfield shift of 0.05 ppm for H-3 (Grebenok et al. 1994), although the magnitude and pattern of shifts depend on the phosphorylation site. Isaac et al. identified 3-acetyl-ecdysone-2-phosphate after hydrolysis of the phosphate group, which was found to release 3-acetyl-ecdysone. The comparison between the NMR data of ecdysone-2-phosphate and those of ecdysone showed that H-2 and H-3 were shifted downfield by 0.52 and 0.22 ppm, respectively (Isaac and Rees 1984). Comparing **bombycosterol** and **bombycosterol 3-phosphate** reveals that H-3 shifts downfield by 0.5 ppm upon phosphorylation (Hiramoto et al. 1988). Both ecdysone 22-phosphate and 2-deoxyecdysone-22-phosphate are present in the eggs of *Schistocerca gregaria.* Once again, the electron-withdrawing nature of the phosphate ester group causes a downfield shift of H-22 (δ_H 4.20–4.22 versus 3.60 for ecdysone) (Isaac et al. 1982). The chemical shift of C-22 was also found to be downfield shifted (δ_C 79.3 vs 75.3 for ecdysone). Peak broadening of the C-22 and C-23 signals was noted due to ^{13}C-^{31}P coupling ($^2J_{COP}$ ca. 4.7 Hz, $^3J_{CCOP}$ ca. 3.1 Hz). Analysis of such phosphate esters in negative ion mode fast atom bombardment mass spectrometry (FAB-MS) revealed ion peaks at *m/z* 79 (PO_3^-) and 97 ($H_2PO_4^-$) (Hiramoto et al. 1988). **Ecdysone 22-*N*⁶-(isopentenyl)-adenosine monophosphate**, isolated from the locust *Locusta migratoria*, represents a distinctive ecdysteroid conjugate (Tsoupras et al. 1983). Its structure was elucidated using mass spectrometry, which revealed fragments indicative of adenosine, N^6-(isopentenyl) adenosine, and phosphoribose, while the C-O-P linkage was reported to be too labile under the acquisition conditions used. Additionally, ^1H NMR showed signals attributable to the extraneous N^6-(isopentenyl)-AMP unit (two vinylic methyl groups, one additional vinyl proton, and two additional aromatic protons), and ^{31}P NMR confirmed phosphate functionality ($\delta_P = 2.0$ ppm). The C-22 carbon resonance was reported to shift downfield by ~6 ppm upon phosphorylation (Tsoupras et al. 1983), consistent with expected phosphate-induced effects (Mantsch and Smith 1972).

2.3 Mass Spectrometry-Based Profiling and Detection

Beside the important role of NMR and chiroptical techniques in characterizing ecdysteroids, MS techniques, often coupled with chromatography (GC-MS or LC-MS), have played a complementary role over the last 40 years. Early work relied on electron impact (EI) MS post-derivatization, while softer atmospheric pressure ionization (API: ESI, APCI) and tandem MS (MS/MS) emerged in the 1990s for underivatized, polar forms. In this chapter, we will focus our survey on how recent advances in mass spectrometry and chemoinformatics have impacted ecdysteroid profiling. We will not describe the use of MS for ecdysteroid structure elucidation in detail (fragmentation patterns for identification), as this has been extensively described elsewhere (Dinan 2001; Destrez et al. 2009; Yusupova et al. 2022).

2.3.1 Computational Mass Spectrometry Applied to Ecdysteroids

Recent developments in computational mass spectrometry, particularly molecular networking (MN) (Watrous et al. 2012), have revolutionized the dereplication and discovery of natural products by visualizing chemical relationships based on MS/MS spectral similarity (Fox Ramos et al. 2019). This is particularly relevant for ecdysteroids because extracts often contain many closely related polyhydroxylated steroids whose LC-MS features overlap, while their MS/MS spectra still retain family-level similarity that MN can exploit.

A representative ecdysteroid-focused example is the study on *Fibraurea recisa* roots by Kakumu et al. (2023), who combined untargeted metabolomics-based molecular networking with MS2LDA motif annotation to prioritize sinapoyl-functionalized metabolites (van der Hooft et al. 2016). Their MN/MS2LDA workflow revealed a characteristic fragment "motif" linked to multiple sinapoylated features, which was then used as a targeting handle for isolation. Guided by this computational prioritization, they isolated two new sinapic acid–ecdysteroid hybrids, 3-*O*-sinapoyl makisterone A, and 2-*O*-sinapoyl makisterone A and elucidated their structures using HR-ESI-QTOF-MS/MS, NMR, and chemical degradation plus optical rotation. Beyond annotation, this study illustrates how MN plus substructure-aware motif mining (MS2LDA) can shift ecdysteroid work from "find-and-identify" to "network-guided discovery," especially for conjugated derivatives (here, sinapate esters) that may be missed by classical targeted screening. A notable application of Feature-Based Molecular Networking (FBMN) (Nothias et al. 2020) was reported by Park et al. for the profiling of *Ajuga spectabilis* (Park et al. 2023). By integrating FBMN with Ion Identity Networking (IIN) (Schmid et al. 2021), the authors constructed a spectral network that grouped known ecdysteroids (such as **20-hydroxyecdysone**, **polypodine B**, and **ajugalactone**) into distinct clusters. This networking strategy highlighted unique nodes connected to the main ecdysteroid cluster, guiding the targeted isolation of two new compounds, **spectasterones A** and **B**. The network topology allowed the researchers to infer structural similarities and prioritize these "outlier" nodes which possessed unique side-chain modifications. Similarly, LC-MS/MS fingerprinting combined with clustering analysis has been employed to map the ecdysteroid content of *Cyanotis arachnoidea*, facilitating the comprehensive annotation of over 50 components in native and commercial extracts (Laczkó et al. 2025). These studies demonstrate that computational workflows such as GNPS (Global Natural Products Social Molecular Networking) can effectively overcome the challenges of traditional targeted analysis by revealing the full "steroidome" of a sample (Chen et al. 2024).

2.3.2 Alternative Ionization Pathways Ecdysteroid Detection

While protonated molecules $[M+H]^+$ are typically targeted in standard LC-MS workflows, alternative ionization pathways involving metal adducts or charge-remote mechanisms offer complementary structural information, particularly for isomer differentiation and side-chain characterization.

2.3.2.1 Charge-Remote Fragmentation (CRF)

Charge-remote fragmentation (CRF) is a gas-phase dissociation mechanism where bond cleavages occur physically remote from the charge site, often driven by high-energy collisions. This behavior is particularly valuable for ecdysteroids conjugated with charged groups, where the charge is sequestered, allowing the steroid skeleton to undergo informative cleavage. Ikeda et al. (1995) demonstrated the utility of CRF for **ecdysteroid 22-phosphates**. In negative ion mode FAB-MS/MS, the $[M-H]^-$ ions of these conjugates underwent charge-remote fragmentation along the side chain, yielding a series of product ions that allowed for the precise localization of the phosphate group and the determination of the side-chain structure (Ikeda et al. 1995). Similar CRF mechanisms have been exploited for steroid sulfates using high-energy MALDI-TOF/TOF, providing diagnostic cleavage at ring junctions and side-chain branch points that are often silent in low-energy CID spectra (Yan et al. 2014).

2.3.2.2 Silver-Cationized Fragmentation

More recently, silver cationization ($[M + Ag]^+$) has emerged as a powerful tool for analyzing steroids that are difficult to protonate or resolve as isomers. Sharma et al. established that silver ions coordinate with steroid olefins and oxygen functionalities, often bridging specific sites (e.g., between C-3 and C-5 or across the side chain) (Sharma and Lanekoff 2023; Sharma et al. 2024). Upon collision-induced dissociation (CID), these silver adducts yield distinctive fragmentation patterns driven by both charge-retention and charge-remote mechanisms (Sharma and Lanekoff 2023; Sharma et al. 2024). Unlike protonated species which often lose water non-specifically, silver-cationized steroids produce diagnostic cross-ring cleavages and side-chain losses that are highly sensitive to stereochemistry. This method has been shown to resolve closely related regioisomers and stereoisomers (such as 3-hydroxy-Δ^5 vs. 3-hydroxy-5α steroids) by providing "fingerprint" spectra unique to each isomer. The application of this technique to ecdysteroids holds significant potential for distinguishing complex side-chain isomers and determining the location of unsaturation without the need for extensive derivatization.

2.4 Conclusions and Future Prospects

Ecdysteroids represent an old and remarkably persistent family of natural products that has, in many ways, accompanied every major evolution in the history of structure elucidation. The early era of ecdysteroid chemistry was sustained by carefully constructed spectroscopic empirical arguments, assembled piece by piece through chemical degradation, strategic derivatization, and synthesis of diagnostic fragments or closely related standards. These approaches were laborious and often indirect, yet they imposed an intellectual rigor that remains instructive.

The modern era has substantially accelerated ecdysteroid characterization through high-field NMR, hyphenated LC–MS/MS, and computational chemistry. These tools now permit rapid navigation of complex natural product mixtures, systematic exploration of conformational landscapes, and statistical ranking of candidate structures. Yet, this progress carries a subtle risk: the ease of generating calculated spectra and automated annotations can encourage researchers to rely on single "best-scoring" predictions without engaging sufficiently with classical diagnostic reasoning or the rich legacy of empirical knowledge. For ecdysteroids, the most reliable approach remains integrative: modern computation should sharpen questions and prioritize hypotheses, while classical strategies, such as targeted derivatization, chiroptical analysis, and rigorous spectroscopic cross-validation, continue to provide decisive constraints that automated workflows alone cannot supply.

The future of ecdysteroid research will depend increasingly on data organization and open science as well as on instrumentation. Curated resources such as Ecdybase demonstrate the value of community-driven knowledge management. From such foundations, ecdysteroid data can be systematically disseminated into broader, machine-actionable platforms (for example, Wikidata via LOTUS workflows (Rutz et al. 2022)) to satisfy FAIR (Findable, Accessible, Interoperable, Reusable) requirements and enable cross-disciplinary integration. This vision also underscores an urgent practical need: comprehensive digitization and standardized structuring of primary experimental evidence for ecdysteroids, including MS/MS (such as GNPS) and NMR (e.g., NP-MRD and nmrXiv)-based spectral databases (Neumann et al. 2024; Wishart et al. 2025). Building these resources would not merely accelerate dereplication and prevent redundant discovery; it would also preserve the hard-won insights of preceding decades, ensuring that future methodological innovations strengthen rather than sever intellectual continuity in ecdysteroid chemistry and natural product sciences.

References

Ajayi TO, Srivedavyasasri R, Nyong EE, Odeniyi MA, Moody JO, Ross SA (2019) Two new phytoecdysteroids from *Sphenocentrum jollyanum* Pierre root. Steroids 150:108456

Amagata T, Minoura K, Numata A (1998) Gymnasterones, novel cytotoxic metabolites produced by a fungal strain from a sponge. Tetrahedron Lett 39:3773–3774

Amagata T, Tanaka M, Yamada T, Doi M, Minoura K, Ohishi H, Yamori T, Numata A (2007) Variation in Cytostatic Constituents of a Sponge-Derived *Gymnascella dankaliensis* by Manipulating the Carbon Source. J Nat Prod 70:1731–1740

Bajguz A, Tretyn A (2003) The chemical characteristic and distribution of brassinosteroids in plants. Phytochemistry 62:1027–1046

Baltaev UA (1995) Rapisterone D, a phytoecdysteroid from *Rhaponticum carthamoides*. Phytochemistry 38:799–800

Baltaev UA, Darmograi VN, Abubakirov NK (1987) Phytoecdysteroids of plants of the genus *Silene*. XIV. Ecdysterone 20-*O*-benzoate from *Silene tatarica*. Chem Nat Compd 23:706–708

Baltayev UA, Galiautdinov IV, Odinokov VN (1999) One-step synthesis of shidasterone 22*S*-analogue from ecdysterone. Mendeleev Commun 9:122–122

Báthori M, Girault J-P, Máthé I, Lafont R (2000) Isolation of 5α- and 5β-dihydrorubrosterone from *Silene otites* L. (Wib). Biomed Chromatogr 14:464–467

Bathori M, Mathe I, Girault J-P, Kalasz H, Lafont R (1998) Isolation and Structural Elucidation of Two Plant Ecdysteroids, Gerardiasterone and 22-Epi-20-hydroxyecdysone. J Nat Prod 61:415–417

Báthori M, Tóth N, Hunyadi A, Márki Á, Zador E (2008) Phytoecdysteroids and anabolic-androgenic steroids-structure and effects on humans. Curr Med Chem 15:75–91

Ben Nejma A, Nguir A, Jannet HB, Hamza MA, Daich A, Othman M, Lawson AM (2015) New septanoside and 20-hydroxyecdysone septanoside derivative from *Atriplex portulacoides* roots with preliminary biological activities. Bioorg Med Chem Lett 25:1665–1670

Blunt J, Lane G, Munro M, Russell G (1979) The Absolute Configuration at C24 of the Ecdysteroids Dacrysterone, Pterosterone and Ponasterone C. Aust J Chem 32:779–782

Bock K, Wiebe L (1973) Effect of oxygen lone pairs on J-1 (C-13H) values in 1,3-dioxanes. Acta Chem Scand 27:2676–2678

Borovikova EB, Baltaev UA (1999) Lesterone, a new phytoecdysteroid from the seeds of *Leuzea carthamoides*. Chem Nat Compd 35:182–183

Bourne PC, Whiting P, Dhadialla TS, Hormann RE, Girault P, Harmatha J, Lafont R, Dinan L (2002) Ecdysteroid 7,9(11)-dien-6-ones as potential photoaffinity labels for ecdysteroid binding proteins. J Insect Sci 2

Bückmann D, Starnecker G, Tomaschko K-H, Wilhelm E, Lafont R, Girault J-P (1986) Isolation and identification of major ecdysteroids from the pycnogonid *Pycnogonum litorale* (Ström) (Arthropoda, Pantopoda). J Comp Physiol B 156:759–765

Buděšínský M, Vokáč K, Harmatha J, Cvačka J (2008) Additional minor ecdysteroid components of Leuzea carthamoides. Steroids 73:502–514

Buděšínský M, Vokáč K, Harmatha J, Cvačka J (2009) Corrigendum to "Additional minor ecdysteroid components of *Leuzea carthamoides*" [Steroids 73 (2008) 502–514]". Steroids 74:1003–1004

Bulcock BW, Chen R, Lacey E, Chooi Y-H, Flematti GR (2024) Ether–Diol Ambiguity: An Inconspicuous Issue in the Structure Elucidation of Oxygenated Natural Products. J Nat Prod 87:2101–2109

Butenandt A, Karlson P (1954) Über die Isolierung eines Metamorphose-Hormons der Insekten in kristallisierter Form. Z Für Naturforschung B 9:389–391

Cai X, Shi Y-M, Pöhlmann N, Revermann O, Bahner I, Pidot SJ, Wesche F, Lackner H, Büchel C, Kaiser M, Richter C, Schwalbe H, Stinear TP, Zeeck A, Bode HB (2017) Structure and Biosynthesis of Isatropolones, Bioactive Amine-Scavenging Fluorescent Natural Products from *Streptomyces* Gö66. Angew Chem Int Ed 56:4945–4949

Calcagno M-P, Camps F, Coll J, Melé E, Sánchez-Baeza F (1996) New phytoecdysteroids from roots of *Ajuga reptans* varieties. Tetrahedron 52:10137–10146

Camps F, Coll J, Cortel A (1982) 29-norsengosterone and 29-norcyasterone, new C-28 phytoecdysteroids from *Ajuga reptans* (Labiatae). Chem Lett 11:1313–1316

Canonica L, Danieli B, Ferrari G, Krepinsky J, Rainoldi G (1973) Structure of calonysterone, an unusually modified phytoecdysone. J Chem Soc Chem Commun 737–738

Castro A, Coll J, Tandrón YA, Pant AK, Mathela CS (2008) Phytoecdysteroids from *Ajuga macrosperma* var. *breviflora* Roots. J Nat Prod 71:1294–1296

Černý V, Buděšínský M, Šorm F (1973) On steroids. CXLVIII. Reactions and stereochemistry of some 2-oxygenated A-homo-5α-cholestane derivatives. Collect Czechoslov Chem Commun 38:565–574

Chen T, Massias J, Bertrand S, Guitton Y, Le Bizec B, Dervilly G (2024) Innovative molecular networking analysis of steroids and characterisation of the urinary steroidome. Sci Data 11:818

Clouse SD (1996) Plant Chromosomes: Brassinosteroids in the spotlight. Curr Biol 6:658–661

Coll J, Reixach N, Sanchez-Baeza F, Casas J, Camps F (1994) New ecdysteroids from *Polypodium vulgare*. Tetrahedron 50:7247–7252

Coll J, Tandrón YA, Zeng X (2007) New phytoecdysteroids from cultured plants of *Ajuga nipponensis* Makino. Steroids 72:270–277

Crouzet S, Maria A, Dinan L, Lafont R, Girault J (2009) Ecdysteroids from *Cyanotis longifolia* Benth. (Commelinaceae). Arch Insect Biochem Physiol 72:194–209

Csábi J, Hsieh T-J, Hasanpour F, Martins A, Kele Z, Gáti T, Simon A, Tóth G, Hunyadi A (2015) Oxidized Metabolites of 20-Hydroxyecdysone and Their Activity on Skeletal Muscle Cells: Preparation of a Pair of Desmotropes with Opposite Bioactivities. J Nat Prod 78:2339–2345

Danieli B, Ferrari G, Krepinsky J, Palmisamo G, Sardini D (1974) The determination of absolute configuration at C-24 of the phytoecdysone makisterone A. J Chem Soc Chem Commun 745–746

De Marino S, Palagiano E, Zollo F, Minale L, Iorizzi M (1997) A novel sulphated steroid with a 7-membered 5-oxalactone B-ring from an Antarctic starfish of the family Asteriidae. Tetrahedron 53:8625–8628

DeMatteo MP, Snyder NL, Morton M, Baldisseri DM, Hadad CM, Peczuh MW (2005) Septanose Carbohydrates: Synthesis and Conformational Studies of Methyl α-D-*glycero*-D-Idoseptanoside and Methyl β-D- *glycero*-D-Guloseptanoside. J Org Chem 70:24–38

Destrez B, Pinel G, Monteau F, Lafont R, Le Bizec B (2009) Detection and identification of 20-hydroxyecdysone metabolites in calf urine by liquid chromatography-high resolution or tandem mass spectrometry measurements and establishment of their kinetics of elimination after 20-hydroxyecdysone administration. Anal Chim Acta 637:178–184

Dinan L (1988) The chemical synthesis of ecdysone 22-long-chain fatty acyl esters in high yield. J Steroid Biochem 31:237–245

Dinan L (2001) Phytoecdysteroids: biological aspects. Phytochemistry 57:325–339

Dinan L, Mamadalieva NZ, Lafont R (2021) Dietary phytoecdysteroids. In: Handbook of dietary phytochemicals. Springer, pp. 1541–1593

Dinan L, Rees HH (1978) Preparation of 3-*epi*-ecdysone and 3-*epi*-20-hydroxyecdysone. Steroids 32:629–638

Dzhukharova MKh, Tashkhodzhaev B, Saatov Z, Abdullaev ND (1993) Phytoecdysteroids of plants of the genus *Silene* XIV. Brahuisterone from *Silene brahuica*. Chem Nat Compd 29:484–489

Eliel EL, Knoeber SMC (1968) Conformational analysis. XVI. 1,3-Dioxanes. J Am Chem Soc 90:3444–3458

Fábián L, Argay G, Kálmán A, Báthori M (2002) Crystal structures of ecdysteroids: the role of solvent molecules in hydrogen bonding and isostructurality. Struct Sci 58:710–720

Fang Y, Ni Z-Y, Yamada T, Wang Y-F, Zhang M-L, Dong M, Cong B, Sauriol F, Huo C-H, Shi Q-W (2010) Two new phytoecdysteroids from the needles of *Taxus canadensis*. Z Für Naturforschung B 65:1401–1405

Fox Ramos AE, Evanno L, Poupon E, Champy P, Beniddir MA (2019) Natural products targeting strategies involving molecular networking: different manners, one goal. Nat Prod Rep 36:960–980

Franco BA, Luciano ER, Sarotti AM, Zanardi MM (2023) DP4+App: Finding the Best Balance between Computational Cost and Predictive Capacity in the Structure Elucidation Process by DP4+. Factors Analysis and Automation. J Nat Prod 86:2360–2367

Fuchino H, Nakamura H, Hakamatsuka T (1997) Two new phytoecdysteroids from the fern *Schizaea dichotoma*. Nat Med 51:491–492

Fujimoto Y, Yamada T, Ikekawa N (1985) Pyridine-induced deshielding of 4-methylene protons for the determination of C-6 stereochemistry of sterols having a 5α, 6-diol moiety. Revision of the C-6 stereochemistry of marine sterol isolated from a sponge, *Dysidea* sp. Chem Pharm Bull (Tokyo) 33:3129–3133

Fukuzawa A, Kumagai Y, Masamune T, Furusaki A, Katayama C, Matsumoto T (1981) Acetylpinnasterol and pinnasterol, ecdysone-like metabolites, from the marine red alga *Laurencia pinnata* Yamada. Tetrahedron Lett 22:4085–4086

Fukuzawa A, Miyamoto M, Kumagai Y, Masamune T (1986) Ecdysone-like metabolites, 14α-hydroxypinnasterols, from the red alga *Laurencia pinnata*. Phytochemistry 25:1305–1307

Fung S, Siddall JB (1980) Stereoselective synthesis of brassinolide: a plant growth promoting steroidal lactone. J Am Chem Soc 102:6580–6581

Garcia M, Girault J-P, Lafont R (1986) Ecdysteroid metabolism in the terrestrial snail *Cepaea nemoralis* (L.). Int J Invertebr Reprod Dev 9:43–58

Giner J-L, Djerassi C (1992) Minor and trace sterols in marine invertebrates 65. 23-Epidihydrocalysterol: a new cyclopropane-containing sponge sterol. Steroids 57:258–261

Giner J-L, Feng J, Kiemle DJ (2016) NMR Tube Degradation Method for Sugar Analysis of Glycosides. J Nat Prod 79:2413–2417

Girault J-P, Báthori M, Kalász H, Mathé I, Lafont R (1996) Sidisterone, a C_{24} Ecdysteroid from *Silene dioica* and *Silene otites*. J Nat Prod 59:522–524

Girault J-P, Bathori M, Varga E, Szendrei K, LaFont R (1990) Isolation and Identification of New Ecdysteroids from the Caryophyllaceae. J Nat Prod 53:279–293

Girault JP, Lafont R (1988) The complete ^{1}H-NMR assignment of ecdysone and 20-hydroxyecdysone. J Insect Physiol 34:701–706

Girault J-P, Lafont R, Varga E, Hajdu Z, Herke I, Szendrei K (1988) Ecdysteroids from *Leuzea carthamoides*. Phytochemistry 27:737–741

Gorelick-Feldman J, Cohick W, Raskin I (2010) Ecdysteroids elicit a rapid Ca2+ flux leading to Akt activation and increased protein synthesis in skeletal muscle cells. Steroids 75:632–637

Govindan M, Abbas SA, Schmitz FJ, Lee RH, Papkoff JS, Slate DL (1994) New Cycloartanol Sulfates from the Alga *Tydemania expeditionis*: Inhibitors of the Protein Tyrosine Kinase pp $^{60v\text{-}src}$. J Nat Prod 57:74–78

Grebenok RJ, Venkatachari S, Adler JH (1994) Biosynthesis of ecdysone and ecdysone phosphates in spinach. Phytochemistry 36:1399–1408

Grove MD, Spencer GF, Rohwedder WK, Mandava N, Worley JF, Warthen Jr JD, Steffens GL, Flippen-Anderson JL, Cook Jr JC (1979) Brassinolide, a plant growth-promoting steroid isolated from *Brassica napus* pollen. Nature 281:216–217

Guerriero A, Traldi P, Pietra F (1986) Gerardiasterone, a new ecdysteroid with a 20, 22, 23, 25-tetrahydroxylated side chain from the mediterranean zoanthid *Gerardia savaglia*. J Chem Soc Chem Commun 40–41

Guibout L, Mamadalieva N, Balducci C, Girault J-P, Lafont R (2015) The minor ecdysteroids from *Ajuga turkestanica*. Phytochem Anal 26:293–300

Guillen PO, Calabro K, Jaramillo KB, Dominguez C, Genta-Jouve G, Rodriguez J, Thomas OP (2018) Ecdysonelactones, Ecdysteroids from the Tropical Eastern Pacific Zoantharian *Antipathozoanthus hickmani*. Mar Drugs 16:58

Guo K, Ren X, Mu R-F, Zhou T-T, Li D, Hu H, Liu Y, Li S-H (2021) Ecdysteroids and spirosterane steroids from the traditional Chinese medicine *Paris polyphylla* var. *yunnanensis*. Phytochem Lett 45:117–120

Hang DTT, Hang NTM, Anh HLT, Nhiem NX, Hue CT, Binh PT, Dat NT, Nam NH, Yen PH, Minh CV, Hung NV, Kiem PV (2015) ^{1} H and^{13} C NMR assignments of new ecdysteroids from *Callisia fragrans*. Magn Reson Chem 53:379–382

Harmatha J, Buděšínský M, Vokáč K (2002) Photochemical transformation of 20-hydroxyecdysone: production of monomeric and dimeric ecdysteroid analogues. Steroids 67:127–135

Hedtmann U, Hobert K, Klintz R, Welzel P, Frelek J, Strangmann-Diekmann M, Klöne A, Pongs O (1989) Side-chain Homologues of 20-Hydroxyecdysone: Synthesis, Configurational Assignment, and Biochemical Characterization. Angew Chem Int Ed Engl 28:1515–1518

Hikino H, Hikino Y, Nomoto K, Takemoto T (1968) Cyasterone, an insect metamorphosing substance from *Cyathula capitata*: structure. Tetrahedron 24:4895–4906

Hikino H, Okuyama T, Konno C, Takemoto T (1975) Carbon-13 nuclear magnetic resonance spectra of phytoecdysones. Chem Pharm Bull (Tokyo) 23:125–132

Hiramoto M, Fujimoto Y, Kakinuma K, Ikekawa N, Ohnishi E (1988) Ecdysteroid conjugates in the ovaries of the silkworm, *Bombyx mori*: 3-phosphates of 2,22-dideoxy-20-hydroxyecdysone and of bombycosterol. Experientia 44:623–625

Hirayama Y, Okuzumi K, Masubuti H, Uekusa H, Girault J-P, Fujimoto Y (2014) Stereochemical Assignment of C-24 and C-25 of Amarasterone A, a Putative Biosynthetic Intermediate of Cyasterone. J Org Chem 79:5471–5477

Hoffmann KH, Bulenda D, Thiry E, Schmid E (1985) Apolar ecdysteroid esters in adult female crickets, *Gryllus bimaculatus*. Life Sci 37:185–192

Horn DHS (1971) Naturally Occurring Insecticides. Marcel Dekker, New York

Hou X-L, Gao Y-Q, Yang J-H, Liu H-W, Bai M-M, Wu Z-H, Li C-H, Tian J-M, Gao J-M (2023) Phytoecdysteroids from *Dianthus superbus* L.: Structures and anti-neuroinflammatory evaluation. Phytochemistry 212:113710

Hu J, Shi X, Mao X, Li H, Chen J, Shi J (2014) Ecdysteroids from the ethanol extract of *Diplopterygium rufopilosum*. Phytochem Lett 8:73–76

Hunyadi A, Gergely A, Simon A, Toth G, Veress G, Bathori M (2007) Preparative-Scale Chromatography of Ecdysteroids of *Serratula wolffii* Andrae. J Chromatogr Sci 45:76–86

Hunyadi A, Herke I, Lengyel K, Báthori M, Kele Z, Simon A, Tóth G, Szendrei K (2016) Ecdysteroid-containing food supplements from *Cyanotis arachnoidea* on the European market: evidence for spinach product counterfeiting. Sci Rep 6:37322

Hunyadi A, Tóth G, Simon A, Mák M, Kele Z, Máthé I, Báthori M (2004) Two New Ecdysteroids from *Serratula wolffii*. J Nat Prod 67:1070–1072

Ikeda M, Fujita T, Naoki H, Naya Y, Mamiya Y, Kamba M, Sonobe H (1995) Charge-remote fragmentation of ecdysteroids conjugated with phosphoric acid. Rapid Commun Mass Spectrom 9:1480–1483

Imai S, Murata E, Fujioka S, Matsuoka T, Koreeda M, Nakanishi K (1970) Structures of stachysterone A, the first natural 27-carbon steroid with a rearranged methyl group and stachysterone B. J Am Chem Soc 92:7510–7512

Isaac RE, Rees HH (1984) Isolation and identification of ecdysteroid phosphates and acetylecdysteroid phosphates from developing eggs of the locust, *Schistocerca gregaria*. Biochem J 221:459–464

Isaac RE, Rose ME, Rees HH, Goodwin TW (1982) Identification of ecdysone-22-phosphate and 2-deoxyecysone-22-phosphate in eggs of the desert locust, *Schistocerca gregaris*, by fast atom bombardment mass spectrometry and NMR spectroscopy. J Chem Soc Chem Commun 249–251

Jayasinghe L, Jayasooriya CP, Hara N, Fujimoto Y (2003a) A pyridine ring-containing ecdysteroid from *Diploclisia glaucescens*. Tetrahedron Lett 44:8769–8771

Jayasinghe L, Jayasooriya CP, Oyama K, Fujimoto Y (2002) 3-Deoxy-1β, 20-dihydroxyecdysone from the leaves of *Diploclisia glaucescens*. Steroids 67:555–558

Jayasinghe L, Kumarihamy BM, Arundathie BS, Dissanayake L, Hara N, Fujimoto Y (2003b) A new ecdysteroid, 2-deoxy-5β, 20-dihydroxyecdysone from the fruits of *Diploclisia glaucescens*. Steroids 68:447–450

Jia A, Li Y, Zhou J, Gao J (2010) Three phytoecdysteroids from *Sagina japonica* and potential biotransforming pathways of japonicone. Chem Nat Compd 46:738–741

Jiang P, Liu Y, Sun Y, Pan J, Guan W, Xu Z, Li X, Wang S, Mei Y, Kuang H, Yang B (2021) Ecdysteroids from the Aerial Parts of *Paris verticillata*. Chem Biodivers 18:e2100239

Jin L, Xue Y, Zong T, Sun J, Zhou W, Jin M, Diao S, Li G (2025) Two new phytoecdysteroids isolated from *Tournefortia sibirica*. Nat Prod Res 1–6

Jizba J, Dolejš L, Herout V, Šorm F (1971) The structure of osladin-the sweet principle of the rhizomes of *Polypodium vulgare* L. Tetrahedron Lett 12:1329–1332

Kakumu Y, Thi Nguyen MT, Mitsunaga T, Hattori H (2023) Identification of Ecdysteroid Sinapate Esters with COX-2 Inhibitory Effects from *Fibraurea recisa* Using Molecular Networking and MS2LDA. J Nat Prod 86:1960–1967

Kalász H, Pongrácz Z, Máthé I, Kálmán A, Argay G (2002) 5-Alpha- and 5-beta-2-deoxyintegristerone A, a 5-alpha and 5-beta isomer pair of ecdysteroids isolated from the *Silene* genus. Biomed Chromatogr 16:373–378

Karlson P, Hoffmeister H, Hummel H, Hocks P, Spiteller G (1965) Zur Chemie des Ecdysons, VI: Reaktionen des Ecdysonmoleküls. Chem Ber 98:2394–2402

Kawahara N, Sekita S, Satake M (1995) Two steroids from *Calvatia cyathiformis*. Phytochemistry 38:947–950

Kawahara N, Sekita S, Satake M (1994) Steroids from *Calvatia cyathiformis*. Phytochemistry 37:213–215

Kayser H, Ertl P, Eilinger P, Spindler-Barth M, Winkler T (2002) Diastereomeric ecdysteroids with a cyclic hemiacetal in the side chain produced by cytochrome P450 in hormonally resistant insect cells. Arch Biochem Biophys 400:180–187

Kim J, Kinghorn AD (1989) Further steroidal and flavonoid constituents of the sweet plant, *Polypodium glycyrrhiza*. Phytochemistry 28:1225–1228

Kim J, Pezzuto JM, Soejarto DD, Lang FA, Kinghorn AD (1988) Polypodoside A, an Intensely Sweet Constituent of the Rhizomes of *Polypodium glycyrrhiza*. J Nat Prod 51:1166–1172

Kitagawa I, Nishino T, Matsuno T, Akutsu H, Kyogoku Y (1978) Structure of holothurin B a pharmacologically active triterpene-oligoglycoside from the sea cucumber *Holothuria leucospilota* Brandt. Tetrahedron Lett 19:985–988

Köck M, Grube A, Seiple IB, Baran PS (2007) The Pursuit of Palau'amine. Angew Chem Int Ed 46:6586–6594

Koreeda M, Nakanishi K (1970) 5β-Hydroxy-ecdysones and a revision of the structure of ponasterone C. J Chem Soc Chem Commun 351–352

Kovganko NV, Sokolov SN (2001) Synthesis of 2α,3α-dihydroxy-Δ4,7-6-ketosteroids, structural analogs of diaulusterols A and B. Chem Nat Compd 37:455–461

Krepinsky J, Findlay JA, Danieli B, Palmisano G, Beynon P, Murakami S (1977) Concerning carbon-13 n.m.r. spectroscopy of 5β-hydroxylated phytoecdysones. Org Magn Reson 10:255–257

Kumpun S, Maria A, Crouzet S, Evrard-Todeschi N, Girault J-P, Lafont R (2011) Ecdysteroids from *Chenopodium quinoa* Willd., an ancient Andean crop of high nutritional value. Food Chem 125:1226–1234

Kuramoto T, Kihira K, Matsumoto N, Hoshita T (1981) Determination of the sulfated position in 5β-bufol sulfate by a carbon-13 nuclear magnetic resonance study. Chem Pharm Bull (Tokyo) 29:1136–1139

Laczkó D, Chu E-L, Chang C-C, Chang F-R, Girst G, Gáti T, Tóth G, Könczöl Á, Hunyadi A (2025) LC-MS/MS Fingerprinting Analysis of *Cyanotis arachnoidea* Extracts: Process-Related Artifacts in Anabolic Food Supplements. ACS Omega https://doi.org/10.1021/acsomega.4c10908

Lafont R, Balducci C, Dinan L (2021) Ecdysteroids. Encyclopedia 1:1267–1302

Lafont R, Bouthier A, Wilson ID (1991) Phytoecdysteroids: structures, occurrence, biosynthesis and possible ecological significance. In: Conf. Incect Chem. Ecolog: Programme and Abstract. Tabor. pp 197–214

Lafont R, Harmatha J, Marion-Poll F, Dinan L, Wilson ID (2002) Ecdybase, a free ecdysteroid database. http://ecdybase.org (accessed 15 January 2026)

Lee J-C, Chang F-R, Chen S-R, Wu Y-H, Hu H-C, Wu Y-C, Backlund A, Cheng Y-B (2016) Anti-Dengue Virus Constituents from Formosan Zoanthid *Palythoa mutuki*. Mar Drugs 14:151

Li M, Zhou P, Wu A (2006) Synthesis of Gymnasterone B, an antitumor steroid from *Gymnascella dankaliensis*. Tetrahedron Lett 47:3409–3412

Liktor-Busa E, Simon A, Tóth G, Báthori M (2008) The first two ecdysteroids containing a furan ring from *Serratula wolffii*. Tetrahedron Lett 49:1738–1740

Ling TJ, Ma WZ, Wei XY (2003) Ecdysteroids from the roots of *Serratula chinensis*. J Trop Subtrop Bot 11:143–147

Liu J, Du Dan SI, Lv H-N, Wu X-F, Li Y, Liu Y-Y, Yu S-S (2010) Application of Dimolybdenum Reagent $Mo_2(OAc)_4$ for Determination of the Absolute Configurations of *vic*-Diols. Chin J Org Chem 30:1270

Liu Y-H, Zhang Q, Zhang D-Z, Fu X-Y, Liu W-Y, Tian J-M (2025) Two undescribed phytoecdysteroids from *Dianthus superbus* L. Phytochem Lett 69:103015

Mamadalieva NZ, Saatov Z, Kachala VV, Shashkov AS (2002) Phytoecdysteroids of plants of the *Silene* genus. 2-Deoxyecdysterone-25-acetate from *Silene wallichiana*. Chem Nat Compd 38:179–181

Mantsch HH, Smith ICP (1972) Fourier-transformed ^{13}C NMR spectra of polyuridylic acid, uridine, and related nucleotides — The use of 31POC13C couplings for conformational analysis. Biochem Biophys Res Commun 46:808–815

Marat K, Templeton JF, Ling Y (1993) NMR study of conformation and configuration in C-20-substituted 5β,14β-pregnanes, 5β-pregn-14-enes and 21-nor-5β,14β-pregnanes. Magn Reson Chem 31:17–22

Maria A, Girault J-P, Saatov Z, Harmatha J, Dinan L, Lafont R (2005) Ecdysteroid glycosides: identification, chromatographic properties, and biological significance. J Chromatogr Sci 43:149–157

Maurer P, Girault J, Larchevêque M, Lafont R (1993) 24-Epi-makisterone a (not makisterone A) is the major ecdysteroid in the leaf-cutting ant *Acromyrmex octospinosus* (reich) (hymenoptera, formicidae: Attini). Arch Insect Biochem Physiol 23:29–35

Mbeng Obame RB, Le Pogam P, Leblanc K, Rharrabti S, Numbi wa Ilunga E, Fontaine V, Gallard J-F, Retailleau P, Ndong JE, N'negué MA, Obiang CS, Ovono Abessolo F, Said-Hassane C, El Kalamouni C, Otogo N'Nang E, Bernadat G, Beniddir MA (2026) Targeted discovery of sesquiterpene indole alkaloids from *Greenwayodendron suaveolens*. Phytochemistry 241:114664

Miller R, Clardy J, Kozlowski J, Mikolajczak K, Plattner R, Powell R, Smith C, Weisleder D, Qi-Tai Z (1985) Phytoecdysteroids of *Diploclisia glaucescens* Seed. Planta Med 51:40–42

Mimaki Y, Sashida Y, Shimomura H (1989) Lipid and steroidal constituents of *Lilium auratum* var. *platyphyllum* and *L. tenuifolium*. Phytochemistry 28:3453–3458

Minli Z, Stout MJ, Kubo I (1992) Isolation of ecdysteroids from *Vitex strickeri* using RLCC and recycling HPLC. Phytochemistry 31:247–250

Miyata Y, Diyabalanage T, Amsler CD, McClintock JB, Valeriote FA, Baker BJ (2007) Ecdysteroids from the Antarctic Tunicate *Synoicum adareanum*. J Nat Prod 70:1859–1864

Nakamura S, Chen G, Nakashima S, Matsuda H, Pei Y, Yoshikawa M (2010) Brazilian Natural Medicines. IV. New Noroleanane-Type Triterpene and Ecdysterone-Type Sterol Glycosides and Melanogenesis Inhibitors from the Roots of *Pfaffia glomerata*. Chem Pharm Bull (Tokyo) 58:690–695

Nakanishi K, Imai S, Murata E, Fujioke S, Matsuoka T, Koreeda M (1970a) Insect hormones. XIX. Structures of stachysterone A, the first natural 27-carbon steroid with a rearranged methyl group, and stachysterone B. J Am Chem Soc 92:7510–7512

Nakanishi K, Koreeda M, Chang ML, Hsu HY (1968) Insect hormones V. The structures of ponasterones B and C. Tetrahedron Lett 9:1105–1110

Nakanishi K, Koreeda M, Goto M (1970b) Insect hormones. XX. Ajugalactone, an insect-molting inhibitor, as tested by the Chilo dipping method. J Am Chem Soc 92:7512–7513

Neumann S, Andres A-C, Bach F, Bender T, Minella CB, Eberl F, Fischer TG, Golub B, Harivyasi SS, Herres-Pawlis S, Huang P-C, Hunold J, Jollife JD, Jung N, Liermann JC, Nainala VC, Razum M, Koepler O, Steinbeck C (2024) Interim Report NFDI4Chem 2023. Res Ideas Outcomes 10:e124977

Nishimoto N, Shiobara Y, Fujino M, Inoue S-S, Takemoto T, De Oliveira F, Akisue G, Akisue MK, Hashimoto G, Tanaka O (1987) Ecdysteroids from *Pfaffia iresinoides* and reassignment of some ^{13}C NMR chemical shifts. Phytochemistry 26:2505–2507

Nishizawa M, Kodama S, Yamane Y, Kayano K, Hatakeyama S, Yamada H (1994a) Synthesis and glycosylation shift of 1, 1'-disaccharides. Chem Pharm Bull (Tokyo) 42:982–984

Nishizawa M, Yamada H, Yamaguchi Y, Hatakeyama S, Lee I-S, Kim J, Kinghorn AD (1994b) Structure Revision of Polypodoside A. Major Sweet Principle of *Polypodium glycyrrhiza*. Chem Lett 23:1555–1558

Nothias L-F, Petras D, Schmid R, Dührkop K, Rainer J, Sarvepalli A, Protsyuk I, Ernst M, Tsugawa H, Fleischauer M, Aicheler F, Aksenov AA, Alka O, Allard P-M, Barsch A, Cachet X, Caraballo-Rodriguez AM, Da Silva RR, Dang T, Garg N, Gauglitz JM, Gurevich A, Isaac G, Jarmusch AK, Kameník Z, Kang KB, Kessler N, Koester I, Korf A, Le Gouellec A, Ludwig M, Martin HC, McCall L-I, McSayles J, Meyer SW, Mohimani H, Morsy M, Moyne O, Neumann S, Neuweger H, Nguyen NH, Nothias-Esposito M, Paolini J, Phelan VV, Pluskal T, Quinn RA, Rogers S, Shrestha B, Tripathi A, van der Hooft JJJ, Vargas F, Weldon KC, Witting M, Yang H, Zhang Z, Zubeil F, Kohlbacher O, Böcker S, Alexandrov T, Bandeira N, Wang M, Dorrestein PC (2020) Feature-based molecular networking in the GNPS analysis environment. Nat Methods 17:905–908

Ochieng C, Ishola I, Opiyo S, Manguro L, Owuor P, Wong K-C (2012) Phytoecdysteroids from the Stem Bark of *Vitex doniana* and Their Anti-Inflammatory Effects. Planta Med 79:52–59

Odinokov VN, Galyautdinov IV, Nedopekin DV, Khalilov LM, Shashkov AS, Kachala VV, Dinan L, Lafont R (2002) Phytoecdysteroids from the juice of *Serratula coronata* L. (Asteraceae). Insect Biochem Mol Biol 32:161–165

Odinokov VN, Kumpun S, Galyautdinov IV, Evrard-Todeschi N, Veskina NA, Khalilov LM, Girault J-P, Dinan L, Maria A, Lafont R (2005) Low-polarity phytoecdysteroids from the juice of *Serratula coronata* L. (Asteraceae). Collect Czechoslov Chem Commun 70:2038–2052

Ohsawa T, Yukawa M, Takao C, Murayama M, Bando H (1992) Studies on constituents of fruit body of *Polyporus umbellatus* and their cytotoxic activity. Chem Pharm Bull (Tokyo) 40:143–147

Ohta S, Guo J-R, Hiraga Y, Suga T (1996) 24-Epi-pterosterone: A novel phytoecdysone from the roots of *Athyrium yokoscense*. Phytochemistry 41:745–747

Okuzumi K, Hara N, Uekusa H, Fujimoto Y (2005) Structure elucidation of cyasterone stereoisomers isolated from *Cyathula officinalis*. Org Biomol Chem 3:1227–1232

Olatunde OZ, Yong J, Lu C (2022) Isolation, structural elucidation of new phytoecdysteroids from *Ajuga decumbens* Thunb. Steroids 186:109089

Otogo N'Nang E, Cauchie G, Retailleau P, Agnandji ST, Gallard J-F, Mouray E, Grellier P, Champy P, Le Pogam P, Beniddir MA (2023) From the Spectroscopic Reassessment of Authentic Alkaloid Samples to the Molecular Networking-Guided Discovery of Criophylline-Related Analogues from *Callichilia inaequalis*. J Nat Prod 86:1202–1210

Park I, Park K, Lee H-S, Hong S-M, Sriramulu DK, Hwang H, Song S, Baek JG, Kim DH, Kim SY, Kwon J, Kwon HC (2023) Ecdysteroids from the Korean Endemic Species *Ajuga spectabilis* with Activities against Glucocorticoid Receptors and 11β-Hydroxysteroid Dehydrogenase Type 1. ACS Omega 8:26191–26200

Perrone A, Masullo M, Bassarello C, Hamed AI, Belisario MA, Pizza C, Piacente S (2007) Sulfated Triterpene Derivatives from *Fagonia arabica*. J Nat Prod 70:584–588

Pinheiro MLB, Wolter Filho W, Da Rocha AI, Porter B, Wenkertt E (1983) Abutasterone, an ecdysone from *Abuta velutina*. Phytochemistry 22:2320–2321

Píš J, Buděšínský M, Vokáč K, Laudová V, Harmatha J (1994) Ecdysteroids from the roots of *Leuzea carthamoides*. Phytochemistry 37:707–711

Pongrácz Z, Báthori M, Tóth G, Simon A, Mák M, Máthé I (2003) 9α,20-Dihydroxyecdysone, a New Natural Ecdysteroid from *Silene italica* ssp. *nemoralis*. J Nat Prod 66:450–451

Ranoux A, Djanashvili K, Arends IWCE, Hanefeld U (2013) 5-Hydroxymethylfurfural Synthesis from Hexoses Is Autocatalytic. ACS Catal 3:760–763

Roth U, König M, Seifert K (1995) Ecdysteroids from *Penstemon venustus*. Phytochemistry 39:941–942

Roussel PG, Turner NJ, Dinan LN (1995) Synthesis of shidasterone and the unambiguous determination of its configuration at C-22. J Chem Soc Chem Commun 933–934

Rudel D, Bathori M, Gharbi J, Girault J-P, Racz I, Melis K, Szendrei K, Lafont R (1992) New Ecdysteroids from *Serratula tinctoria*. Planta Med 58:358–364

Russell G, Fenemore P, Horn D, Middleton E (1972) Insect Moulting Hormones: The Phytoecdysones of *Dacrydium intermedium*. Aust J Chem 25:1935–1941

Russell G, Fraser J (1973) Insect Moulting Hormones: Dacrysterone, a New Phytoecdysone From *Dacrydium intermedium*. Aust J Chem 26:1805–1807

Russell GB, Greenwood DR, Lane GA, Blunt JW, Munro MHG (1981) 2-deoxy-3-epiecdysone from the fern *Blechnum vulcanicum*. Phytochemistry 20:2407–2410

Russell GB, Horn DHS, Middleton EJ (1971) New phytoecdysones from *Dacrydium intermedium*. J Chem Soc Chem Commun 71

Rutz A, Sorokina M, Galgonek J, Mietchen D, Willighagen E, Gaudry A, Graham JG, Stephan R, Page R, Vondrášek J, Steinbeck C, Pauli GF, Wolfender J-L, Bisson J, Allard P-M (2022) The LOTUS initiative for open knowledge management in natural products research. elife 11:e70780

Saatov Z, Abdullaev ND, Gorovits MB, Abubakirov NK (1984) Phytoecdysteroids of plants of the genus *Silene*. VI. α-Ecdysone 22-sulfate — A new ecdysteroid from *Silene brahuica*. Chem Nat Compd 20:441–444

Saatov Z, Gorovits MB, Abdullaev ND, Abubakirov NK (1987) Phytoecdysteroids of plants of the genus *Silene* XII. 5α-Ecdysterone 22-*O*-benzoate from *Silene scabrifolia*. Chem Nat Compd 23:563–565

Saleem M, Musaddiq S, Riaz N, Zubair M, Ashraf M, Nasar R, Jabbar A (2013) Ecdysteroids from the flowers of *Aerva javanica*. Steroids 78:1098–1102

Samek Z, Buděšínský M (1979) *In situ* reactions with trichloroacetyl isocyanate and their application to structural assignment of hydroxy compounds by ^{1}H NMR spectroscopy. A general comment. Collect Czechoslov Chem Commun 44:558–588

Sarker SD, Dinan LN, Girault J-P, Lafont R, Waterman PG (1996) Punisterone [(20*R*,24*S*)-25-Deoxy-11α,20,24-trihydroxyecdysone]: A New Phytoecdysteroid from *Blandfordia punicea*. J Nat Prod 59:789–793

Schmid R, Petras D, Nothias L-F, Wang M, Aron AT, Jagels A, Tsugawa H, Rainer J, Garcia-Aloy M, Dührkop K, Korf A, Pluskal T, Kameník Z, Jarmusch AK, Caraballo-Rodríguez AM, Weldon KC, Nothias-Esposito M, Aksenov AA, Bauermeister A, Albarracin Orio A, Grundmann CO, Vargas F, Koester I, Gauglitz JM, Gentry EC, Hövelmann Y, Kalinina SA, Pendergraft MA, Panitchpakdi M, Tehan R, Le Gouellec A, Aleti G, Mannochio Russo H, Arndt B, Hübner F, Hayen H, Zhi H, Raffatellu M, Prather KA, Aluwihare LI, Böcker S, McPhail KL, Humpf H-U, Karst U, Dorrestein PC (2021) Ion identity molecular networking for mass spectrometry-based metabolomics in the GNPS environment. Nat Commun 12:3832

Searle PA, Molinski TF (1995) 4-Dehydroecdysterone, a New Ecdysteroid from the Zoanthid *Parazoanthus* sp. J Nat Prod 58:264–268

Sharma VV, De Chavez D, Slade SE, Lanekoff I (2024) Enhancing isomer specificity in mass spectrometry by combining silver ion adduction and ion mobility. Talanta Open 10:100373

Sharma VV, Lanekoff I (2023) Revealing Structure and Localization of Steroid Regioisomers through Predictive Fragmentation Patterns in Mass Spectrometry Imaging. Anal Chem 95:17843–17850

Shigemori H, Sato Y, Kagata T, Kobayashi J (1999) Palythoalones A and B, New Ecdysteroids from the Marine Zoanthid *Palythoa australiae*. J Nat Prod 62:372–374

Simon A, Pongrácz Z, Tóth G, Mák M, Máthé I, Báthori M (2004) A new ecdysteroid with unique 9β-OH and four other ecdysteroids from *Silene italica* ssp. *nemoralis*. Steroids 69:389–394

Simon A, Tóth G, Liktor-Busa E, Kele Z, Takács M, Gergely A, Báthori M (2007) Three new steroids from the roots of *Serratula wolffii*. Steroids 72:751–755

Simon A, Tóth N, Tóth G, Kele Z, Groska J, Báthori M (2009) Ecdysteroids from *Silene viridiflora*. Helv Chim Acta 92:753–761
Singh SB, Thakur RS (1982) Structure and stereochemistry of paristerone, a novel phytoecdysone from the tubers of *Paris polyphylla*. Tetrahedron 38:2189–2194
Songsheng H, Guoliang W, Kemin X (1982) The Isolation and Identification of Phytoecdysones from *Dacrydium pierrei* Hickel. J Integr Plant Biol 24
Stoldt M, Porzel A, Adam G, Brandt W (1997) Side Chain Conformation of the Growth-Promoting Phytohormones Brassinolide and 24-Epibrassinolide. Magn Reson Chem 35:629–636
Suksamrarn A, Jankam A, Tarnchompoo B, Putchakarn S (2002a) Ecdysteroids from a *Zoanthus* sp. J Nat Prod 65:1194–1197
Suksamrarn A, Kumpun S, Yingyongnarongkul B (2002b) Ecdysteroids of *Vitex scabra* Stem Bark. J Nat Prod 65:1690–1692
Suksamrarn A, Promrangsan N, Chitkul B, Homvisasevongsa S, Sirikate A (1997) Ecdysteroids of the root bark of *Vitex canescens*. Phytochemistry 45:1149–1152
Suksamrarn A, Promrangsan N, Jintasirikul A (2000) Highly oxygenated ecdysteroids from *Vitex canescens* root bark. Phytochemistry 53:921–924
Suksamrarn A, Sommechai C, Charulpong P, Chitkul B (1995) Ecdysteroids from *Vitex canescens*. Phytochemistry 38:473–476
Sun Y, Yasukawa K (2008) New anti-inflammatory ergostane-type ecdysteroids from the sclerotium of *Polyporus umbellatus*. Bioorg Med Chem Lett 18:3417–3420
Tabekoueng GB, Akak CM, Happi GM, Langat MK, Frese M, Stammler H-G, Neumann B, Azebaze AGB, Waffo AFK, Wansi JD, Lenta BN, Sewald N, Vardamides JC, Nkengfack AE (2020) The chemistry of the West and Central African *Penianthus zenkeri* Diels (Menispermaceae). Phytochem Lett 38:12–18
Takács M, Simon A, Liktor-Busa E, Báthori M, Zsila F, Bikádi Z, Horváth P, Veress G, Gergely A, Tóth G (2010) Structure and stereochemistry of novel ecdysteroids from the roots of *Serratula wolffii*. Magn Reson Chem 48:386–391
Takemoto T, Nomoto K, Hikino Y, Hikino H (1968) Structure of capitasterone, a novel C_{29} insect-moulting substance from *Cyathula capitata*. Tetrahedron Lett 9:4929–4932
Tóth G, Herke I, Gáti T, Vágvölgyi M, Berkecz R, Parfenova LV, Ueno M, Yokoi T, Nakagawa Y, Hunyadi A (2021) A Commercial Extract of *Cyanotis arachnoidea* Roots as a Source of Unusual Ecdysteroid Derivatives with Insect Hormone Receptor Binding Activity. J Nat Prod 84:1870–1881
Tóth G, Santa-Maria AR, Herke I, Gáti T, Galvis-Montes D, Walter FR, Deli MA, Hunyadi A (2023) Highly Oxidized Ecdysteroids from a Commercial *Cyanotis arachnoidea* Root Extract as Potent Blood–Brain Barrier Protective Agents. J Nat Prod 86:1074–1080
Tóth N, Simon A, Tóth G, Kele Z, Hunyadi A, Báthori M (2008) 26-Hydroxylated Ecdysteroids from *Silene viridiflora*. J Nat Prod 71:1461–1463
Tsoupras G, Luu B, Hoffmann JA (1983) A Cytokinin (Isopentenyl-adenosyl-mononucleotide) Linked to Ecdysone in Newly Laid Eggs of *Locusta migratoria*. Science 220:507–509
Tsubuki M, Takada H, Katoh T, Miki S, Honda T (1996) Synthesis and structural elucidation of a new zooecdysteroid gerardiasterone. Tetrahedron 52:14515–14532
van der Hooft JJJ, Wandy J, Barrett MP, Burgess KEV, Rogers S (2016) Topic modeling for untargeted substructure exploration in metabolomics. Proc Natl Acad Sci 113:13738–13743
Ványolós A, Simon A, Tóth G, Polgár L, Kele Z, Ilku A, Mátyus P, Báthori M (2009) C-29 Ecdysteroids from *Ajuga reptans* var. *reptans*. J Nat Prod 72:929–932
Vokáč K, Buděšínský M, Harmatha J (2002) Minor ecdysteroid components of *Leuzea carthamoides*. Collect Czechoslov Chem Commun 67:124–139
Vokáč K, Buděšínský M, Harmatha J, Píš J (1998) New ergostane type ecdysteroids from fungi. Ecdysteroid constituents of mushroom *Paxillus atrotomentosus*. Tetrahedron 54:1657–1666
Wang P, Li S, Ownby S, Zhang Z, Yuan W, Zhang W, Scott Beasley R (2009) Ecdysteroids and a sucrose phenylpropanoid ester from *Froelichia floridana*. Phytochemistry 70:430–436

Wang Q-H, Yang L, Jiang H, Wang Z-B, Yang B-Y, Kuang H-X (2011) Three New Phytoecdysteroids Containing a Furan Ring from the Roots of *Achyranthes bidentata* Bl. Molecules 16:5989–5997
Wang X-J, Zhang Q, Peng Y-R, Li L, Qu J, Liu Y-B, Xu S, Ma S-G, Li Y, Zou Z-M, Wang R-B, Yu S-S (2019) Two azafluoranthene alkaloids and a phytoecdysone from the stems of *Cyclea barbata*. J Asian Nat Prod Res 21:217–226
Wang Z, Li H, Li Y, Nie K, Yang X, Bai X, Li Y, Huang X (2024) Phytosteroids from roots of *Psammosilene tunicoides* WC Wu et CY Wu. Rec Nat Prod 18:290–295
Watrous J, Roach P, Alexandrov T, Heath BS, Yang JY, Kersten RD, van der Voort M, Pogliano K, Gross H, Raaijmakers JM, Moore BS, Laskin J, Bandeira N, Dorrestein PC (2012) Mass spectral molecular networking of living microbial colonies. Proc Natl Acad Sci 109:E1743–E1752
Williams DE, Ayer SW, Andersen RJ (1986) Diaulusterols A and B from the skin extracts of the dorid nudibranch *Diaulula sandiegensis*. Can J Chem 64:1527–1529
Wishart DS, Sajed T, Pin M, Poynton EF, Goel B, Lee BL, Guo AC, Saha S, Sayeeda Z, Han S, Berjanskii M, Peters H, Oler E, Gautam V, Jordan T, Kim J, Ledingham B, Tretter ZM, Koller JT, Shreffler HA, Stillwell LR, Jystad AM, Govind N, Bade JL, Sumner LW, Linington RG, Cort JR (2025) The Natural Products Magnetic Resonance Database (NP-MRD) for 2025. Nucleic Acids Res 53:D700–D708
Wu P, Xie H, Tao W, Miao S, Wei X (2010) Phytoecdysteroids from the rhizomes of *Brainea insignis*. Phytochemistry 71:975–981
Yamada H, Nishizawa M (1995) Synthesis and Structure Revision of Intensely Sweet Saponin Osladin. J Org Chem 60:386–397
Yamada H, Nishizawa M, Katayama C (1992) Osladin, a sweet principle of *Polypodium vulgare*. Structure revision. Tetrahedron Lett 33:4009–4010
Yan Y, Ubukata M, Cody RB, Holy TE, Gross ML (2014) High-Energy Collision-Induced Dissociation by MALDI TOF/TOF Causes Charge-Remote Fragmentation of Steroid Sulfates. J Am Soc Mass Spectrom 25:1404–1411
Yang L, Teng L, Huang Y, Tang M, Liao W, Fu Q (2022) Undescribed ecdysteroids and phenolic glycosides from the roots of *Cyathula officinalis* Kuan and their anti-inflammatory activity in LPS-induced RAW 264.7 macrophages in vitro. Phytochemistry 196:113101
Yi J, Luo Y, Li B, Zhang G (2004) Phytoecdysteroids and glycoceramides from *Eriophyton wallchii*. Steroids 69:809–815
Ying-Ying W, Jia-Yuan LI, Chang-Liang YAO, Yang YU, Shuai YAO, Min GAO, Shi-Fei WU, Wen-Long WEI, Qi-Rui BI, De-An GUO (2022) Deep chemical identification of phytoecdysteroids in *Achyranthes bidentata* Blume by UHPLC coupled with linear ion trap-Orbitrap mass spectrometry and targeted isolation. Chin J Nat Med 20:551–560
Yokosuka A, Inomata M, Yoshizawa Y, Iguchi T, Mimaki Y (2021) Bufadienolides and ecdysteroids from the whole plants of *Helleborus Niger* and their cytotoxicity. J Nat Med 75:393–402
Yokota T, Arima M, Takahashi N (1982) Castasterone, a new phytosterol with plant-hormone potency, from chestnut insect gall. Tetrahedron Lett 23:1275–1278
Yokota T, Kim SK, Fukui Y, Takahashi N, Takeuchi Y, Takematsu T (1987) Brassinosteroids and sterols from a green alga, *Hydrodictyon reticulatum*: configuration at C-24. Phytochemistry 26:503–506
Yuan R, Wang H, Liu Z, Jia L, Ma Y, Yan R, Yin S, Tu W, Hu M, Ding L, Wu X (2025) Clerodane Diterpenoids and Ecdysteroids From the Fruits of *Tinospora sagittata* (Oliv.) Gagnep. Chem Biodivers e01502
Yusupova UY, Ramazonov NS, Syrov VN, Sagdullaev SS (2022) Phytoecdysteroids: Properties, Biological Activity and Applications. Springer, Singapore
Zeng Q, Chen Y, Wang J, Shi X, Che Y, Chen X, Zhong W, Zhang W, Wei X, Wang F, Zhang S (2022) Diverse Secondary Metabolites from the Coral-Derived Fungus *Aspergillus hiratsukae* SCSIO 5Bn1003. Mar Drugs 20:150
Zheng S-Z, Yang H-P, Ma X-M, Shen X-W (2004) Two new Polyporusterones from *Polyporus umbellatus*. Nat Prod Res 18:403–407

Zhou W-W, Lin W-H, Guo S-X (2007) Two new polyporusterones isolated from the sclerotia of *Polyporus umbellatus*. Chem Pharm Bull (Tokyo) 55:1148–1150

Zhu L, Zhang G, Chen L, Wang S, Li P, Li L (2009) A new ecdysteroside from *Lygodium japonicum* (Thunb.) Sw. J Nat Med 63:215–219

Zhylitskaya H, Litvinovskaya R, Drach S, Khripach V (2011) Synthesis of sidisterone, a phytoecdysteroid from *Silene dioica* (L.) Clairv. Tetrahedron Lett 52:5267–5269

Zughdani M, Yusufoğlu HS, Ekiz G, Linden A, Çalış İ (2020) Ecdysteroids from the underground parts of *Rhaponticum acaule* (L.) DC. Phytochemistry 180:112530

Chapter 3
Metabolic and Hormonal Modulation by Phytoecdysteroids in Mammals

René Lafont

3.1 Introduction—The Early Studies

The story began with the isolation from silkworm pupae of the first ecdysteroid, the insect molting hormone ecdysone (Butenandt and Karlson 1954). Investigations into the effects of ecdysteroids in mammals started in 1963, i.e., before steroids related to insect molting hormones were discovered in plants. These precursor experiments were performed in a context of hormonal heterophylly, i.e., investigations on the possible effects of insect hormones on mammals and vice versa (Burdette and Coda 1963). The authors used a mouse liver preparation (a 14,000 x g supernatant) for in vitro incorporation of ^{14}C-leucine and observed a dose-dependent stimulation of protein synthesis when ecdysone was added to this preparation (20–200 μM, assuming that one "Calliphora unit" is equivalent to 0.1 μg of ecdysone).

The discovery of ecdysteroids in plants (Nakanishi et al. 1966) allowed the availability of much larger amounts of these molecules and suggested a possible use of these molecules as endocrine disruptors toward insect pests in agronomy. Accordingly, investigations were undertaken to check the safety of these molecules for mammals/humans and, although they were indeed shown to be nontoxic (oral LD_{50} > 9 g/kg [Matsuda et al. 1970]; see Chap. 7 in this volume), they were not devoid of pharmacological activity.

Their stimulatory effect on protein synthesis (= anabolic effect) was confirmed (Okui et al. 1968; Uchiyama and Otaka 1974) (Fig. 3.1); this effect did not depend on increased transcription (it was not abolished by actinomycin D) and operated at the translational level.

Ecdysteroids (in practice mainly 20-hydroxyecdysone [20E], the most abundant natural representative) also displayed a hypoglycemic effect in diabetic ani-

R. Lafont (✉)
Sorbonne University, Paris Cedex 05, France

V. N. Syrov et al. (eds.), *Phytoecdysteroids II*,
https://doi.org/10.1007/978-981-95-8423-9_3

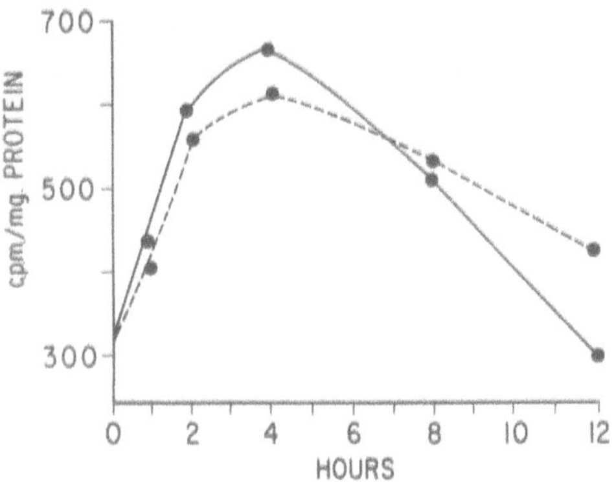

Fig. 3.1 Time course of the stimulation of amino acid incorporation after the in vivo injection of 20E (— 0.5 mg/kg, --- 5 mg/kg). Experiment used a cell-free fraction for protein synthesis in vitro: an S20 fluid prepared from mouse liver was incubated with a ^{14}C Chlorella hydrolysate (amino acid mixture) (after Uchiyama and Otaka 1974)

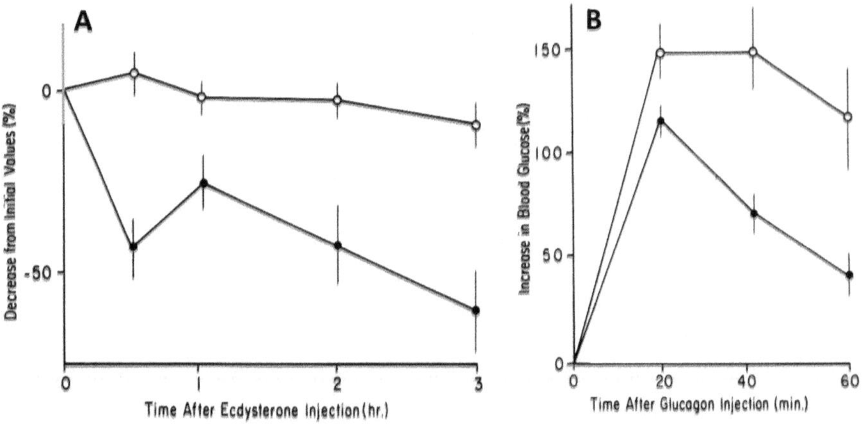

Fig. 3.2 A. Effect of a single dose of ecdysterone (20E) on hyperglycemia induced by alloxan in mice (o Control, • 20E). B. Effect of a single dose of 20E on hyperglycemia induced by glucagon in mice—20E was given 1 h before injection of glucagon (o Control, • 20E). (Reproduced from Uchiyama and Yoshida 1974)

mals (Yoshida et al. 1971; Uchiyama and Yoshida 1974; Fig. 3.2) and a hypocholesterolemic effect in rabbits fed a high-cholesterol diet (Matsuda et al. 1974; Fig. 3.3). All those effects were consistent with worldwide use of ecdysteroid-rich plants in traditional medicines (Sláma and Lafont 1995; Báthori and Pongrácz 2005).

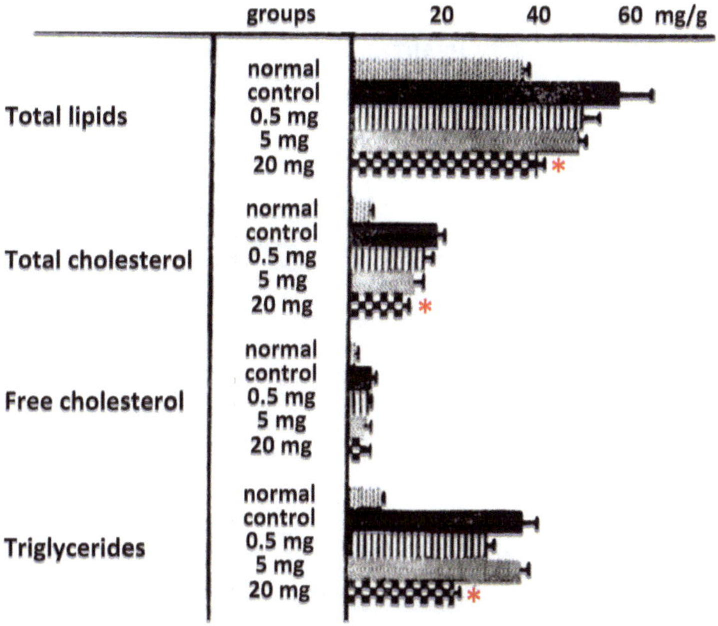

Fig. 3.3 Effect of ecdysterone on liver lipids in cholesterol-fed rabbits. Animals in each group were fed a diet containing 1% cholesterol together with 0, 0.5, 5, and 20 mg of ecdysterone per body per day for 12 weeks. Increases in total lipid, total cholesterol, and triglyceride levels were inhibited by oral administration of ecdysterone. The asterisk signals significant differences from control (modified after Matsuda et al. 1974)

3.2 Toward Understanding the Mechanisms of Action of Phytoecdysteroids

Vertebrate steroid hormones classically bind to specific nuclear receptors that operate as transcriptional factors, and this is also the case for ecdysteroids in insects. Additionally, steroids may display rapid effects by binding to membrane receptors, and this also applies to insects (Zhao 2020).

Binding studies performed with vertebrate steroid receptors using competition of ecdysteroids with radiolabeled vertebrate steroid hormones proved rather unsuccessful, although some displacement was observed (Báthori et al. 2008; Seidlova-Wuttke et al. 2010a, b).

First attempts to characterize early effects of 20E were undertaken by Catalán et al. (1980), who injected 20E intraperitoneally into mice (0.5 mg/kg) and measured the changes of cyclic nucleotide levels in several organs. They observed a transitory decrease in cAMP and increase in cGMP, suggesting a membrane effect (Fig. 3.4).

The cyclic nucleotide mechanism was further investigated by Mykhaylyk et al. (2001). These authors treated rat erythrocytes in vitro with 20E immobilized at the

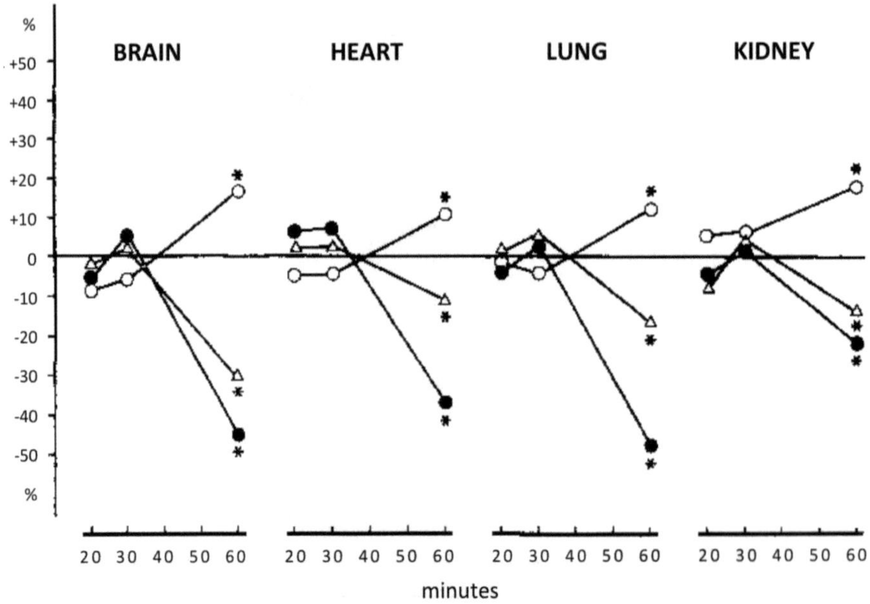

Fig. 3.4 Variations in the cAMP-protein kinase system in several organs of the mouse after administration of ecdysterone. -•-•-•- cAMP; -o-o-o cGMP;-Δ-Δ-Δ- cAMP-dependent protein kinase. Changes are expressed as a percentage of the control values (= 100%). Asterisks indicate statistically significant changes with respect to the control levels ($P < 0.001$) (Catalán et al. 1980)

surface of nanodispersed magnetite particles (10^{-15}–10^{-8} M) and then measured its effects on the cyclic nucleotide system, and they confirmed the decrease in cAMP and the increase in cGMP (Fig. 3.5).

More recently, Gorelick-Feldman et al. (2010) provided good evidence for a dual membrane effect on a GPCR coupled with phospholipase C and a calcium channel. The resulting increase in free Ca^{2+} activates PI3K/Akt, which results in the observed dose-dependent increase in protein synthesis by C2C12 myoblast cells (Fig. 3.6). These effects resemble those on a GPCR-type 20E receptor in insects (Cai et al. 2014).

The involvement of estrogen hormone receptor(s) (ERs) was then proposed (Parr et al. 2015). This was established by the similarity of estradiol (E2) and 20E effects on muscle cells in vitro, both being suppressed by an estrogen hormone receptor β (ERβ) antagonist. In silico docking experiments of 20E with ERα and ERβ showed multiple interactions, especially with ERβ, which is known to be involved in the anabolic effects of E2 (Fig. 3.7).

These observations raise several comments: (1) "binding" does not necessarily mean "activation" rather than "inhibition," whereas lack of binding means "no effect"; (2) direct binding studies of 20E to unspecified ERs (Báthori et al. 2008; Seidlova-Wuttke et al. 2010a) or specifically to ERα (Lafont et al. 2021a) were negative; (3) 20E did not increase uterus weight of castrated female rats, a specific

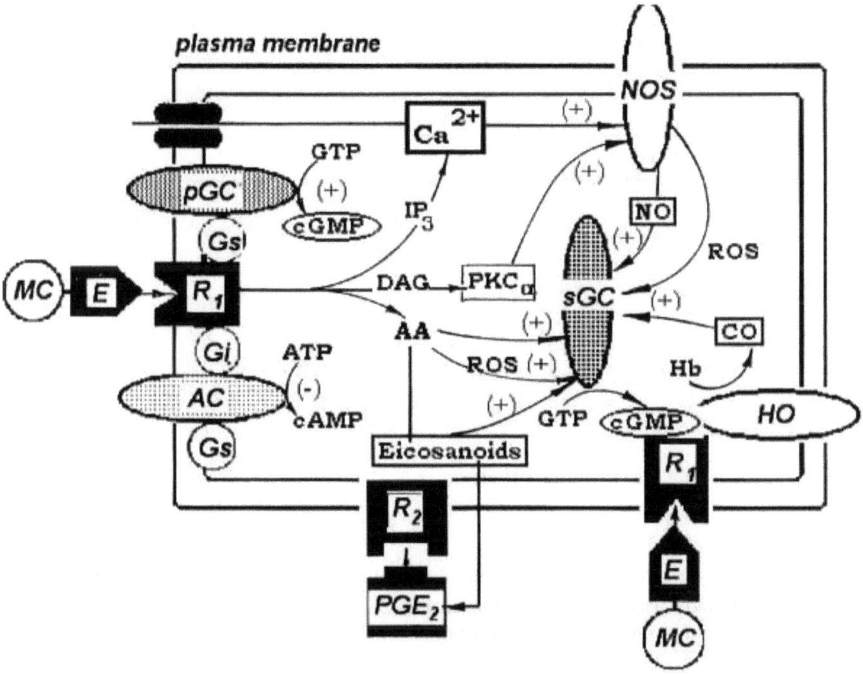

Fig. 3.5 Membrane action of microparticulate-bound 20E activates GC and inhibits AC. Hypothetical scheme for molecular mechanisms of receptor-mediated stimulation of cyclic GMP formation in erythrocytes. E, C27-steroid; MC, magnetic carrier; R1 and R2, receptors; AC, adenylate cyclase; pGC, particulate or membranous guanylate cyclase; sGC, soluble guanylate cyclase; Gs and Gi, G-protein stimulating and inhibiting subunits; NOS, NO synthase; HO, heme oxygenase; PKCα; α-isoform of protein kinase C; cGMP, cyclic 3′,5′ guanosine monophosphate; GTP, guanosine triphosphate; cAMP, cyclic adenosine monophosphate; ATP, adenosine triphosphate; NO, nitric oxide; IP3, inositol 3,4,5-triphosphate; DAG, 1,2-diacylglycerol; PGE2, prostaglandin E2; CO, carbon monoxide; Hb, hemoglobin; ROS, reactive oxygen species. (reproduced from Mykhaylyk et al. 2001)

effect of ERα activation (Seidlova-Wuttke et al. 2010b); (4) another in silico study concluded the absence of 20E binding to ERα and ERβ (Lapenna et al. 2015).

In our opinion, we should remain cautious with the conclusions of in silico docking experiments. Thus, recently, Tariq et al. (2025) found a very strong binding of 20E to androgen receptor (AR), ERα and ERβ, with a higher affinity of 20E compared with their respective endogenous ligands, testosterone and estradiol, which seems difficult to believe, as, for instance, protein synthesis stimulation in vitro requires much higher concentrations of 20E than of E2. Furthermore, the highest binding affinity for ERs was found for ERβ.

A remaining question was: Are we faced with a nuclear or a membrane estrogen receptor? This question is justified by the fact that ERβ signaling seems to be mainly non-nuclear (Maneix et al. 2015) and that rapid cell signaling involves palmitoylated

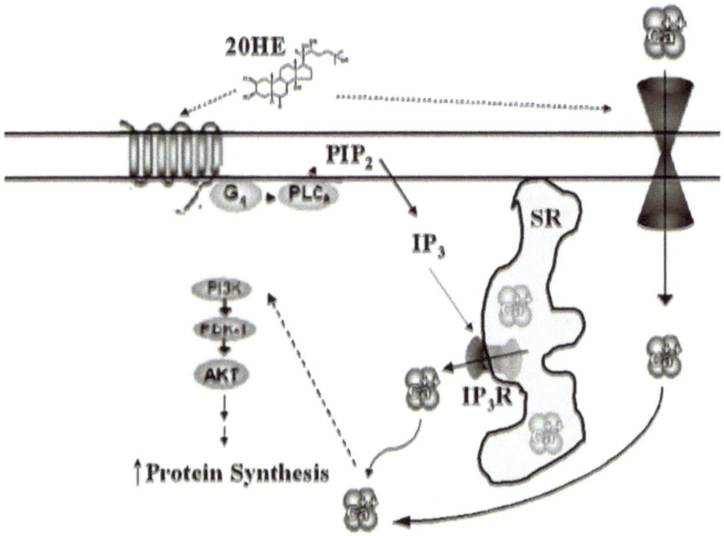

Fig. 3.6 20E elicits a free Ca^{2+} increase by two mechanisms (1) through activation of a GPCR coupled to PLC and (2) a membrane calcium channel. This activates the PI3K/Akt cascade leading to increased protein synthesis (reproduced from Gorelick-Feldman et al. 2010)

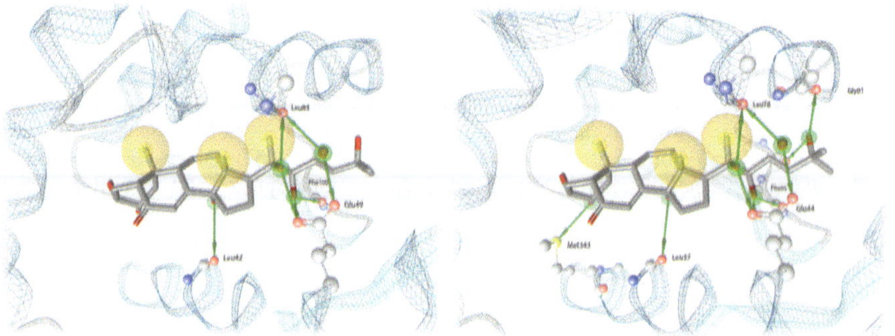

Fig. 3.7 In silico binding of 20E to estrogen receptors. ERα (PDB 3UUD, left) and ERβ (PDB 3OLL, right) with in silico docked ecdysterone (Parr et al. 2015). ERβ forms additional hydrogen bonds with Met343 to a hydroxyl group of ecdysterone and with the protein backbone to the terminal alkyhydroxyl group. Arrows indicate hydrogen bonds (green: donor, red: acceptor), while yellow spheres show hydrophobic interactions (reproduced with permission from Parr et al. 2015)

nuclear receptors anchored to the inner side of the cell membrane (Marino and Ascenzi 2006).

More recently, a new membrane mechanism was proposed based on the similarity of the pleiotropic effects of 20E and of angiotensin (1–7), which activates the protector arm of the renin-angiotensin-aldosterone system (RAAS) through the Mas receptor, a GPCR (Lafont et al. 2021a). Firstly, the activity of 20E persisted when it was covalently bound to albumin, thus establishing its binding to a cell membrane receptor. The effects of both 20E and Ang(1–7), i.e., the inhibition of myostatin gene expression and the production of NO, were abolished in the presence of A779, a Mas antagonist, and after treating target cells with a Mas siRNA in order to reduce Mas receptor concentrations. On this basis, we have tried to combine these data with those previously established with ERβ antagonists.

E2 also inhibited myostatin gene expression, but its effect was not abolished by A779. By contrast, A779 abolished the vasodilatory effect of E2, thus establishing crosstalk between Mas and ER (as proposed by Sobrino et al. 2017). But which ER subtype? The involvement of GPR30, a GPCR receptor of E2, was excluded, given the lack of effect of specific agonists or antagonists (Lafont et al. 2021a). Interestingly, 17αE2, an E2 epimer that does not bind nuclear estrogen receptors, was able to inhibit myostatin expression. Moreover, it was shown that a diarylheptanoid compound enhanced myoblast differentiation through a membrane estrogen receptor, and its effect was abolished by an inhibitor of palmitoylation (Tipbunjong et al. 2019).

The combination of all those data enabled the proposal of a mechanism involving both Mas and a membrane-anchored ER to allow the crosstalk to proceed (Fig. 3.8).

These data did not establish the direct binding of 20E to Mas. It could indeed bind to another GPCR that would then heterodimerize with Mas (in insects, 20E binds to a dopamine receptor [Lark et al. 2017]). Nevertheless, a very recent in silico docking experiment proposed such a binding with a rather low affinity (Lu et al. 2025). A second article also reported a high binding affinity of ecdysone to MasR (Quiroga et al. 2025).

Of course, these findings do not exclude other potential targets for 20E; thus, high concentrations of 20E (30–300 μM) can inhibit glutamate receptors (Okada et al. 1998; Mukai et al. 2002), an effect consistent with a small binding observed in an off-target assay (Lafont et al. 2021a). In silico docking experiments have

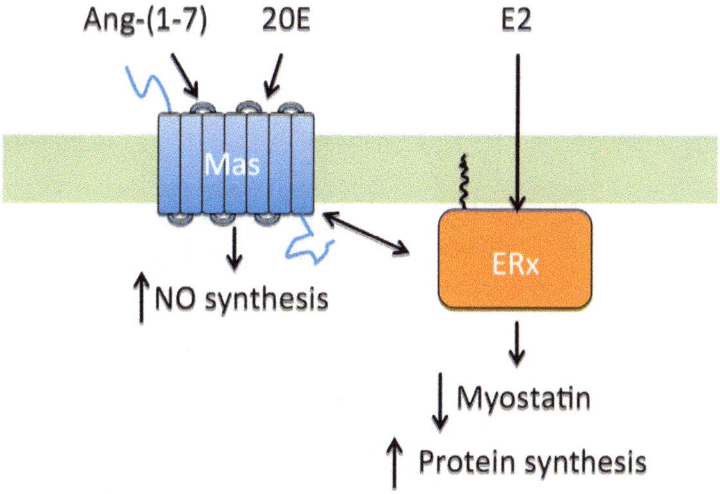

Fig. 3.8 Proposed mechanism of myostatin gene control by angiotensin(1–7), 20E and E2. According to this scheme, myostatin and protein synthesis would be controlled directly by ER and indirectly by Mas, while NO synthesis would be controlled directly by angiotensin(1–7) (Sampaio et al. 2007) and 20E (Omanakuttan et al. 2016), and indirectly by E2 (Sobrino et al. 2017). (Reproduced from Lafont et al. 2021a)

proposed many different targets, including not only nuclear and membrane receptors, but also various enzymes (Mamadalieva et al. 2022), and this will thus deserve additional studies in the future.

3.3 Multiple Transduction Mechanisms Involved by Mas Activation

Subsequent to Mas activation, several transduction mechanisms are activated, and this is also observed when treating C2C12 cells with 20E (Serova et al. 2024). The Mas receptor indeed not only activates a wide range of transduction mechanisms but also forms heterodimers with angiotensin II AT1R receptor (and inhibit it) and with bradykinin B2R receptor (McFall et al. 2020). Moreover, its transduction mechanisms may also depend on the ligand nature—peptidic vs. nonpeptidic (Tirupula et al. 2014) (Fig. 3.9).

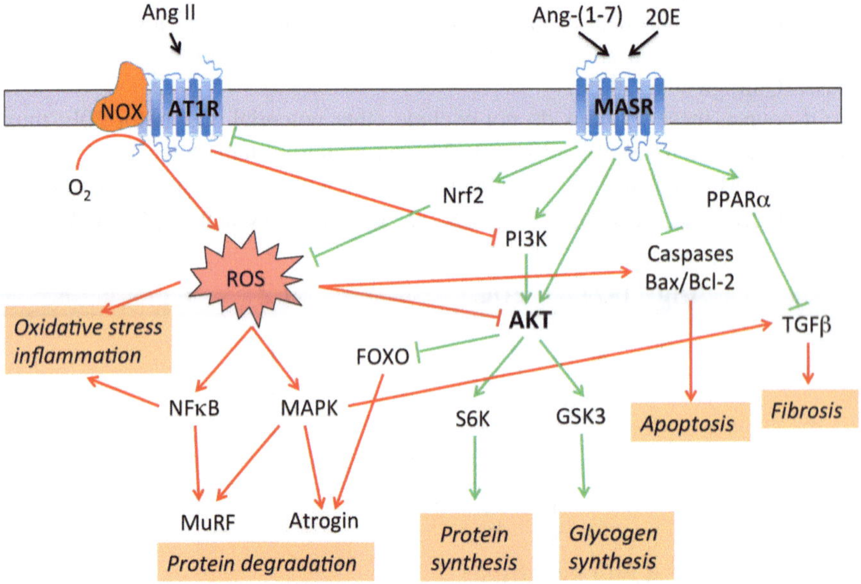

Fig. 3.9 The multiple transduction mechanisms of the Mas receptor, which are supposed to apply to both angiotensin-(1–7) and 20E. (Drawn after Cabello-Verrugio et al. 2015; Sigurta et al. 2016; McFall et al. 2020). AKT: protein kinase B; Bax/Bcl-2: Bcl-2-associated X protein; FOXO: Forkhead box class O transcription factor; GSK3: glycogen synthase kinase 3; MAPK: mitogen-activated protein kinase; NFκB: nuclear factor kappa-B; NOX: NADPH oxidase; Nrf2: nuclear factor erythroid 2–related factor 2; PI3K: phosphoinositide 3-kinase; PPARα: peroxisome proliferator-activated receptor alpha; ROS: reactive oxygen species; S6K: S6 kinase; TGFβ: transforming growth factor β. Red arrows indicate deleterious effects, and green arrows indicate beneficial effects

3.4 The Pleiotropic Effects of Ecdysteroids

Ecdysteroids have general antioxidative, anti-inflammatory, and antifibrotic effects. As their receptors are expressed in most/all tissues, they control plenty of organs, and their effects are very similar to those of angiotensin-(1–7), the endogenous ligand of the Mas receptor. This similarity also applies to many estradiol effects (Lafont et al. 2021a) (Tables 3.1 and 3.2).

These multiple effects of ecdysteroids on mammals/humans have already been described in many reviews (Sláma and Lafont 1995; Lafont and Dinan 2003; Báthori and Pongrácz 2005; Dinan and Lafont 2006, 2015; Báthori et al. 2008; Dinan 2009; Cahliková et al. 2011; Laekeman & Vlietinck 2013; Bajguz et al. 2015; Vasiliev et al. 2015; Akhmetkarimova and Temirgaziev 2021; Das et al. 2021; Dinan et al. 2021; Lafont et al. 2021b; Yusupova et al. 2021; Molokhova et al. 2022; Todorova et al. 2024).

While there has been a substantial increase in ecdysteroid consumption in the form of dietary supplements, there are still a limited number of clinical studies to assess the health claims for humans.

We will consider the present status of some recent preclinical/clinical developments (for a more exhaustive historical list of nutritional and clinical trials, refer to Lafont and Dinan 2003; Dinan et al. 2021 for historical studies, and to Lafont et al. 2021b more specifically for anabolic studies on muscles).

The first pharmacokinetic study of 20E in humans is recent (Dioh et al. 2023) and showed that the oral bioavailability is rather low, and thus large amounts (≥ 1 g) of 20E were required to reach micromolar peak concentrations in plasma. Such experiments provide an estimation of 20E doses to be administered and of the range of "physiological" concentrations to use in vitro. This will be discussed more extensively in Chap. 7, this volume.

Table 3.1 The general effects of 20E and related ecdysteroids

Effect	References (non-exhaustive)
Anti-fibrotic	Hung et al. (2012)
Anti-inflammatory	Kurmukov and Syrov (1988); Mhashilkar et al. (2016); Xia et al. (2016); Fang et al. (2017); Bhardwaj et al. (2019); Song et al. (2019); Yang et al. (2021)
Anti-oxidant	Osynskaya et al. (1992); Kuz'menko et al. (1999); Dai et al. (2002); Feng et al. (2012); Gao et al. (2012); Hu et al. (2012); Si et al. (2014); Liu et al. (2022)
Anti-stress	Syrov et al. (2014); Ramazonov et al. (2020); Baev et al. (2022)
Anti-apoptotic	Pan et al. (2016); Liu et al. (2025)

Table 3.2 The pleiotropic effects of 20E and related ecdysteroids

Organ	Effects	References
Skeletal muscles	Anabolic (young sportsmen) Anti-atrophic Anti-proteolytic Protection against DMD, SMA	Chermnykh et al. (1988); Syrov (2000); Emirova et al. (2004); Tóth et al. (2008); Gorelick-Feldman (2009); Gorelick-Feldman et al. (2008, 2010); Lawrence (2012); Parr et al. (2014); Hirunsai et al. (2016); Zwetsloot et al. (2022); Bézier et al. (2025); Blitek et al. (2025)
Adipose tissues	Reduces fat accumulation Hypolipidemic/ hypocholesterolemic	Syrov et al. (1983); Kizelsztein et al. (2009); Seidlova-Wuttke et al. (2010b); Foucault et al. (2012)
Nervous tissue	Neuroprotective	Wu et al. (2001); Mukai et al. (2002); Luo et al. (2009); Liu et al. (2011); Hu et al. (2012); Wu et al. (2017)
	Anti-nociceptive	Sarpong et al. (2016); Archana et al. (2022); Frachet et al. (2025); Tuleuov et al. (2025)
	Anti-Alzheimer, anti-Parkinson	Yang et al. (2010); Pan et al. (2016); Xu et al. (2018); Lim et al. (2020); Qu et al. (2022); Rong et al. (2024); Xing et al. (2024, 2025); Gholipour et al. (2025)
Circulatory system	Cardioprotective	Kurmukov and Ermishina (1991); Korkach et al. (2007); Xia et al. (2013a); Buniam et al. (2020)
	Vasorelaxant	Dongmo et al. (2014); Phungphong et al. (2017)
	Hematopoïesis activator	Syrov et al. (1997); Abdullaevna (2018)
Liver	Detox, NASH, anti-steatosis Reduction of cholestasis	Shakhmurova et al. (2010a); Xia et al. (2013b); Agoun et al. (2021); Syrov et al. (2021a); Kang and Li (2022); Buniam et al. (2023)
Pancreas	β-cell protective	Sundaram et al. (2012b); Wang et al. (2017); Kuchkarova et al. (2020); Syrov et al. (2021a, b)
Lungs	Lung protective, surfactant production, COVID-19	Wu et al. (1998); Li et al. (2013); Xia et al. (2016); Song et al. (2019); Lobo et al. (2024)
Kidneys	Protective, anti-fibrotic	Syrov et al. (1992); Zou et al. (2010); Hung et al. (2012)
Stomach	Anti-ulcerous	Shakhmurova et al. (2010b); Zhou et al. (2010)
Bones, cartilages	Anti-osteoporotic Glucocorticoids counteraction	Gao et al. (2008); Kapur et al. (2010); Seidlova-Wuttke et al. (2010b); Dai et al. (2015, 2017); Tang et al. (2021, 2023); Xie et al. (2025)
Skin and hair	Photoprotective, wound repair, Anti-psoriatic, anti-alopecia	Detmar et al. (1994); Ishida et al. (1999); Syrov and Khushbaktova (1996); Hou et al. (2007); Li et al. (2008); Nsimba et al. (2008); Zhegn et al. (2008); Ho et al. (2015); Ramazanov et al. (2017)
Diabetes	Anti-diabetic, glycemia reduction, insulin sensibilization	Yoshida et al. (1971); Kizelsztein et al. (2009); Sundaram et al. (2012a, b); Naresh Kumar et al. (2013); Mallek et al. (2018); Syrov et al. (2021b)
Cancer	Anti-cancer adjuvant, proapoptotic	Konovalova et al. (2002); Martins et al. (2012, 2015); Csábi et al. (2016); Hunyadi et al. (2017); Romaniuk-Drapała et al. (2021); Savchenko et al. (2021, 2025); Makhamov et al. (2025)

3.5 Effects of 20E/Ecdysteroids on Muscles

The effects on muscles have been especially investigated, and significant improvements of muscle mass, muscle strength, and physical performances have been established. Most studies were performed with young sportsmen receiving 20E with intensive physical training eventually associated with protein/amino acids supplementation. Those early studies provided the background data for the marketing of 20E-based dietary supplements. More recent nutritional clinical studies were performed (Isenmann et al. 2019) with the aim of prohibiting the use of ecdysteroids as "doping substances." Analytical methods were developed to detect their unauthorized use by analyzing sportsmen's urine (Kraiem et al. 2021).

These beneficial effects of 20E on muscles have been recently investigated in pathological contexts, i.e., the age-related degradation of muscle mass and performances (sarcopenia) and two genetic muscle degenerative diseases such as Duchenne Muscular Dystrophy (DMD) and Spinal Muscular Atrophy (SMA).

Sarcopenia corresponds to a progressive reduction of muscle mass/strength which may progressively lead to frailty and disability. It results from an imbalance of muscle protein synthesis and degradation, a reduction of muscle renewal capacity (satellite cell senescence), and the progressive death of motoneurons. At first, an observational study was undertaken over 6 months in order to appreciate the kinetics of degradation of performances in selected patients over 65y (SARA-OBS study, NCT03021798 2020, Fielding et al. 2025a). Then the effect of BIO101 (a pharmaceutical-grade preparation of 20E, 175 or 350 mg twice daily) was investigated over 6–9 months and demonstrated an improvement of walking mobility with the higher dose only (SARA-INT, NCT03452488 2024; Fielding et al. 2025b). A confirmatory phase 3 study is expected to start in 2026.

Genetic myopathies result from gene defects that result in a progressive degradation of muscles. Of course, 20E cannot be expected to cure such pathologies, but it could possibly slow the muscle degradation process. Moreover, when exon-skipping treatments with ASOs (antisense oligonucleotides) are being developed to counteract the genetic problem, it makes sense to try the effects of a cotreatment with 20E. These investigations have been conducted in animal models and have given very encouraging results.

Duchenne muscular dystrophy (DMD) is a muscle-degenerative disease due to the dystrophin (a muscle structural protein) gene defect, resulting in the formation of a poorly functional truncated protein. ASO treatment can restore the formation of a quasi-dystrophin. Experiments with genetically modified mice showed that both single treatments improved functional outcomes (distance run and maximum speed in the treadmill exhaustion test). Moreover, improvements in specific force and a decrease in force drop after eccentric contraction were observed only with the combination therapy (Blitek et al. 2025).

Spinal Muscular Atrophy (SMA) is a neuromuscular disease linked with a defect of SMN1 (survival of motor neuron 1) gene results in progressive paralysis and finally death when respiratory muscles are concerned. In severe SMA mice, 20E alone partially protected muscles and motor neurons independently of SMN production. When combined with ASO treatment, the improvements were significantly enhanced (Bézier et al. 2025).

In both DMD and SMA, the results of cotreatments ASO+20E are good enough to envisage future clinical developments.

3.6 20E and Pulmonary Diseases

Lungs are directly faced to both chemical pollutants and a plenty of invasive microorganisms (viruses, bacteria, yeasts …). Microorganisms in particular can compromise lung function (inflammation, edema, …) and trigger sepsis-induced Acute Lung Injury (ALI) or Acute Respiratory Distress Syndrome (ARDS), which represent a major cause of death in intensive care units (Xu et al. 2024).

The renin–angiotensin system is once again directly involved in the control of lung damage (classical arm—angiotensin II) or repair (protective arm—angiotensin-(1–7)) (Tan et al. 2018; Margalhães et al. 2019). The different levels of control are shown in Fig. 3.10.

Given the high similarity of 20E and angiotensin-(1–7) effects, it was expected that 20E would have a protective role in the lungs, and indeed this has been the case in lung ALI evoked by exposure of mice to bacterial lipopolysaccharides (LPS) (Song et al. 2019). 20E can reduce lung edema (Xiong et al. 2006) and reduce the

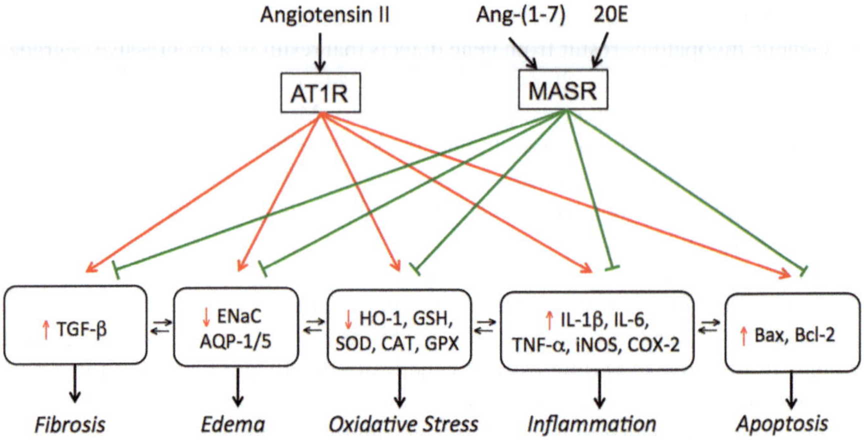

Fig. 3.10 Angiotensin II and angiotensin-(1–7) play opposite roles on lungs and control many pathological developments. AQP-1/5: aquaporins-1/5; Bax/Bcl-2: Bcl-2-associated X protein; CAT: catalase; COX-2: cyclooxygenase-2; ENaC: epithelial sodium channel; GPX: glutathione peroxidase; GSH: glutathione; HO-1: heme oxygenase-1; IL-1β: interleukin-1β; IL-6: interleukin-6; SOD: superoxide dismutase; TGFβ: transforming growth factor β.; TNFα: tumor necrosis factor α

production of inflammatory cytokines, thus preventing the occurrence of a "cytokine storm" (Xia et al. 2016). It can also increase the production of lung surfactant protein A (SP-A) (Li 2013; Li et al. 2013). This finding is particularly relevant because surfactant functions as an opsonin by binding to microorganism surfaces and stimulating the innate immune system. This has been particularly shown with several viruses (Benne et al. 1995; Ghildyal et al. 1999).

The COVID-19 pandemic has provided the opportunity to perform a clinical trial in order to test the efficacy of 20E on patients developing a severe form of the infection (NCT04472728 2023). The use of 20E in this case seems particularly relevant given the fact that the virus targets ACE2, a peptidase involved in angiotensin-(1–7) synthesis, and is therefore expected to impair the protective arm of the renin–angiotensin system; thus, providing an exogenous Mas activator was expected to represent a most accurate treatment (Latil et al. 2021). A clinical project was therefore designed (Dioh et al. 2021), and a multicenter trial was performed using BIO101 (300 mg twice daily) with hospitalized patients developing a severe form of pneumonia with impaired respiration. This treatment proved effective in preventing the need for intubation or preventing death (Lobo et al. 2024). However, the rapid development of RNA vaccines and the appearance of less severe virus mutants did not allow the further development of this 20E treatment, although it had proved more effective than many antiviral substances (Camelo et al. 2021, 2025).

3.7 Ecdysteroids for the Treatment of Infections/Parasitoses

This section will be devoted to the possible protection provided by 20E/ecdysteroids against pathogenic bacteria, fungi (yeasts), or parasitic protozoa or helminths. Rather than an effect on the host to enhance its defense mechanisms, it may be understood as a direct toxic effect of 20E against the pathogens/parasites themselves (Dinan and Lafont 2022). This is, however, debatable (see the above data on the production of SP-A, an antiviral agent).

Concerning bacteria, most studies performed with pure ecdysteroids have been negative, even if extracts of the plants containing them were positive (Shirshova et al. 2006; Mamadalieva et al. 2013; Yusupova et al. 2020). Ecdysteroids are nevertheless active when combined with antibiotics on resistant bacterial strains (Kim et al. 2009; Smirnova et al. 2016).

Helminths (Platyhelminths and Nemathelminths) contain ecdysteroids, which would control some physiological processes (growth and reproduction) (Franke and Käuser 1989). It is therefore expected that their exposure to (high) exogenous ecdysteroids would perturb their physiology and possibly kill them (Spindler et al. 1986; Franke and Käuser 1989).

Given the low oral bioavailability of ecdysteroids in mammals, this approach would be particularly suitable for intestinal parasites. Indeed, a positive result was obtained against hymenolepiasis (Table 3.3), although the 20E dose used was rather low (Makhmudova 2012), and this result deserves additional studies with higher doses.

Table 3.3 Antiparasite studies using 20E

Type	Class	Daily dose	Duration	Effect	References
Giardiasis	Protozoa	5 mg 3–4x	10 days	68.7% parasites eliminated	Osipova et al. (2002)
Lambliasis	Protozoa	5 mg 4x	10 days	Most parasites eliminated	Islamova et al. (2010)
Hymenolepiasis	Cestodes	5 mg 3x	2 weeks	36.4% parasites eliminated	Makhmudova (2012)
Giardiasis	Protozoa	100 mg 2x	10 days	Most parasites eliminated	Toychiev et al. (2022)

Parasitic protozoa deserve special attention. Already in 1960, Cleveland et al. (1960) described the effect of exogenous ecdysteroids on the sexual reproduction of insect protozoan parasites. Studies on Babesia gibsoni, an intraerythrocytic parasite causing hemolytic anemia in dogs (Subeki et al. 2005), showed some in vitro susceptibility to 20E, but this required high concentrations (100 μg/mL). Even worse, experiments with *Toxoplasma gondii* showed a stimulation by 20E of parasite proliferation (Dzitko et al. 2015).

The case of giardiasis appears more interesting, as it concerns an intestinal protozoan parasite. Indeed, small-scale studies (Osipova et al. 2002; Islamova et al. 2010) gave positive results. More recently, Toychiev et al. (2022) used much higher doses of 20E and obtained improved results, which enabled a larger clinical study (NCT04827537 2025). This study will enroll 200 male participants (aged 18–27) recruited from among water sport athletes, who are very frequently faced with this parasitosis. It will compare the antiprotozoal activity of 20E and metronidazole to a placebo. For this specific use, the low oral bioavailability of 20E allows it to remain at high concentration within the intestinal tract, which is an advantage for being active against the parasites.

3.8 Summary

20-Hydroxyecdysone (and several related phytoecdysteroids) displays pleiotropic beneficial effects on mammals.

The mechanism of their action appears to involve at least a partnership of the Mas receptor of angiotensin-(1–7) and of a membrane-associated estrogen receptor.

On the basis of the effects of many ecdysteroid-containing medicinal plants and the effects observed for the pure molecule in many animal models, 20-hydroxyecdysone has been successfully tested in several clinical trials and represents a promising candidate medicine.

References

Abdullaevna SG (2018) Influence of phytoecdysteroids to erythropoiesis, leukopoiesis and titer of antibodies in experimental animals. Int J Sci Res Pub 8(6): 58–61. https://doi.org/10.29322/IJSRP.8.6.2018.p7809

Agoun H, Semiane N, Mallek A, Bellahreche Z, Hammadi S, Madjreb M, Abdlalli M, Khalkhal A, Dahmani Y (2021) High-carbohydrate diet-induced metabolic disorders in Gerbillus tarabuli (a new model of non-alcoholic fatty liver disease). Protective effects of 20-hydroxyecdysone. Arch Physiol Biochem 127(2):127–135. https://doi.org/10.1080/13813455.2019.1621350

Akhmetkarimova ZhS, Temirgaziev BS (2021) Phytoecdysteroids: Chemical structure and biological activity. Euras J Appl Biotech 2021(2):3–22. *(In Russian)*

Archana J, Annapurna A, Devayani P (2022) Neuroprotective role of Tinospora cordifolia extract in streptozotocin induced neuropathic pain. Braz J Pharm Sci 58:e18501. https://doi.org/10.1590/s2175-97902020000118501

Baev, A. Y., Charishnikova, O. S., Khasanov, F. A., Nebesnaya, K. S., Makhmudov, A. R., Rakhmedova, M. T., Khushbaktova, Z. A., Syrov, V. N. and Levitskaya, Y. V. (2022) Ecdysterone prevents negative effect of acute immobilization stress on energy metabolism in rat liver mitochondria. J Steroid Biochem Molec Biol 219 :106066. https://doi.org/10.1016/j.jsbmb.2022.106066

Bajguz A, Bakala I, Talarek M (2015) Chapter 5: Ecdysteroids in plants and their pharmacological effects in vertebrates and Humans. In: Studies in Natural Product Chemistry, Ed. Atta-ur-Rahman, FRS, 45, pp.121–145, Elsevier B.V.

Báthori M, Pongrácz Z (2005) Phytoecdysteroids—From isolation to their effects on humans. Curr Med Chem 12:153–172. https://doi.org/10.2174/0929867053363450

Báthori M, Tóth N, Hunyadi A, Márki Á, Zádor E (2008) Phytoecdysteroids and anabolic-androgenic steroids—Structure and effects on Humans. Curr Med Chem 15:75–91. https://doi.org/10.2174/092986708783330674

Benne CA, Kraaijeveld CA, Van Strijp JAG, Brouwer E, Harmsen M, Verhoef J, Van Golde LMG, Van Iwaarden JF (1995) Interactions of surfactant protein A with influenza a viruses: Binding and neutralization. J Infect Dis 171(2):335–341. https://doi.org/10.1093/infdis/171.2.335

Bézier C, Cottin S, Hashemi PN, El Khoury M, Clerc Z, Balducci C, Sapaly D, Weill L, Lafont R, Veillet S, Dilda PJ, Charbonnier F, Latil M, Biondi O (2025) Drug candidate BIO101 for spinal muscular atrophy as monotherapy or combined with the antisense oligonucleotide ASO-10-27. J Cachexia Sarcopenia Muscle 16(5):e70104. https://doi.org/10.1002/jcsm.70104

Bhardwaj M, Mamadalieva NZ, Chauhan AK, Kang SC (2019) α-Ecdysone suppresses inflammatory responses via the Nrf2 pathway in lipopolysaccharide-stimulated RAW 264.7 cells. Int Immunopharmacol 73:405–413. https://doi.org/10.1016/j.intimp.2019.05.038

Blitek M, Gastaldi C, Doisy M, Le Coz O, Tensorer T, Garcia L, Goyenvalle A (2025) Combined 20-hydroxyecdysone and antisense-mediated exon skipping improve functional outcomes in a mouse model of Duchenne Muscular Dystrophy. Nucleic Acid Ther 35(2):55–67. https://doi.org/10.1089/nat.2024.0085

Buniam J, Chukijrungroat N, Rattanavichit Y, Surapongchai J, Weerachayaphom J, Bupha-Intr T, Saengsirisuwan V (2020) 20-Hydroxyecdysone ameliorates metabolic and cardiovascular dysfunction in high-fat-high-fructose-fed ovariectomized rats. BMC Complement Med Ther 20: 140. https://doi.org/10.1186/s12906-020-02936-1

Buniam J, Chansela P, Weerachayaphum J, Saengsirisuwan V (2023) Dietary supplementation with 20-hydroxyecdysone ameliorates hepatic steatosis and reduces white adipose tissue mass in ovariectomized rats fed a high-fat, high-fructose diet. Biomedicines 11(7):2071. https://doi.org/10.3390/biomedicines11072071

Burdette WJ, Coda RL (1963) Effect of ecdysone on incorporation of C^{14}-leucine into hepatic protein in vitro. *Proc Soc Exp Biol Med* 112:216–217. https://doi.org/10.3181/00379727-112-27996

Butenandt A, Karlson P (1954) Über die Isolierung eines Metamorphose-Hormons der Insekten in kristallisierter Form. Z Naturforsch 9b:389–391.

Cabello-Verrugio C, Morales MG, Rivera JC, Cabrera D, Simon F (2015) Renin-angiotensin system: An old player with novel functions in skeletal muscle. Med Res Rev 35(3):437–463. https://doi.org/10.1002/med.21343

Cahliková L, Macáková K, Chlebek J, Host'álková A, Kulkhánková A, Opletal L (2011) Ecdysterone and its activity on some degenerative diseases. Nat Prod Commun 6(5):707–718.

Cai MJ, Dong DJ, Wang Y, Liu PC, Miu W, Wang JX (2014) G-protein-coupled receptor participates in 20-hydroxyecdysone signaling on the plasma membrane. Cell Commun Signal 12:9. https://doi.org/10.1186/1478-811X-12-9

Camelo S, Latil M, Agus S, Dioh W, Veillet S, Lafont R, Dilda PJ (2021) A comparison between virus- versus patients- centred therapeutic attempts to reduce COVID-19 mortality. Emerg Microbes Infect 10(1):2256–2263. https://doi.org/10.1080/22221751.2021.2006579

Camelo S, Dioh W, Teixeira JP, Busse LW, Nair G, Plantefeve G, Morelot-Panzini C, Lobo SM, Self WH, Collins SP, Van Maanen R, Veillet S (2025) Modulation of the renin-angiotensin system against COVID-19: A path forward? Int J Infect Dis 154:107867. https://doi.org/10.1016/j.ijid.2025.107867

Catalán RE, Martinez, AM, Aragonés MD (1980) Heterophylic action of ecdysterone. Effect on the cAMP-protein kinase system in mice. Naturwissenschaften 67:520. https://doi.org/10.1007/BF01047642

Cleveland LR, Burker Jr AW, Karlson P (1960) Ecdysone induced modifications in the sexual cycles of the protozoa of *Cryptocercus*. J Protozool 7(3):229–239.

Csábi J, Martins A, Sinka I, Molnár J, Zupkó I, Tóth G, Tillekeratne LVM, Hunyadi, A (2016) Synthesis and in vitro evaluation of the antitumor potential and chemo-sensitizing activity of fluorinated ecdysteroid derivatives. MedChemComm 7:2282–2289. https://doi.org/10.1039/c6md00431h

Chermnykh NS, Shimanovsky NL, Shutko GV, Syrov VN (1988) Effects of methandrostenolone and ecdysterone on physical endurance of animals and protein metabolism in the skeletal muscles. *Farmakologiya i Toksikologiya* 6: 57–62. [in Russian]

Dai JQ, Cai YJ, Shi YP, Zhang YH, Liu ZL, Yang L, Li Y (2002) Antioxidant activity of ecdysteroids from *Serratula strangulata*. Chinese J Chem 20(5):497–501.

Dai W, Zhang H, Zhong ZA, Jiang L, Chen H, Lay YA, Kot A, Ritchie RO, Lane NE, Yao W (2015) β-Ecdysone augments peak bone mass in mice of both sexes. Clin Orthop Relat Res 473(8):2495–2504. https://doi.org/10.1007/s11999-015-4246-5

Dai WW, Wang LB, Lin GQ, Wu HJ, Zhang J, Wang CL, Wei YJ, Lee JH, Lay YE, Yao W (2017) Beta-ecdysone protects mouse osteoblasts from glucocorticoid-induced apoptosis *in vitro*. Planta Med 83(11):888–894. https://doi.org/10.1055/s-0043-107808

Das N, Mishra SK, Bishayee A, Ali ES, Bishayee A (2021) The phytochemical, biological, and medicinal attributes of phytoecdysteroids: an updated review. Acta Pharm Sin B 11(7):1740–1766. https://doi.org/10.1016/j.apsb.2020.10.012

Detmar M, Dumas M, Bonté F, Meybeck A, Orfanos CE (1994) Effects of ecdysterone on the differentiation of normal human keratinocytes in vitro. Eur J Dermatol 4:558–562.

Dinan L (2009) The Karlson lecture. Phytoecdysteroids : what use are they ? Arch Insect Biochem Physiol 72(3): 126–141. https://doi.org/10.1002/arch.20334

Dinan L, Lafont R (2006) Effects and applications of arthropod steroid hormones (ecdysteroids) in mammals. J Endocrinol 191:1–8. https://doi.org/10.1677/joe.1.06900

Dinan L, Lafont R (2015) Phytoecdysteroids occurrence, distribution, biosynthesis, metabolism, mode of action and applications: Developments from 2005 to 2015. Pharmaceutical Bulletin (Kazakhstan) (3–4):9–37.

Dinan L, Lafont R (2022) Ecdysteroids as defensive chemicals. Adv Insect Physiol 63:107–154.

Dinan L, Dioh W, Veillet S, Lafont R (2021) 20-Hydroxyecdysone, from plant extracts to clinical use: Therapeutic potential for the treatment of neuromuscular, cardio-metabolic and respiratory diseases. Biomedicines, 9:492. https://doi.org/10.3390/biomedicines9050492

Dioh W, Chabane M, Tourette C, Azbekyan A, Morelot-Panzini C, Hajjar LA, Lins M, Nair GB, Whitehouse T, Mariani J, Latil M, Camelo S, Lafont R, Dilda PJ, Veillet S, Agus S (2021) Testing the efficacy and safety of BIO101, for the prevention of respiratory deterioration, in patients with COVID-19 pneumonia (COVA study): a structured summary of a study protocol for a randomised controlled trial. Trials 22(1):42. https://doi.org/10.1186/s13063-020-04998-5

Dioh W, Tourette C, Del Signore S, Daudigny L, Dupont P, Dilda PJ, Lafont R, Veillet S (2023) A phase I study for safety and pharmacokinetics of BIO101 (20-hydroxyecdysone) in healthy young and older adult adults. J Cachexia Sarcopenia Muscle 14:1259–1273. https://doi.org/10.1002/jcsm.13195

Dongmo AB, Nkeng-Efouet PA, Devkota KP, Wegener JW, Sewald N, Wagner H, Vierling W (2014) Tetra-acetylajugasterone a new constituent of *Vitex cienkowskii* with vasorelaxant activity. Phytomedicine 21:787–792. https://doi.org/10.1016/j.phymed.2014.02.009

Dzitko, K., Grzybowski, M. M., Pawełczyk, J., Dziadek, B., Gatkowska, J., Stączek, P. and Długońska, H. (2015) Phytoecdysteroids as modulators of the *Toxoplasma gondii* growth rate in human and mouse cells. Parasit Vectors 8:422. https://doi.org/10.1186/s13071-015-1019-7

Emirova LR, Rozhkova EA, Panyushkin VV, Mirzoyan RS, Seifulla NR (2004) The effect of cytamines and their combinations with ecdystene, apilak, vitamax and essentiale on the work capability of high-rank athletes. Eksp klin Farmakol (Exp clin Pharmacol) 67(3):66–68. *(In Russian)*

Fang L, Li J, Zhou J, Wang X, Guo L (2017) Isolation and purification of three ecdysteroids from the stems of *Diploclisia glaucescens* by high-speed and their anti-inflammatory activities. Molecules 22(8):1310. https://doi.org/10.3390/molecules22081310

Feng CY, Huang XR, Qi MX, Tang SW, Guo N, Hu YH, Chen S, Ke FJ (2012) Effects of ecdysterone on the expression of NF-κB p65 in H_2O_2 induced oxidative damage of human lens epithelial cells. Zhongguo Zhong Xi Yi Jie He Za Zhi (In Chinese Journal of Integrated Traditional and Western Medicine) 32(1):76–79. *(In Chinese)*

Fielding RA, Rolland Y, Bruyère O, Desvarieux M, Donini LM, Incalzi RA, Muscaritoli M, Tchalia A, Bonnefoy M, Rondanelli M, Van Maanen R, Mariani J, Margalev C, Del Signore S, Tourette C, Dioh W, Veillet S (2025a) Characterizing sarcopenia and sarcopenic obesity in patients aged 65 years and over, at risk of mobility disability: a multicenter observational trial (SARA-OBS). BMC Geriatr 25(1):590. https://doi.org/10.1186/s12877-025-05895-9

Fielding RA, Dao MM, Cannon K, Desvarieux M, Miller SS, Gimness MP, Brandon DM, Villareal DT, Bruyère O, Bautmans I, Rickner K, Perry R, Kritchevsky SB, Musi N, Chehade JM, Kirsten JL, Gielen E, Pickrell P, Dilda P, Lafont R, Margalev C, Rolland Y, Del Signore S, Mariani J, Agus S, Tourette C, Dioh W, van Maanen R, Veillet S (2025b) BIO101 in sarcopenic seniors at risk of mobility disability: Results of a double-blind randomized interventional phase 2b trial. J Cachexia Sarcopenia Muscle, 16(2):e13750. https://doi.org/10.1002/jcsm.13750

Foucault AS, Mathé V, Lafont R, Even P, Dioh W, Veillet S, Tomé D, Huneau D, Hermier D, Quignard-Boulangé A (2012) Quinoa extract enriched in 20-hydroxyecdysone protects mice from diet-induced obesity and modulates adipokines expression. Obesity 20:270–277. https://doi.org/10.1038/oby.2011.257

Frachet, S., Danigo, A., Latil, M., Dilda, P. J., Bessaguet, F., Richard, L., Sturtz, F., Magy, L. and Demiot, C. (2025) Neuroprotective effect of Mas activation by BIO101 in vincristine-induced small fiber-neuropathy. J Peripher Nerv Syst 30:e70055. https://doi.org/10.1111/jns.70055

Franke S, Käuser G (1989) Occurrence and hormonal role of ecdysteroids in non-arthropods. In: Ecdysone—from Chemistry to mode of action, J. Koolman Ed., pp. 296–307. Georg Thieme Verlag, Stuttgart.

Gao L, Cai G, Shi X (2008) β-Ecdysterone induces osteogenic differentiation in mouse mesenchymal stem cells and relieves osteoporosis. Biol Pharm Bull 31(12):2245–2249. https://doi.org/10.1248/bpb.31.2245

Gao S, Tang Z, Chen T, Li F, Shan Y, Feng H (2012) Ecdysterone protects brain injury against free radical in rats. Di-San Junyi Daxue Xuebao (Journal of Third Military Medical University) 34(23):2357–2359. *(In Chinese)*

Ghildyal R, Hartley C, Varrasso A, Meanger J, Voelker RD, Anders EM, Mills J (1999) Surfactant protein A binds to the fusion glycoprotein of respiratory syncytial virus and neutralizes virion infectivity. J Infect Dis 180(6):2009–2013. https://doi.org/10.1086/315134

Gholipour P, Komaki A, Parsa H, Ralezani M (2025) Ecdysterone and high-intensity interval training mitigate Alzheimer's pathology in rats: impacts on depression, synaptic plasticity, and neuroinflammation. Mol Neurobiol 62(11):13933–13957. https://doi.org/10.1007/s12035-025-05168-x

Gorelick-Feldman J, MacLean D, Ilic N, Poulev A, Lila MA, Cheng D, Raskin I (2008) Phytoecdysteroids increase protein synthesis in skeletal muscle cells. J Agric Food Chem 56:3532–3537. https://doi.org/10.1021/jf073059z

Gorelick-Feldman JI. 2009. Phytoecdysteroids – Understanding their anabolic activity. PhD Thesis, Rudgers University. https://rucore.libraries.rutgers.edu/rutgers-lib/25806/PDF/1/play/

Gorelick-Feldman J, Cohick W, Raskin I (2010) Ecdysteroids elicit a rapid Ca^{2+} flux leading to Akt activation and increased protein synthesis in skeletal muscle cells. Steroids 70:632–637. https://doi.org/10.1016/j.steroids.2010.03.008

Hirunsai M, Yimlamai T, Suksamrarn A (2016) Effect of 20-hydroxyecdysone on proteolytic regulation in skeletal muscle atrophy. In Vivo 30: 869–878. https://doi.org/10.21873/invivo.11007

Ho R, Teai T, Meybeck A, Raharivelomanana P (2015) UV-protective effects of phytoecdysteroids from *Microsorum grossum* extracts on human dermal fibroblasts. Nat Prod Commun 10(1):33–36.

Hou L, Wu X, Wang WJ (2007) Ecdysterone promotes wound healing in rabbits. Nan Fang Yi Ke Da Xue Xue Bao (J South Med Univ) 27(3):312–314. *(in Chinese)*

Hu J, Luo CX, Chu WH, Shan YA, Qian ZM, Zhu G, Yu YB, Feng H (2012) 20-Hydroxyecdysone protects against oxidative stress-induced neuronal injury by scavenging free radicals and modulating NF-κB and JNK pathways. PLoS ONE 7(12):e50764. https://doi.org/10.1371/journal.pone.0050764

Hung, T. J., Chen, W. M., Liu, S. F., Liao, T. N., Lee, T.C., Chuang, L. Y., Guh, J. Y., Hung, C. Y., Hung, H. J., Chen, P. Y., Hsieh, P. F. and Yang, Y. L. (2012) 20-Hydroxyecdysone attenuates TGF-β1-induced renal cellular fibrosis in proximal tubule cells. J Diabetes Complications 26(6):463–469. https://doi.org/10.1016/j.jdiacomp.2012.06.014

Hunyadi A, Csábi J, Martins A, Molnár J, Balázs A, Tóth G (2017) Backstabbing P-gp: Side-chain cleaved ecdysteroid 2,3-dioxolanes hyper-sensitize mdr cancer cells to doxorubicin without efflux inhibition. Molecules, 22 (2), 199. https://doi.org/10.3390/molecules22020199

Ishida, H., Inaoka, Y., Shibatani, J., Fukushima, M. and Tsuji, K. (1999) Studies of the active substances in herbs used for hair treatment. II. Isolation of hair regrowth substances, acetosyringone and polyporusterone A and B, from *Polyporus umbellatus* Fries. Biol Pharm Bull 22:1189–1192. https://doi.org/10.1248/bpb.22.1189

Islamova JI, Syrov VN, Khushbaktova A, Osipova SO (2010) The efficacy of ecdystene versus metronidazole in the treatment of lambliasis. Med Parazitol (Mosk) (2):14–17. *(In Russian)*

Isenmann E, Ambrosio G, Joseph JF, Mazzarino M, de la Torre X, Zimmer P, Kazlauskas R, Goebel C, Botrè F, Diel P, Parr MC (2019) Ecdysteroids as non-conventional anabolic agent: performance enhancement by ecdysterone supplementation in humans. Archives of Toxicology 93:1807–1816. https://doi.org/10.1007/s00204-019-02490-x

Kang P, Li S (2022) Makisterone A attenuates experimental cholestasis by activating the farnesoid X receptor. Biochem Biophys Res Commun 623:162–169. https://doi.org/10.1016/j.bbrc.2022.07.030

Kapur P, Wuttke W, Jarry H, Seidlova-Wuttke D (2010) Beneficial effects of β-ecdysone on the joint, epiphyseal cartilage tissue and trabecular bone in ovariectomized rats. Phytomedicine 17:350–355. https://doi.org/10.1016/j.phymed.2010.01.005

Kim, E. S., Jeong, I. S., Kim, J. H., Park, C., Kim, S. M., Kim, J. K., Lee, K. M., Lee, S. H., So, H. and Park, R. (2009) Synergistic effects of the combination of 20-hydroxyecdysone with ampicillin and gentamicin against methicillin-resistant *Staphylococcus aureus*. J Microbiol Biotechnol 19(12):1576–1581. https://doi.org/10.4014/jmb.0903.03015

Kizelsztein, P., Govorko, D., Komarnytsky, S., Evans, A., Wang, Z., Cefalu, W. T. and Raskin, I. (2009) 20-Hydroxyecdysone decreases weight and hyperglycemia in a diet-induced obesity mice model. Am J Physiol Endocrinol Metab 296:E433-E439. https://doi.org/10.1152/ajpendo.90772.2008

Konovalova NP, Mitrokhin IuI, Volkova LM, Sidorenko LI, Todorov IN (2002) Ecdysterone modulates antitumor activity of cytostatics and biosynthesis of macromolecules in tumor-bearing animals. Biology Bulletin (6):650–658.

Korkach YuP, Kotsiuruba AV, Psryslazhna OD, Mohyl'nyts'ka LD, Sahach VF (2007) NO-dependent mechanisms of ecdysterone protective action on the heart and vessels in streptozotocin-induced diabetes mellitus in rats. Fiziol Zh 53(5):3–8. *(In Ukrainian)*

Kraiem S, Al-Jaber MY, Al-Mohammed H, Al-Menhali AS, Al-Thani NJ, Helaleh M, Samsam W, Touil S, Georgakopoulas C, Bouabdallah S, Mohamed-Ali V, Al Maadheed M (2021) Analytical strategy for the detection of ecdysterone and its metabolites in vivo in uPA(+/+)-SCID mice with humanized liver, human urine samples and estimation of prevalence of its use in anti-doping samples. Drug Test Anal 113:1341–1353. https://doi.org/10.1002/dta.3032

Kuchkarova LS, Rokhimova SO, Syrov VN (2020) Effect of turkesterone on the pancreas histology and function in diabetic rats. Int J Current Res Rev 12(21): 1–5. https://doi.org/10.31782/IJCRR.2020.12216

Kurmukov AG, Ermishina OA (1991) The effect of ecdysterone on experimental arrhythmias and changes in the hemodynamics and myocardial contractility induced by coronary artery occlusion. Farmakol Toksikol 54(1):27–29. *(In Russian)*

Kurmukov AG, Syrov VN (1988) Anti-inflammatory properties of ecdysterone. Meditsin Zh Uzbek (10):68–70. *(In Russian)*

Kuz'menko AI, Morozova RP, Nikolenko IA, Donchenko GV (1999) Antioxidant effect of 20-hydroxyecdysone in a model system. Ukrainskii Biokhim Zh 71(3):35–38. *(In Ukrainian)*

Laekeman G, Vlietinck A (2013) Phytoecdysteroids: Phytochemistry and pharmacological activity. In: Natural Products, K.G. Ramawat & J.M. Mérillon eds, Springer-Verlag Berlin, pp. 3827–3849.

Lafont R, Dinan L (2003) Practical uses for ecdysteroids in mammals including humans: an update. J Insect Sci 3:7 https://doi.org/10.1093/jis/3.1.7

Lafont R, Serova M, Didry-Barca B, Raynal S, Guibout L, Dinan L, Veillet S, Latil M, Dioh W, Dilda PJ (2021a) 20-Hydroxyecdysone activates the protective arm of the RAAS via the Mas receptor. J Mol Endocrinol 68(2):77–87. https://doi.org/10.1530/JME-21-0033

Lafont R, Balducci C, Dinan L (2021b) Ecdysteroids. Encyclopedia 1:1267–1302. https://doi.org/10.3390/encyclopedia1040096

Lapenna S, Gemen R, Wollgast J, Worth A, Maragkoudakis P, Caldeira S (2015) Assessing herbal products with health claims. Crit Rev Food Sci Nutr 55:1918–1928. https://doi.org/10.1080/10408398.2012.726661

Lark A, Kitamoto T, Martin JR (2017) Modulation of neuronal activity in the Drosophila mushroom body by DopEcR, a unique dual receptor from ecdysone and dopamine. Biochim Biophys Acta - Mol Cell Res 1864 (10), pp. 1578–1588. https://doi.org/10.1016/j.bbamcr.2017.05.015

Latil M, Camelo S, Veillet S, Lafont R, Dilda PJ (2021) Developing new drugs that activate the protective arm of the renin-angiotensin system as a potential treatment for respiratory failure in COVID-19 patients. Drug Discov Today 26(5):1311–1318. https://doi.org/10.1016/j.drudis.2021.02.010

Lawrence MM (2012) *Ajuga turkestanica* as a countermeasure against sarcopenia and dynapenia. Ms thesis, Appalachian State University. https://files01.core.ac.uk/download/pdf/149240702.pdf

Li J (2013) Effect of ecdysterone on the expression of TNF-α mRNA and SP-A in the lung tissue of rats with acute lung injury [Dissertation thesis]. Southern Medical University, China.

Li J, Wu X, Zhang J, Wu X, Gao D, Shen T, Gu C (2013) Effect of ecdysterone on the expression of toll-like receptor 4 and surfactant protein A in lung tissue of rats with acute lung injury. Infect Inflamm Rep 14(1):22–26. https://doi.org/10.3969/j.issn.1672-8521.2013.01.007 *(In Chinese)*

Li, Y.L., Wu, X., Zhang, C., Zhenggao, Y. and Wang, W. (2008) Effect of ecdysterone paste on wound healing in rabbits. J Xinxiang Med Coll 25(2):116–118. *(In Chinese)*

Lim HS, Moon BC, Lee J, Choi G, Park G (2020) The insect molting hormone 20-hydroxyecdysone protects dopaminergic neurons against MPTP-induced neurotoxicity in a mouse model of Parkinson's disease. Free Radic Biol Med 159:23–36. https://doi.org/10.1016/j.freeradbiomed.2020.07.010

Liu Y, Liu Q, Chen D, Matsuura A, Xiang L, Qi J (2022) Inokosterone from *Gentiana rigescens* Franch extends the longevity of yeast and mammalian cells via antioxidative stress and mitophagy induction. Antioxidants (Basel) 11:214. https://doi.org/10.3390/antiox11020214

Liu Y, Chang AB, Liu YR, Tang ZS, Song ZX, Fan XH, Zhou M, Wang XC (2025) Achyranthes Bidentata Radix mitigates neuronal apoptosis via regulation of JUN and CREB in the cAMP signaling pathway. Phytomedicine 145:157022. https://doi.org/10.1016/j.phymed.2025.157022

Liu Z, Chen Y, Chen Z, Tang W, Zhu G, Wang X, Feng H (2011) Effect of ecdysterone on the nervous lesions of rabbits acquired after subarachnoid hemorrhage. Med J Chin PLA 36(12):1351–1353. *(In Chinese)*

Lobo SM, Plantefève G, Nair G, Cavalcante AJ, de Moraes NF, Nunes E, Barnum O, Stadnik CMB, Lima MP, Lins M, Hajjar LA, Lipinski C, Islam S, Ramos F, Simon T, Martinot JB, Guimard T, Desclaux A, Lioger B, Neuenschwander FC, De Souza Paolino B, Amin A, Acosta SA, Dilling DF, Cartagena EE, Snyder B, Devaud E, Berselli Marinho AKB, Beato PMM, De Wit S, Selvan V, Gray J, Fernandez R, Pourcher V, Maddox L, Kay R, Azbekyan A, Chabane M, Tourette C, Esmeraldino LE, Dilda PJ, Lafont R, Mariani J, Camelo S, Rabut S, Agus S, Veillet S, Dioh W, van Maanen R, Morelot-Panzini C (2024) Oral 20-hydroxyecdysone (BIO101), a MAS receptor activator, reduces the combined proportion of respiratory failures or early deaths in adults with severe COVID-19. Findings of a randomized, placebo-controlled, phase 2/3 trial (COVA). EClinicalMedicine 68: 102383. https://doi.org/10.1016/j.eclinm.2023.102383

Lu T, Jia W, Wang Y, Ma Y, Du F, Gao H, Song C, Li H, Jin X, Liu C, Jin, H, Lin Y (2025) Therapeutic potential of 20E in pulmonary atrial hypertension: involvement of Mas receptor and PI3/Akt pathway. BioRXIV preprint, https://doi.org/10.1101/2025.04.03.647130

Luo H, Luo C, Zhang Y, Chi L, Li L, Chen K (2009) Effect of ecdysterone on injury of lipid peroxidation following focal cerebral ischemia in rats. Zhongguo Yaoye (Pharmaceutical Research) 18(15):12–13. *(In Chinese)*

Makhamov MM, Abduvikhidov DI, Razzokov JI, Sharipov AT (2025) Repurposing 20-hydroxyecdysone for cancer therapy: computational evidence for aquaporin-5 inhibition. MHAPP 5(10):521. https://doi.org/10.5281/zenodo.17343845

Makhmudova LB (2012) Experience in using ecdisten in the treatment of hymenolepiasis. Med Parazitol (Moskva) (1):45–47. *(In Russian)*

Mallek A, Movassat J, Ameddah S, Liu J, Semiane N, Khalkhal A, Dahmani Y (2018) Experimental diabetes induced by streptozotocin in the desert gerbil, *Gerbillus gerbillus*, and the effects of short-term 20-hydroxyecdysone administration. Biomed Pharmacother 102:354–361. https://doi.org/10.1016/j.biopha.2018.03.070

Mamadalieva NZ, El-Readi MZ, Ovidi E, Ashour ML, Hamoud R, Sagdullaev SS, Azimova SS, Tiezzi A, Wink M (2013) Antiproliferative, antimicrobial and antioxidant activities of the chemical constituents of *Ajuga turkestanica*. Phytopharmacology 4(1):1–18.

Mamadalieva NZ, Hussain H, Mollica A, Zengin G, Mamadalieva RZ, Elhady SS, Fadil SA, Ashour ML, Youssef FS (2022) Ecdysteroids as potent enzyme inhibitors and verification of their activity using in vitro and in silico docking studies. Life (Basel), 12(6):824. https://doi.org/10.3390/life12060824

Maneix L, Antonson P, Humire P, Rochel-Maia S, Castaneda J, Omoto Y, Kim HJ, Warner M, Gustafsson JA (2015) Estrogen receptor beta exon 3-deleted mouse: The importance of non-ERE pathways in ERbeta signaling. Proc Natl Acad Sci USA 112:5135–5140. https://doi.org/10.1073/pnas.1504944112

Margalhães GS, Campagnole-Santos MJ, da Glória Rodrigues-Machado M (2019) Lung, In: Angiotensin-(1-7), RAS Santos (ed), Springer Nature Switzerland, pp 131–152.

Marino M, Ascenzi P (2006) Steroid hormone rapid signaling: the pivotal role of S-palmitoylation. IUBMB Life 58:716–719. https://doi.org/10.1080/15216540601019485

Martins A, Tóth N, Ványolós A, Béni Z, Zupkó J, Báthori M, Hunyadi A (2012) Significant activity of ecdysteroids on the resistance to doxorubicin in mammalian cancer cells expressing the human ABCB1 transporter. J Med Chem 55(11):5034–5043. https://doi.org/10.1021/jm300424n

Martins A, Sipos P, Dér K, Csábi J, Miklos W, Berger W, Zalatnai A, Amaeral L, Molnár J, Szabó-Révész P, Hunyadi A (2015) Ecdysteroids sensitize MDR and Non-MDR cancer cell lines to doxorubicin, paclitaxel, and vincristine, but tend to protect them from Cisplatin. Biomed Res Int 2015:895360. https://doi.org/10.1155/2015/895360

Matsuda H, Kawaba T, Yamamoto Y (1970) Pharmacological studies of insect metamorphosing steroids from Achyranthis radix. Folia Pharm Jap 66(5):551–563.

Matsuda H, Kawaba T, Yamamoto Y, Ogawa S 1974. Effect of ecdysterone on experimental atherosclerosis in rabbits. Folia Pharm Jap 70:325–339.

McFall A, Nicklin SA, Work LM (2020) The counter regulatory axis of the renin angiotensin system in the brain and ischemic stroke: Insight from preclinical stroke studies and therapeutic potential. Cellular Signal 76:109809. https://doi.org/10.1016/j.cellsig.2020.109809

Mhashilkar AS, Adapa SR, Jiang RHY, Williams SA, Zaky W, Slatko BE, Luck AN, Moorhead AR, Unnash TR (2016) Phenotype and molecular analysis of the effect of 20-hydroxyecdysone on the human filarial parasite *Brugia malayi*. Int J Parasitol 46(5–6):333–341. https://doi.org/10.1016/j.ijpara.2016.01.005

Molokhova EI, Lipin DE, Sorokina YuV (2022) The main directions of pharmaceutical development of phytoecdysteroid preparations (review). Issues of Biological, Medical and Pharmaceutical Chemistry 25(8):3–9. https://doi.org/10.29296/25877313-2022-08-01 *(In Russian)*

Mukai S, Mishima HK, Shoge K, Shinya M, Ishihara K, Sasa M (2002) Existence of ionotropic glutamate receptor subtypes in cultured rat retinal ganglion cells obtained by the magnetic cell sorter method and inhibitory effects of 20-hydroxyecdysone, a neurosteroid, on the glutamate response. Jpn J Pharmacol 89(1):44–52. https://doi.org/10.1254/jjp.89.44

Mykhaylyk OM, Kotsyuruba AV, Buchanevich OM, Korduban AM, Mengel EF, Gulaya NM (2001) Signal transduction of erythrocytes after specific binding of ecdysterone and cholesterol immobilized on nanodispersed magnetite. J Magnetism Magnetic Materials 225:226–234.

Nakanishi K, Koreeda N, Sasaki S, Chang ML, Hsu HY (1966) The structure of ponasterone A, an insect moulting hormone from the leaves of Podocarpus nakaii Hay. Chem Comm 1966:915–917.

Naresh Kumar R, Sundaram R, Shanthi P, Sachdanandam P (2013) Protective role of 20-OH ecdysone on lipid profile and tissue fatty acid changes in streptozotocin induced diabetic rats. Eur J Pharmacol 698:489–498. https://doi.org/10.1016/j.ejphar.2012.10.016

NCT03021798 (2020) An OBServational Clinical Trial (SARA-OBS) in Sarcopenia and Sarcopenic Obesity in Patients Aged 65 Years and Over (SARA-OBS). Completed. https://clinicaltrials.gov/study/NCT03021798

NCT03452488 (2024) A Double-blind, Placebo-controlled, Randomized INTerventional Clinical Trial (SARA-INT). Completed. https://clinicaltrials.gov/study/NCT03452488

NCT04472728 (2023) Testing the Efficacy and Safety of BIO101 for the Prevention of Respiratory Deterioration in COVID-19 Patients (COVA). Terminated. https://clinicaltrials.gov/study/NCT04472728

NCT04827537 (2025) Anti-protozoal Effect of Steroid Hormone 20-hydroxyecdysone. Active, not recruiting. https://clinicaltrials.gov/study/NCT04827537

Nsimba RY, Kikuzaki H, Konishi Y (2008) Ecdysteroids act as inhibitors of calf skin collagenase and oxidative stress. J Biochem Mol Toxicol 22(4):240–250. https://doi.org/10.1002/jbt.20234

Okada M, Ishihara K, Sasa M, Izumi R, Yajin K, Harada Y (1998) Enhancement of GABA-mediated inhibition of rat medial vestibular nucleus neurons by the neurosteroid 20-hydroxyecdysone. Acta Oto-Laryngol 118(1):11–16. https://doi.org/10.1080/00016489850155062

Okui S, Otaka T, Uchiyama M, Takemoto T, Hikino H, Ogawa S, Nishimoto N (1968) Stimulation of protein synthesis in mouse liver by insect-moulting steroids. Chem Pharm Bull 16:384–387.

Omanakuttan, A, Bose C, Pandurangan N, Kumar GB, Banerji A, Nair BG (2016) Nitric oxide and ERK mediates regulation of cellular processes by ecdysterone. Exp Cell Res 346:167–175. https://doi.org/10.1016/j.yexcr.2016.07.019

Osipova SO, Islamova ZhI, Syrov VN, Badalova S, Khushbaktova ZA (2002) Ecdysten in the treatment of giardiasis. Med Parazitol (Mosk), (1):29–33. *(In Russian)*

Osynskaya LF, Saad LM, Kholodova YuD (1992) Antiradical properties and antioxidative activity of ecdysterone. Ukr Biokhim Zh 64(1):114–117. *(In Ukrainian)*

Pan Z, Niu Y, Liang Y, Zhang X, Dong M (2016) β-Ecdysterone protects SH-SY5Y cells against 6-hydroxydopamine-induced apoptosis via mitochondria-dependent mechanism: involvement of p38MAPK-p53 signaling pathway. Neurotox Res 30(3):453–466. https://doi.org/10.1007/s12640-016-9631-7

Parr MK, Zhao P, Haupt O, Tchoukouegno Ngueu S, Hengevoss J, Fritzemeier KH, Piechotta M, Schlörer N, Muhn P, Zheng WY, Xie MY, Diel P (2014) Estrogen receptor beta is involved in skeletal muscle hypertrophy induced by the phytoecdysteroid ecdysterone. Mol Nutr Food Res 58:1861–1872. https://doi.org/10.1002/mnfr.201300806

Parr MK, Botrè F, Naß A, Hengevoss J, Diel P, Wolber G (2015) Ecdysteroids: a novel class of anabolic agents? Biol Sport 32(2):169–173. https://doi.org/10.5604/20831862.1144420

Phungphong S, Kijtawornrat A, Chaidhuang S, Saengsirisuwan V, Bupha-Intr T (2017) 20-Hydroxyecdysone attenuates cardiac remodeling in spontaneously hypertensive rats. Steroids 126:79–84. https://doi.org/10.1016/j.steroids.2017.08.004

Qu L, Zhang Y, Dong H, Liu D, Chen P, Zhang X (2022) In vivo protection study of β-ecdysterone against MPTP-induced Parkinson's disease. China Medical Herald 19(35): 4–7. https://doi.org/10.20047/j.issn1673-7210.2022.35.01 *(In Chinese)*

Quiroga I, Cardenas-Garcia M, Hernandez-Linares MG, Guerrero-Luna, G, Flores-Manuel F (2025) Theoretical-experimental analysis to elucidate the mechanism of action of novel anabolic agents. Molecules 30:4486. https://doi.org/10.3390/molecules30224486

Ramazanov NSh, Bobayev ID, Yusupova UY, Aliyeva NK, Egamova FR, Yuldasheva NKh, Syrov VN (2017) Phytoecdysteroids-containing extract from *Stachys hissarica* plant and its wound-healing activity. *Natural Product Research* 31(5): 593–597. https://doi.org/10.1080/14786419.2016.1205058

Ramazonov N, Yusupova U, Egamova F, Syrov V (2020) Anti stress activity of phytoecdysteroids isolated from aerial part of *Silene claviformis*. Chem Sci Int J 29(7):9–17. https://doi.org/10.9734/CSJI/2020/v29i730190

Romaniuk-Drapała A, Lisiak N, Totón E, Matysiak A, Nawrot J, Nowak G, Kaczmarek M, Rybczynska M, Rubis B (2021) Proapototic and proautophagic activity of 20-hydroxyecdysone in breast cancer cells in vitro. Chem Biol Interact 342:109479. https://doi.org/10.1016/j.cbi.2021.109479

Rong, H., Li X, Yu C, Wang W, Li L, Xu T, Ding Y, Su Y, Wang X (2024) Targeting the Nrf2-dependent mechanism of β-ecdysterone in attenuating the motor dysfunction in the MPTP/Pro-induced Parkinson's disease mice model. Folia Neuropathol 62(1):83–95. https://doi.org/10.5114/fn.2023.133039

Sampaio WO, Dos Santos RAS, Faria-Silva R, de Mata Machado LT, Schiffrin EL, Touyz RM (2007) Angiotensin-(1-7) through receptor Mas mediates endothelial nitric oxide synthase activation via Akt-dependent pathways. Hypertension 49(1):185-192. https://doi.org/10.1161/01.HYP.0000251865.35728.2f

Sarpong FM, Armah FA, Amponsah IK, Atchoglo PK (2016) Antinociceptive ecdysteroids and other constituents of *Palisota hirsuta* K. Schum (Commelinaceae). J Appl Pharm Sci 6(10):147–153. https://doi.org/10.7324/JAPS.2016.601020

Savchenko RG, Nové M, Spengler G, Hunyadi A, Parfenova LV (2021) In vitro adjuvant antitumor activity of various classes of semi-synthetic poststerone derivatives. Bioorg Chem 106:104485. https://doi.org/10.1016/j.bioorg.2020.104485

Savchenko RG, Mescheryakova ES, Shuvalov OY, Parfenova LV (2025) Synthesis and cytotoxic activity of triphenylphosphonium salts of ecdysteroids, lithocholic acid ans cholesterol. Nat Prod Res 26:1–7. https://doi.org/10.1080/14786419.2025.2592114

Seidlova-Wuttke D, Christel D, Kapur P, Nguyen BT, Jarry H, Wuttke W (2010a) β-Ecdysone has bone protective but no estrogenic effects in ovariectomized rats. Phytomedicine 17(11):884–889. https://doi.org/10.1016/j.phymed.2010.03.021

Seidlova-Wuttke D, Ehrhardt C, Wuttke W (2010b) Metabolic effects of 20-OH-ecdysone in ovariectomized rats. J Steroid Biochem Mol Biol 119(3–5):121–126. https://doi.org/10.1016/j.jsbmb.2010.01.006

Serova M, Didry-Barca B, Deloux R, Foucault AS, Veillet S, Lafont R, Dilda PJ, Latil M (2024) BIO101 stimulates myoblast differentiation and improves muscle function in adult and old mice. J Cachexia Sarcopenia Muscle 15(1):55–66. https://doi.org/10.1002/jcsm.13326

Shakhmurova GA, Khushbaktova ZA, Syrov VN (2010a) Estimation of hepatoprotective and immunocorrecting effects of the sum of phytoecdysteroids from *Silene viridiflora* in experimental animals treated with tetrachlormethan. O'zbekiston Biologiya Jurnali, (5):16–20.

Shakhmurova GA, Syrov, VN, Khushbaktova ZA (2010b) Immunomodulating and antistress activity of ecdysterone and turkesterone under immobilization-induced stress conditions in mice. Pharm Chem J 44(1):7–9. *(In Russian)*

Shirshova, T. I., Politova, N. K., Burtseva, S. A., Beshlei, I. V. and Volodin, V. V. (2006) Antimicrobial activity of natural ecdysteroids from *Serratula coronata* and their acyl derivatives. Pharm Chem J 40(5):268–271.

Si X, Ma Z, Chen Y, Huang L, Feng W (2014) Protective effects of 20-hydroxyecdysone on H_2O_2-induced cytotoxicity in human neuroblastoma SH-SY5Y cells. J Chin Pharm Sci 23(1):33–38. https://doi.org/10.5246/jcps.2014.01.005

Sigurta A, Zambelli V, Bellani G (2016) Renin-angiotensin system in ventilator-induced diaphragmatic dysfunction: Potential protective role of angiotensin (1-7). Med Hypotheses 94:132–137. https://doi.org/10.1016/j.mehy.2016.07.013

Sláma K, Lafont R (1995) Insect hormones—ecdysteroids: their presence and action in vertebrates. Eur J Entomol 92:355–377.

Smirnova G, Bekmatemykh K, Oktyabrsky ON (2016) The effect of 20-hydroxyecdysone on the susceptibility of *Escherichia coli* to different antibiotics. J Appl Microbiol 121(6):1511–1518. https://doi.org/10.1111/jam.13304

Sobrino A, Vallejo S, Novella S, Lazaro-Franco M, Mompeon A, Bueno-Beti C, Walthe T, Sanchez-Ferrer C, Peiro C, Hermenegildo C (2017) Mas receptor is involved in the estrogen-receptor induced nitric oxide-dependent vasorelaxation. Biochem Pharmacol 129:67–72. https://doi.org/10.1016/j.bcp.2017.01.012

Song G, Xia XC, Zhang K, Yu R, Li B, Li M, Yu X, Zhang J, Xue S (2019) Protective effect of 20-hydroxyecdysterone against lipopolysaccharides-induced acute lung injury in mice. J Pharm Drug Res 2(3):109–114.

Spindler KD, Spindler-Barth M, Mehlhorn H (1986) Effects of the juvenile hormone antagonist precocene II and the moulting hormone 20-OH-ecdysone on *Litomosoides carinii* and *Dipetalonema viteae*. Z Parasitenkd 72(6):837–841. https://doi.org/10.1007/BF00925106

Subeki, Matsuura H, Takahashi K, Yamasaki M, Yamato O, Maede Y, Katakura K, Suzuki M, Trimurningsih, Chairul, Yoshihara T (2005) Antibabesial activity of protoberberine alkaloids and 20-hydroxyecdysone from *Arcangelisia flava* against *Babesia gibsoni* in culture. J Vet Med Sci 67(2):223–227. https://doi.org/10.1292/jvms.67.223

Sundaram R, Naresh R, Shanthi P, Sachdanandam, P (2012a) Efficacy of 20-OH-ecdysone on hepatic key enzymes of carbohydrate metabolism in streptozotocin induced diabetic rats. Phytomedicine 19(8–9):725–729. https://doi.org/10.1016/j.phymed.2012.02.019

Sundaram R, Naresh R, Shanthi P, Sachdanandam P (2012b) Ameliorative effect of 20-OH ecdysone on streptozotocin induced oxidative stress and β-cell damage in experimental hyperglycemic rats. Process Biochem 47:2072–2080. https://doi.org/10.1016/j.procbio.2012.07.025

Syrov VN (2000) Comparative experimental investigations of the anabolic activity of ecdysteroids and steranabols. Pharm Chem J 34(4):193–197.

Syrov, V.N., Khushbaktova ZA, Abzalova MKh, Sultanov MB. (1983) On the hypolipidemic and antiatherosclerotic action of phytoecdysteroids. Dokl Akad Nauk Resp Uzbek (9):44–45. *(In Russian)*

Syrov VN, Khushbaktova ZA, Nabiev AN (1992) An experimental study of the hepatoprotective properties of phytoecdysteroids and nerobol in carbon tetrachloride - induced liver injury. Eksp Klin Farmakol 55(3):61–65. *(In Russian)*

Syrov VN, Khushbaktova ZA (1996) Wound-healing effect of ecdysteroids. Doklady Akademii Nauk Respubliki Uzbekistan, (12):47–50.

Syrov VN, Nasyrova SS, Khushbaktova ZA (1997) The results of experimental study of phytoecdysteroids as erythropoiesis stimulators in laboratory animals. Eksp Klin Farmakol 60(3):41–44. *(In Russian)*

Syrov VN, Islamova ZhI, Egamova FR, Iuldasheva NKh, Khushbaktova ZA (2014) Stress-protective properties of phytoecdysteroids. Eksp Klin Farmakol 77(7):35–38.

Syrov VN, Khushbaktova ZA, Yusupova, SM, Shakhmurova GA (2021a) A hepatoprotective influence of ecdisten under conditions of alcohol intoxication in rats. Annals of the Romanian Society for Cell Biology 25(6):20392–20398.

Syrov VN, Shakhmurova GA, Yudalsheva NK, Yusupova SM, Khushbaktova ZA (2021b) Pharmacological correction of ecdisten of hyperglycemic states, morphological changes in the pancreas and immunobiological status of the organism of rats with alloxan diabetes. European Science Review (5–6):3–4. https://doi.org/10.29013/ESR-21-5.6-3-8

Tan WSD, Liao W, Zhou S, Mei D, Wong WSF (2018) Targeting the renin-angiotensin system as novel therapeutic strategy for pulmonary diseases. Curr Opin Pharmacol 40:9–17. https://doi.org/10.1016/j.coph.2017.12.002

Tang Y, Mo Y, Xin D, Xiong Z, Zeng L, Luo G, Cao Y (2021) Regulation of osteoblast autophagy based on PI3K/AKT/mTOR signaling pathway study on the effect of β-ecdysterone on fracture healing. J Orthop Surg Res 16(1):719. https://doi.org/10.1186/s13018-021-02862-z

Tang, Y., Zhou, K., Mo, Y., Xin, D., Xiong, Z. and Yu, X. (2023) An Injectable sustained release hydrogel of hyaluronic acid loaded with β-ecdysterone ameliorates cartilage damage in osteoarthritis via activating autophagy. Adv Therap 6:2300101. https://doi.org/10.1002/adtp.202300101

Tariq SS, Sardar M, Shafiq M, Heinz H, Nur-e-Alam M, Ahmed A, Ul-Haq Z (2025) Molecular insights into the role of estrogen receptor beta in ecdysterone mediated anabolic activity. PLoS One 20(6): e0320865. https://doi.org/10.1371/journal.pone.0320865

Tipbunjong C, Khuituan P, Kitiyanant Y, Suksamrarn A, Pholpramool C (2019) Diarylheptanoid 1-(4-hydroxyphenyl)-7-phenyl-(6E)-6-hepten-3-one enhances C2C12 myoblast differentiation by targeting membrane estrogen receptors and activates Akt-mTOR and p38 MAPK-NF-kappaB signaling axes. Journal of Natural Medicines 73(4):735–744. https://doi.org/10.1007/s11418-019-01322-7

Tirupula KC, Desnoyer R, Speth RC, Karnik SS (2014) Atypical signaling and functional desensitization response of Mas receptor to peptide ligands. PLoS One 9(7):e103520. https://doi.org/10.1371/journal.pone.0103520

Todorova V, Ivanova S, Chakarov D, Kraev K, Ivanov K (2024) Ecdysterone and turkesterone—Compounds with prominent potential in sport and healthy nutrition. Nutrients 16(9):1382. https://doi.org/10.3390/nu16091382

Tóth N, Szabó A, Kacsala P, Héger J, Zádor E (2008) 20-Hydroxyecdysone increases fiber size in a muscle-specific fashion in rat. Phytomedicine 15(9):691–698. https://doi.org/10.1016/j.phymed.2008.04.015

Toychiev A, Kerimov F, Islamova J, Shaislamova M, Mirzakhmedov M, Davis N, Syrov, V, Osipova S (2022) Antiprotozoal effect of steroid hormone 20-hydroxyecdysone in giardosis patients. Ann Parasitol 68(3):605–615. https://doi.org/10.17420/ap6803.467

Tuleuov B, Abulyaissova L, Temirgaziyev B, Kassymova M, Almagambetov K, Melsov M, Kozhanova A, Drasar P, Adekenov S (2025) Synthesis, anti-inflammatory and analgesic activity of the supramolecular complex of 3-epi-2-deoxyecdysone and prospects for its medical applications. Fitoterapia 187:106952. https://doi.org/10.1016/j.fitote.2025.106952

Uchiyama M, Otaka T (1974) Phytoecdysones and protein metabolism in mammalia. In Burdette WJ Ed., Invertebrate Endocrinology and Hormonal Heterophylly, pp. 375–400, Springer-Verlag.

Uchiyama M, Yoshida T (1974) Effect of ecdysterone on carbohydrate and lipid metabolism. In Burdette WJ Ed., Invertebrate Endocrinology and Hormonal Heterophylly, pp. 401–416, Springer-Verlag.

Vasiliev AS, Abdrashitova (Polomeyeva) NYu, Udut VV (2015) Ecdysteroids and their biological activity Rastit Res 51(2):229–259. *(In Russian)*

Wang JJ, Jin H, Zheng SL, Xia P, Cai Y, Ni XJ (2017) Phytoecdysteroids from *Ajuga iva* act as a potential antidiabetic agent against alloxan-induced diabetic male albino rats. Biomed Pharmacother 96:480–488. https://doi.org/10.1016/j.biopha.2017.10.029

Wu J, Gao L, Shang L, Wang G, Wei N, Chu T, Chen S, Zhang Y, Huang J, Wang J, Lin R (2017) Ecdysterones from *Rhaponticum carthamoides* (Willd.) Iljin reduce hippocampal excitotoxic cell loss and upregulate mTOR signaling in rats. Fitoterapia, 119:158–167. https://doi.org/10.1016/j.fitote.2017.03.015

Wu X, Jiang Y, Fan S, Wang R, Xiang M, Niu H, Li T (1998) Effects of ecdysterone on rat lung reperfusion injury. Zhongguo Yaolixue Tongbao (In Chinese Pharm Bull), 14:256–258. *(In Chinese)*

Wu, X, Yao, Y, Zhuo Y (2001) Protective effect and mechanism of ecdysterone on rat acute cerebral injury. Zhongguo Yaoxue Zazhi 36(7):454–457. *(In Chinese)*

Xia X, Zhang Q, Liang G, Lu S, Yang Y, Tian Y (2013a) Role of 20-hydroxyecdysone in protecting rats against diabetic cardiomyopathy. Zhonghua Laonian Xin Nao Xuegangbing Zazhi 15(4): 412–415. *(In Chinese)*

Xia X, Zhang Q, Wang Z, Gui G, Liang G, Liu R (2013b) Protective effect of 20-hydroxyecdysone on diabetic hepathopathy of rats. Xiandai Yufang Yixue, 40(21):4031–4034. *(In Chinese)*

Xia XC, Tang NY, Xue SP, Wang XY, Ma YH, Wang WN, Liu RZ (2016) Effects of 20-hydroxyecdysone on expression of inflammation cytokines in acute lung injury mice. Xiandai Yufang Yixue (Modern Preventive Medicine) 43(5):870–874. *(In Chinese)*

Xie A, Wang L, Zhang Y, Zhang S, Cao J, Wang C, Wu H, Dai W (2025) Beta-ecdysone protects osteocytes from excess glucocorticoids via Akt1-mediated regulation of Connexin43. Steroids, 224:109695. https://doi.org/10.1016/j.steroids.2025.109693

Xing, H., Chen, G., Raza, F., Zafar, H., Xing, L., Li, Q., Li, R., Xu, S. and Chen, D. (2024) The protective effects of ecdysterone on cognitive impairment through regulating Akt/GSK-3β/Nrf2 signaling pathway and oxidative stress in cognitive mice model and Aβ-induced cell toxicity. Frontiers in Bioscience (Landmark Ed), 29(3): 109. https://doi.org/10.31083/j.fbl2903109

Xing H, Sun X, Xing L, Li Q, Puyang M, Liu L, Chen D (2025) ROS-responsive ecdysterone nanodelivery system ameliorates Alzheimer's disease via synergistic modulation of oxidative stress and amyloid-β deposition. Chem Engin J 522:167717. https://doi.org/10.1016/j.cej.2025.167717

Xiong J, Wu X, Wang W, Hou L (2006) Study on role of ecdysterone on endotoxin-induced pulmonary injury. Guangdong Yixue 27(5):661–663. *(In Chinese)*

Xu, T., Niu, C., Zhang, X. and Dong, M. (2018) β-Ecdysterone protects SH-SY5Y cells against β-amyloid-induced apoptosis via c-Jun-terminal kinase- and Akt-associated complementary pathways. Lab Invest 98(4):489–499. https://doi.org/10.1038/s41374-017-0009-0

Xu Y, Xin J, Sun Y, Wang X, Sun L, Zhao F, Niu C, Liu S (2024) Mechanisms of sepsis-induced acute lung injury and advancements of natural small molecules in its treatment. Pharmaceuticals (Basel) 17(4):472. https://doi.org/10.3390/ph17040472

Yang L, Friedemann T, Pan J (2021) Ecdysterone attenuates the development of radiation-induced oral mucositis in rats at early stage. Radiat Res 196(4):366–374. https://doi.org/10.1667/RADE-21-00042.1

Yang SF, Zhang X, Sun XS, Ling TJ, Feng Y, Du XY, Zhao M, Yang Y, Wang L, Liu RT (2010) Diverse ecdysterones show different effects on amyloid-β42 aggregation but all uniformly inhibit amyloid-β42-induced cytotoxicity. J Alzheimers Dis 22(1):107–117. https://doi.org/10.3233/JAD-2010-100621

Yoshida T, Otaka T, Uchiyama M, Ogawa S (1971) Effect of ecdysterone on hyperglycemia in experimental animals. Biochem Pharmacol 20:3263–3268.

Yusupova UYu, Sasmakov SA, Usmanov DA, Ramazonov NSh, Azimova SS, Sagdullaev SSh (2020) Phytoecdysteroids from *Silene claviformis* and their antibacterial and antifungal activities. Pol J Natur Sc 35(3):313–321.

Yusupova UYu, Ramazonov, NS, Syrov VN, Sagdullaev SS (2021) Phytoecdysteroids—Properties, Biological Activity and Applications. Springer Nature, Singapore. ISBN 978-981-16-6711-4.

Zhao XF (2020) G protein-coupled receptors function at cell membrane receptors for the steroid hormone 20-hydroxyecdysone. Cell Commun Signal 18(1):146. https://doi.org/10.1186/s12964-020-00620-y

Zhegn GY, Wu X, Li YL, Zhang JH, Wang WJ (2008) Preparation and dose-effect analysis of ecdysterone cream for promoting wound healing. Nan Fang Yi Ke Da Xue Xue Bao (J South Med Univ) 28(5):828–831. *(In Chinese)*

Zhou Y, Wu X, Liao J, Wu C, Zhang Y, Zhang Z (2010) Effect of ecdysterone on the healing of gastric ulcer in model rats. Zhonggu Yaofang 21(25):2332–2335. *(In Chinese)*

Zou D, Xu Z, Cao L, Chen Q (2010) Effects of ecdysterone on early stage diabetic nephropathy in streptozotocin-induced diabetic rats. Zhongguo Xinyao Yu Lingchuan Zazhi 29(11):842–846. *(In Chinese)*

Zwetsloot KA, Shanely RA, Godwin JS, Hodgman CF (2022) Phytoecdysteroids accelerate recovery of skeletal muscle function following in vivo eccentric contraction-induced injury in adult and old muscles. Front Rehabil Sci 2:757789. https://doi.org/10.3389/fresc.2021.757789

Chapter 4
Stress-Protective and Adaptogenic Roles of Phytoecdysteroids

Vladimir Nikolayevich Syrov

Abstract The purpose of this chapter is to summarize experimental evidence supporting the efficacy, safety, and adaptogenic potential of phytoecdysteroids (PEs), including ecdysterone, turkesterone, and cyasterone. These compounds have been shown to be effective, nontoxic, and powerful natural adaptogens that enhance stress resistance through synergistic metabolic actions. A key feature distinguishing phytoecdysteroids from androgenic anabolic steroids is their lack of interaction with classical steroid hormone receptors. Despite this, they exert strong anabolic and homeostasis-stabilizing effects under diverse stress conditions, including hepatotoxic injury, oxidative stress, hypoxia, catecholamine intoxication, and physical exhaustion. Phytoecdysteroids attenuate stress-induced disturbances in protein and nitrogen metabolism. They normalize total protein and albumin levels in blood and restore urea synthesis rates. At the mitochondrial level, phytoecdysteroids improve lipid redox balance by reducing lipid peroxidation products and normalizing lactate--to-pyruvate ratios. These changes reflect the restoration of lipid metabolism and hepatobiliary function, including normalization of the bile acid-to-cholesterol balance. At the systemic level, phytoecdysteroids preserve myocardial structure under catecholaminergic stress, enhance physical performance, stimulate tissue repair, and promote mineralization during bone regeneration. Neuroendocrine studies further demonstrate their central antistress activity, including wake-promoting effects and a reduction in sedative-induced sleep duration, which clearly distinguishes them from classical anabolic agents. Across various experimental models, phytoecdysteroids showed efficacy comparable to—and in several cases exceeding—that of the anabolic drug Nerobol and established adaptogens such as Eleutherococcus and Saparal. Importantly, these effects were achieved with an excellent pharmacological safety profile, underscoring the translational potential of phytoecdysteroids as multifunctional adaptogenic and regenerative agents.

V. N. Syrov (✉)
Department of Pharmacology and Toxicology, Institute of the Chemistry of Plant Substances, Academy of Science of the Republic of Uzbekistan, Tashkent, Uzbekistan

V. N. Syrov et al. (eds.), *Phytoecdysteroids II*,
https://doi.org/10.1007/978-981-95-8423-9_4

Keywords Phytoecdysteroids · Adaptogen · Anabolic metabolism · Protein biosynthesis · Nerobol · Eleutherococcus · Saparal

4.1 Introduction

Stress is one of the most universal biological phenomena, reflecting the fundamental relationship between organisms and their environment. In the classical sense, stress is defined as a nonspecific response of an organism to any demand placed upon it. This concept, formulated and systematically developed in the mid-twentieth century by Hans Selye, brought a revolutionary shift to biology and medicine, directing attention not only to specific pathological conditions but also to the general mechanisms of adaptation and resistance (Selye 1936; Gusol et al. 2022). Selye described general adaptation syndrome (GAS) as a three-phase response of the organism to stressors, including the alarm reaction, the stage of resistance, and the stage of exhaustion. Energy resources are adaptively mobilized in these reactions: changes in the hypothalamic–pituitary–adrenal axis (HPA), metabolism, and immune functions enable organisms to maintain homeostasis under adverse conditions. However, in the vast majority of cases, prolonged or excessive stress leads to maladaptation, resulting in the pathological conditions of cardiovascular diseases, liver dysfunction, immunosuppression, and neurodegeneration (Shakhmurova et al. 2013).

The concept of adaptogens emerged within the Soviet scientific school during the 1940s–1960s. It was initially proposed by N. V. Lazarev and subsequently developed by I. I. Brekhman and I. V. Dardymov. Adaptogens were defined as plant--derived substances that enhance the nonspecific resistance of the organism to a wide range of adverse factors while exerting a normalizing effect irrespective of the nature of the stressor. The classical adaptogenic plants include ginseng (*Panax ginseng*), Siberian ginseng (*Eleutherococcus senticosus*), and golden root (*Rhodiola rosea*). These plants have been widely studied for their ability to improve physical performance, increase resistance to the development of fatigue and hypoxia, and exert positive effects on immune and endocrine functions. The main bioactive constituents identified in these plants include saponins, phenolic glycosides, and lignans, which display pleiotropic effects on stress-related physiology. Despite the efficacy of classical adaptogens, these agents are not without limitations. Their active compounds often lack high specificity and display variable biological activities, while their molecular mechanisms of action remain only partially understood. Furthermore, their use has at times been associated with undesirable stimulation of the cardiovascular or central nervous system. In response to these limitations, researchers over the past several decades have actively sought new classes of natural adaptogens with more favorable pharmacological profiles (Syrov 1996). It was against this background that the discovery of phytoecdysteroids marked a new stage in studying the adaptogens. PEs are polyhydroxylated steroids that were first identified as analogs of insect molting hormones, the ecdysteroids. Unexpectedly, it was discovered that these compounds are present in a wide variety of plant species, including the genera *Rhaponticum, Ajuga*, *Silene*, etc. (Krasnov et al. 1976; Baltaev and Abubakirov 1987; Usmanov et al. 1971; Dzhukarova et al. 1995; Saatov et al. 1993).

In plants, their biological function is thought to be primarily defensive; they may prevent herbivory by interfering with insect development and reproduction (Usmanov et al. 2019, 2025). However, studies in vertebrates demonstrate that phytoecdysteroids exert a wide range of physiological effects (Yusupova et al. 2019, 2020, 2021, 2022a, b, 2025). This represented an unexpected finding, since it showed that steroid-like compounds are capable of exerting profound physiological effects without adverse side effects typical for anabolic-androgenic steroids, such as Nerobol (Bazulko and Rogozkin 1973). On this basis, interest in phytoecdysteroids developed along two principal directions. First, their considerable pharmacological potential attracted attention, as they influence fundamental metabolic processes—such as protein biosynthesis, carbohydrate and lipid metabolism, and mitochondrial function—allowing correction of stress-induced metabolic disturbances without hormonal toxicity. Second, their structural homology with insect hormones highlighted their value for comparative physiology, providing a unique model for studying the evolution of steroid signaling and receptor selectivity across biological kingdoms. Further investigations revealed that phytoecdysteroids occupy a distinct position among natural adaptogens due to several unique properties. Most notably, their activity is characterized by a pronounced anabolic–metabolic orientation. Unlike classical adaptogens, which act primarily through the central nervous system or endocrine regulation, phytoecdysteroids directly influence protein metabolism by enhancing protein biosynthesis in the liver, skeletal muscle, and myocardium. In addition, they improve nitrogen balance, stimulate ribosomal activity, and normalize serum protein fractions (Fedorova et al. 2020; Solovyova et al. 2021). Simultaneously, they have been shown to exhibit protective properties against different types of experimental stressors, including adrenaline-induced myocardial damage, toxic hepatitis induced by CCl_4, oxidative stress, hypoxia, and fatigue conditions, emphasizing their role in systemic adaptation mechanisms. Particular attention has been given to the exceptionally high safety profile of phytoecdysteroids. Preclinical studies consistently demonstrate their extremely low toxicity, with LD_{50} values in rodents exceeding 6–9 g/kg. To date, no teratogenic or mutagenic effects have been identified (Bajguz et al. 2015). This contrasts markedly with traditional anabolic steroids, associated with hepatotoxicity, cardiovascular risks, and endocrine disturbances. Compared with other well-known adaptogens, including Eleutherococcus or Saparal, phytoecdysteroids demonstrated a more pronounced adaptogenic potential, which was particularly realized in an enhancement of physical performance and prevention of fatigue onset (Ramazonov et al. 2020). Finally, the therapeutic perspectives of phytoecdysteroids far exceed the frames of the classical concept of adaptogen. Phytoecdysteroids have been studied as hepatoprotective (Yusupova et al. 2022a, b; Sidorova et al. 2021), cardioprotective (Fedorov et al.

2009; Politova et al. 2005) agents, antioxidants (Arif et al. 2022), and bioactive compounds stimulating the processes of tissue regeneration (Ramazanov et al. 2016). Thus, phytoecdysteroids represent a distinct class of natural adaptogens whose properties fully meet the criteria of the classical definition of stress-protective agents while substantially extending it. Their ability to combine metabolic activation with systemic protective effects supports the consideration of phytoecdysteroids as promising candidates for further pharmacological development, particularly in the treatment of stress-related pathologies and in performance and recovery medicine (Syrov et al. 2014).

4.2 Biological Basis of Stress Protection and Adaptogenic Effects

The adaptogenic and stress-protective effects of phytoecdysteroids arise from their cumulative actions on key components of cellular and systemic metabolism. There are many experimental studies that have successfully shown the universal metabolic properties of ecdysterone, turkesterone, and other analogous compounds. Phytoecdysteroids enhance anabolic processes, correct disturbances in carbohydrate and lipid metabolism, support the maintenance of electrolyte balance, and preserve mitochondrial function. An important component of their activity is a pronounced antioxidant effect, which reduces oxidative damage and contributes to the preservation of tissue structure. Taken together, these effects provide a solid physiological basis for the adaptogenic properties of phytoecdysteroids. The following section outlines the major metabolic mechanisms underlying their stress-protective activity.

4.2.1 Protein Metabolism and Anabolic Functions of Phytoecdysteroids

One of the distinctive features of phytoecdysteroids is their strong impact on protein biosynthesis, which is the reason for their anabolic activity and also their stress-protective properties. These ideas were confirmed by the data of our own experiments intended to study the anabolic effects of phytoecdysteroids on male rats. The experimental data showed that all the substances of this group induced an increase in body weight, although the degree of this increase significantly differed among individual compounds (Table 4.1). Indeed, the treatment of animals with the 2,3-monoacetonide of ecdysterone, diacetonide of ecdysterone, ecdysterone-22-O-benzoate, 2-deoxyecdysterone, α-ecdysone, 2-deoxy-α-ecdysone, sileneosides E and C, integristerone A, and other analogs induced a rather strong increase in body weight of rats within 10 days. At the same time, substances such as ecdysterone,

Table 4.1 Effects of phytoecdysteroids and anabolic steroids on body weight gain in male rats depending on age and hormonal status (M ± m, n = 10)

Groups of animals	Mature adults (210–250 g)		Immature (70–80 g) Intact		Mature adults (210–250 g), gonadectomized	
Investigated compounds	Body weight gain in rats					
	$mg{\cdot}g^{-1}{\cdot}day^{-1}$	% of control	mg/g/day	% of control	mg/g/day	% of control
1	2	3	4	5	6	7
Control	5.2 ± 0.6	–	15.3 ± 1.5	–	12.5 ± 1.1	–
Ecdysterone	7.9 ± 0.8*	51.9	27.9 ± 3.7*	81.7	18.1 ± 1.4*	44.8
Viticosterone E	7.3 ± 0.6*	40.4	24.6 ± 2.5*	60.8	16.2 ± 1.1*	29.6
Ecdysterone triacetate	7.1 ± 0.5*	36.5	25.2 ± 1.7*	64.7	15.9 ± 1.1*	27.2
Ecdysterone tetraacetate	6.8 ± 0.56	30.8	23.9 ± 2.3*	56.2	15.2 ± 1.4	21.6
2,3-Ecdysterone monoacetonide	–	–	20.1 ± 1.2*	31.4	14.3 ± 1.9	14.4
Ecdysterone diacetonide	–	–	17.3 ± 0.67	13.1	13.5 ± 1.6	8.0
Oxime of ecdysterone	5.8 ± 0.6	11.5	18.5 ± 0.74	20.9	14.6 ± 1.8	16.8
Ecdysterone-22-O-benzoate	5.9 ± 0.7	13.5	21.9 ± 1.4*	43.1	14.9 ± 1.6	19.2
Ecdysterone-22-O-benzoate-2,3-monoacetonide	5.9 ± 0.6	13.5	19.5 ± 1.1*	27.4	14.4 ± 1.8	15.2
2-Deoxyecdysterone	6.3 ± 0.5	21.2	20.7 ± 1.1*	35.3	14.9 ± 1.1	19.2
α-Ecdysone	6.3 ± 0.5	21.2	19.5 ± 1.2*	27.4	14.1 ± 1.3	12.8
2-Deoxy-α-ecdysone	6.1 ± 0.5	17.3	18.9 ± 0.6*	23.5	14.5 ± 1.3	16.0
2-Deoxy-α-ecdysone-22-O- acetate	5.9 ± 0.3	13.5	18.2 ± 0.9	18.9	13.6 ± 1.6	8.8
Sileneoside A	8.1 ± 0.6*	55.8	28.8 ± 2.4*	88.2	18.7 ± 1.2*	49.6
Sileneoside E	6.6 ± 0.4	26.9	22.4 ± 1.5*	46.4	15.3 ± 1.5	22.4
Sileneoside C	6.3 ± 0.7	21.1	21.5 ± 1.7*	40.5	14.9 ± 1.6	19.2
Cyasterone	7.9 ± 0.6*	51.9	30.2 ± 2.4*	97.4	18.9 ± 2.7*	51.2
Cyasterone triacetate	7.6 ± 0.6*	46.1	27.5 ± 2.2*	79.7	17.8 ± 1.8*	42.4
22-acetylcyasterone	7.7 ± 0.7*	48.1	28.4 ± 1.8*	85.6	18.3 ± 2.1*	46.4
Turkesterone	8.5 ± 0.6*	63.5	33.9 ± 3.0*	121.6	20.8 ± 2.8*	66.4
Turkesterone tetraacetate	8.3 ± 0.6*	59.6	31.5 ± 2.4*	105.8	19.2 ± 2.2*	53.6
Polypodine B	8.0 ± 0.5*	53.8	28.4 ± 2.8*	85.6	18.5 ± 2.3*	48.0
Integristerone A	6.5 ± 0.5	25.0	21.1 ± 2.1*	37.9	15.3 ± 1.9	22.4
24 (28)-dehydromacisterone A	7.8 ± 0.8*	50.0	26.8 ± 2.8*	75.2	17.5 ± 1.4*	40.0
Methylandrostenediol	7.9 ± 0.5*	51.9	27.1 ± 2.5*	77.1	23.0 ± 2.9*	84.0
Nerobol	8.2 ± 0.5*	57.6	32.7 ± 3.7*	113.7	27.8 ± 3.4*	122.4

Note: *The significance of the differences compared with the control ($p < 0.05$)

sileneoside A, cyasterone, polypodine B, and others were found to possess an anabolic activity that is equivalent to or even exceeds that of methylandrostenediol (in this case, the data were obtained on intact animals). Of particular interest is the phytoecdysteroid turkesterone, which, according to experimental data, demonstrates efficacy comparable to that of Nerobol, one of the most effective and widely used anabolic steroids. The increase in body weight observed in animals treated with phytoecdysteroids, similar to that induced by steroidal anabolics, was associated with enhanced protein synthesis from the early stages of treatment. This conclusion is supported by the experimental data presented in Fig. 4.1 and Table 4.2.

According to these data, treatment of animals with phytoecdysteroids—using ecdysterone, sileneoside A, and turkesterone as model compounds—produced effects similar to those of Nerobol. Specifically, it increased the incorporation of ^{14}C-labeled amino acids into proteins of various organs and tissues of male rats and elevated total protein content without significant changes in the relative distribution of protein fractions. These changes ultimately resulted in an increase in tissue body weight. At the same time, it should be emphasized that the degree of this increase significantly differed between young and adult male rats treated with phytoecdysteroids. In this regard, a similarity regarding the mode of action also occurred with steroidal anabolic agents, despite their different mechanisms of action. In addition, it should be noted that, under the influence of both phytoecdysteroids and steroidal anabolics, the gain in body weight was less pronounced in adult, sexually mature animals, as compared to intact immatures, due to a reduced stimulation of protein biosynthesis induced by representatives of both substances. As is clear from Fig. 4.1 and Table 4.3, a limited stimulation of protein synthesis induced by phytoecdysteroids occurred in all tissues investigated, as is evident from a moderate increase in labeled amino acids incorporated into protein of the liver, heart, kidneys, and anterior tibial muscles. At the same time, however, under these conditions, Nerobol elicited a more significant increase in protein radioactivity specifically in skeletal muscles (by 22%; $p < 0.05$). Thus, as a result, a moderate and relatively even increase occurred in total protein weights and organ weights in rats injected with phytoecdysteroids, while Nerobol elicited a preferential, selective increase specifically in weights of anterior tibial muscles (by 10.9%; $p < 0.05$), suggesting a greater tissue specificity exhibited by its anabolic activity in these experiments.

However, all of these effects were found to be significantly greater in young whole rats. Proteins of this group exhibited an adjusted increase in radioactivity per mg of protein per min due to the effects of ecdysterone, sileneoside A, and turkesterone by 18.0, 21.7, and 34.0%, respectively, in the liver; 18.0, 25.0, and 37.9% in the heart; 13.2, 22.4, and 26.4% in the kidney; and 17.5, 25.2, and 31.0% in skeletal muscles. Protein biosynthesis in this group was also markedly stimulated by Nerobol, particularly in skeletal muscle and kidneys, where protein radioactivity increased by 35.9% and 28.0%, respectively. Increases were also observed in the liver and heart, with protein radioactivity rising by 19.0% and 20.0%, respectively. The more pronounced activation of protein biosynthesis in immature rats treated with the tested compounds resulted in a greater increase in organ protein content, which was accompanied by an increase in organ mass. In growing rats treated with

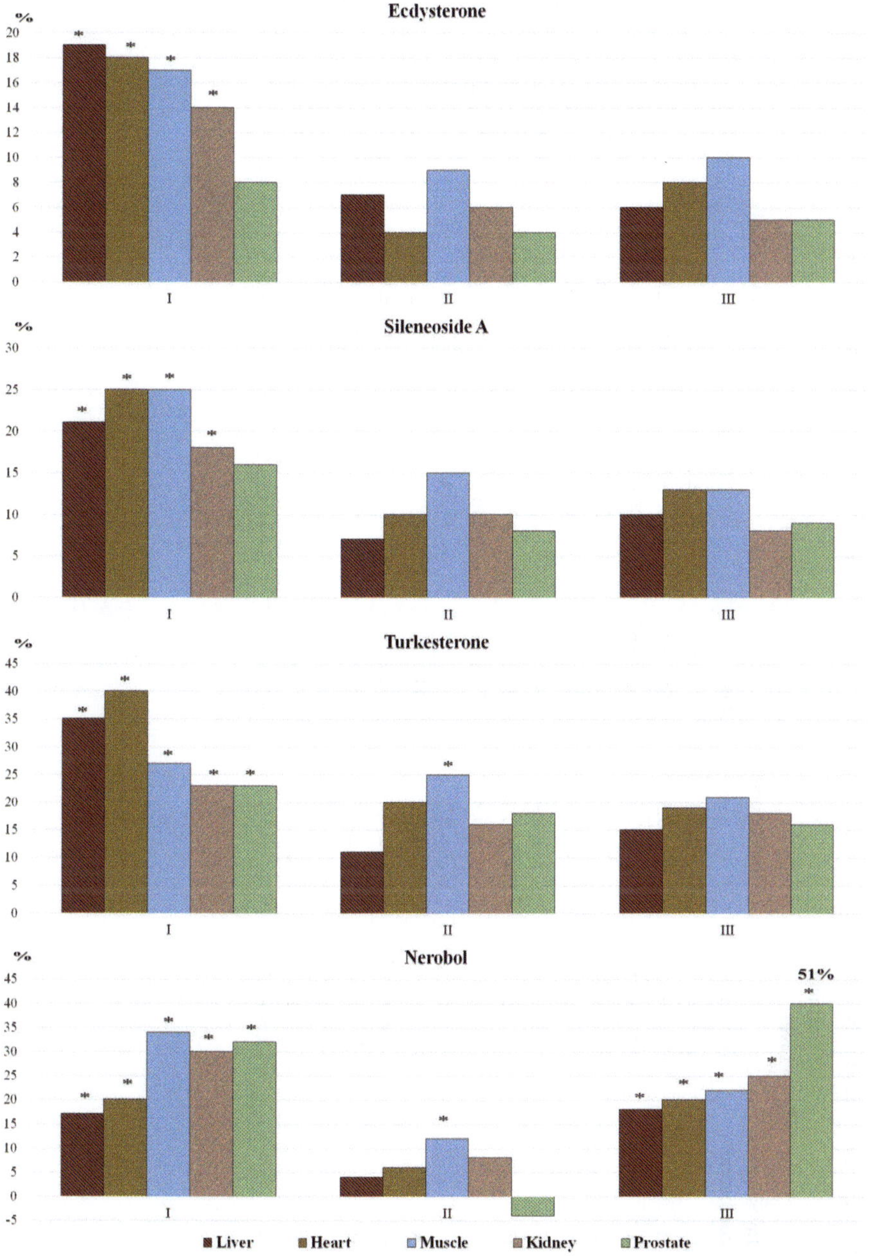

Fig. 4.1 Effect of ecdysterone, sileneoside A, turkesterone, and Nerobol on the intensity of incorporation of ^{14}C amino acids into proteins of internal organs and skeletal muscle (imp/min/mg) in male rats, depending on age and hormonal status of the body (in % of control), n = 6–8. I—immature intact rats; II—sexually mature rats; III—immature neutered rats. The asterisk indicates the reliability of the differences compared to the control (p < 0,05)

Table 4.2 Effects of phytoecdysteroids and Nerobol on organ mass and protein content in male rats according to age and hormonal status (M ± m, n = 6–8)

Groups of animals	Investigated compounds	Treatment					
		Liver			Heart		
		Weight (mg)	Protein content		Weight (mg)	Protein content	
			%	mg		%	mg
Adult sexually mature rats	C	8650 ± 258	17,4 ± 0,4	1511 ± 72	763 ± 31	16,3 ± 1,1	124 ± 9,3
	E	8925 ± 250	17,5 ± 0,3	1558 ± 45	863 ± 31*	16,5 ± 0,3	142 ± 3,8
	S	9283 ± 353	17,5 ± 0,6	1625 ± 67	906 ± 24*	17,0 ± 0,8	153 ± 8,6*
	T	9325 ± 249	18,0 ± 1,1	1676 ± 98	923 ± 29*	17,0 ± 0,7	157 ± 10,7*
	N	8591 ± 445	17,3 ± 0,4	1479 ± 66	815 ± 41	16,3 ± 0,4	132 ± 6,9
Immature intact rats	C	4408 ± 263	17,0 ± 0,4	746 ± 33	358 ± 21	16,9 ± 0,6	60,1 ± 2,0
	E	5425 ± 335*	17,2 ± 0,4	927 ± 44*	429 ± 19*	17,0 ± 0,3	73,8 ± 4,1*
	S	5525 ± 283*	17,5 ± 0,8	971 ± 71*	436 ± 21*	17,0 ± 0,9	74,7 ± 5,3*
	T	5786 ± 373*	17,7 ± 0,6	1021 ± 74*	481 ± 18*	17,3 ± 0,7	83,9 ± 5,9*
	N	5191 ± 217*	17,3 ± 0,6	893 ± 23*	427 ± 17*	17,0 ± 0,1	72,6 ± 3,3*
Immature gonadectomized rats	C	4366 ± 197	17,2 ± 0,5	752 ± 39	333 ± 22	16,6 ± 0,3	55,2 ± 2,8
	E	5183 ± 403	17,4 ± 0,3	902 ± 76	387 ± 16	16,6 ± 0,6	64,0 ± 3,0
	S	5190 ± 493	17,5 ± 0,4	901 ± 74	391 ± 17	16,6 ± 0,5	64,8 ± 3,6
	T	5210 ± 479	17,6 ± 0,4	914 ± 88	402 ± 27	16,7 ± 0,5	67,4 ± 5,8
	N	5225 ± 329*	17,7 ± 0,4	922 ± 53*	410 ± 22*	16,8 ± 0,2	69,1 ± 4,2*

Adult sexually mature rats	C	807 ± 51	15,9 ± 0,7	127 ± 7,7	422 ± 7,8	16,5 ± 1,2	69,7 ± 4,8
	E	912 ± 66	16,2 ± 0,4	147 ± 10,6	477 ± 16,8*	17,4 ± 0,5	82,8 ± 2,1*
	S	914 ± 47	16,3 ± 0,6	149 ± 11,2	482 ± 12,2*	17,3 ± 0,8	83,4 ± 3,8*
	T	928 ± 25	16,3 ± 0,6	152 ± 9,1	492 ± 12,3*	17,4 ± 0,9	85,5 ± 4,2*
	N	817 ± 14	16,1 ± 0,6	130 ± 4,2	468 ± 8,7*	17,5 ± 0,4	82,1 ± 1,5*
Immature intact rats	C	391 ± 24	16,1 ± 0,2	63,2 ± 4,1	169 ± 9,0	17,3 ± 0,2	29,3 ± 1,8
	E	475 ± 15*	16,4 ± 0,4	78,2 ± 3,7*	202 ± 10,5*	17,4 ± 0,2	35,1 ± 1,8*
	S	489 ± 15*	16,4 ± 0,3	79,9 ± 1,8*	209 ± 10,8*	17,4 ± 0,3	36,4 ± 1,9*
	T	500 ± 15*	16,8 ± 0,3	83,8 ± 3,2*	220 ± 14,1*	18,1 ± 0,7	39,5 ± 1,5*
	N	518 ± 19*	16,2 ± 0,2	84,0 ± 3,7*	226 ± 13,9*	17,8 ± 0,5	40,2 ± 2,7*
Immature gonadectomized rats	C	363 ± 18	15,7 ± 0,3	57,3 ± 3,3	147 ± 7,7	16,9 ± 0,3	25,0 ± 1,7
	E	418 ± 25	16,4 ± 0,5	68,4 ± 4,3	168 ± 2,7*	17,3 ± 0,2	29,2 ± 0,5*
	S	419 ± 24	16,3 ± 0,5	68,1 ± 4,2	173 ± 3,4*	17,3 ± 0,4	30,0 ± 0,9*
	T	421 ± 24	16,4 ± 0,4	68,8 ± 4,1	178 ± 4,8*	17,3 ± 0,3	30,9 ± 0,9*
	N	441 ± 25*	16,4 ± 0,3	72,9 ± 3,5*	184 ± 11,4*	17,2 ± 0,5	31,9 ± 2,3*

Table 4.2 (continued)

Groups of animals	Investigated compounds	Treatment			
		The ventral prostate			Seminal vesicles
		Weight (mg)	Protein content %	Protein content mg	Weight (mg)
Adult mature rats	C	388 ± 20	11,3 ± 0,4	43,6 ± 2,1	1284 ± 65
	E	409 ± 28	11,4 ± 0,4	46,4 ± 3,4	1395 ± 46
	S	415 ± 25	11,3 ± 0,3	47,1 ± 3,7	1436 ± 62
	T	421 ± 26	11,7 ± 0,7	49,5 ± 4,5	1458 ± 44
	N	318 ± 23*	11,9 ± 0,3	35,4 ± 2,6*	1096 ± 33*
Immature intact rats	C	24,7 ± 3,5	11,5 ± 0,4	2,8 ± 0,4	21,3 ± 2,9
	E	29,8 ± 2,2	11,6 ± 0,6	3,4 ± 0,1	27,8 ± 3,5
	S	32,2 ± 2,1	11,7 ± 0,6	3,8 ± 0,4	32,2 ± 3,7*
	T	34,0 ± 2,2*	11,7 ± 0,5	3,9 ± 0,2*	42,2 ± 3,4*
	N	36,7 ± 2,7*	11,9 ± 0,4	4,4 ± 0,4*	45,6 ± 6,8*
Immature gonadectomized rats	C	8,6 ± 0,8	9,3 ± 0,8	0,70 ± 0,04	10,8 ± 1,3
	E	9,5 ± 1,1	8,9 ± 0,8	0,79 ± 0,07	11,5 ± 1,1
	S	9,6 ± 0,6	8,7 ± 0,4	0,84 ± 0,07	11,7 ± 0,7
	T	10,0 ± 0,9	8,8 ± 0,8	0,87 ± 0,09	12,0 ± 1,2
	N	18,5 ± 1,9*	10,4 ± 1,2	1,8 ± 0,2*	24,8 ± 2,7*

Note: C—control, E—ecdysterone, S– sileneoside A, T—turkesterone, N—Nerobol, *-the significance of differences compared with the control ($p < 0,05$)

Table 4.3 Effect of phytoecdysteroids and Nerobol on organ mass and protein content in female rats depending on age and hormonal status (M ± m, n = 6–8)

Groups of animals	Investigated compounds	Treatment					
		liver			Heart		
		Weight (mg)	Protein content		Weight (mg)	Protein content	
			%	mg		%	mg
Adult mature rats	C	6366 ± 468	17,6 ± 0,6	1119 ± 95	669 ± 28	16,9 ± 0,5	113 ± 6,3
	E	7445 ± 443	17,5 ± 0,6	1294 ± 53	710 ± 38	17,3 ± 0,3	122 ± 4,4
	T	7468 ± 162	17,7 ± 0,3	1328 ± 42	761 ± 26*	17,3 ± 0,4	131 ± 6,6
	N	7366 ± 329	17,4 ± 0,6	1273 ± 52	702 ± 30	17,2 ± 0,3	120 ± 5,0
Immature intact rats	C	3925 ± 306	17,7 ± 0,4	687 ± 41	347 ± 19	16,4 ± 0,8	56,9 ± 4,2
	E	4850 ± 126	18,0 ± 0,7	867 ± 19*	405 ± 21	16,7 ± 0,5	67,3 ± 3,5
	T	4983 ± 280	17,9 ± 0,5	893 ± 47*	415 ± 21*	16,9 ± 0,5	70,2 ± 2.4*
	N	5025 ± 115	17,9 ± 0,6	898 ± 32*	412 ± 19*	17,1 ± 0,4	70,5 ± 4,2*
Immature gonadectomized rats	C	4628 ± 246	18,2 ± 0,5	836 ± 33	362 ± 34	16,8 ± 0,4	60,8 ± 5,7
	E	5528 ± 375	17,7 ± 0,3	974 ± 58	417 ± 28	16,8 ± 0,7	69,1 ± 1,9
	T	5486 ± 295*	18,4 ± 0,5	1009 ± 58*	420 ± 20	16,7 ± 0,6	69,9 ± 3,8
	N	5170 ± 441	18,4 ± 0,6	942 ± 68	414 ± 21	16,9 ± 0,5	69,9 ± 1,8

Groups of animals	Investigated compounds	Treatment					
		Kidney			Tibialis anterior muscle		
		Weight (mg)	Protein content		Weight (mg)	Protein content	
			%	mg		%	mg
Adult mature rats	C	653 ± 39	16,6 ± 0,5	107 ± 3,8	369 ± 14,4	17,3 ± 0,5	63,9 ± 2,6
	E	721 ± 33	16,8 ± 0,4	121 ± 6,0	380 ± 18,9	17,4 ± 0,3	66,0 ± 2,6
	T	734 ± 16	16,7 ± 0,6	123 ± 6,3	423 ± 17,5*	17,9 ± 0,8	75,0 ± 4,2*
	N	733 ± 24	16,4 ± 0,4	120 ± 5,1	373 ± 14,1	17,9 ± 0,3	66,8 ± 2,1
Immature intact rats	C	365 ± 7,2	16,3 ± 0,2	59,4 ± 0,4	138 ± 2,8	17,1 ± 0,3	23,6 ± 0,2
	E	445 ± 22*	16,3 ± 0,4	72,5 ± 4,2*	163 ± 8,8*	17,2 ± 0,6	28,0 ± 1,5*
	T	489 ± 15*	16,4 ± 0,3	79,9 ± 1,8*	169 ± 3,5*	17,3 ± 0,5	29,4 ± 1,3*
	N	451 ± 33*	16,4 ± 0,4	73,4 ± 3,7*	172 ± 8,6*	17,5 ± 0,9	30,0 ± 1,9*

(continued)

Table 4.3 (continued)

Groups of animals	Investigated compounds	Treatment					
		Kidney			Tibialis anterior muscle		
		Weight (mg)	Protein content		Weight (mg)	Protein content	
			%	mg		%	mg
Immature gonadectomized rats	C	401 ± 29	16,6 ± 0,4	66,7 ± 5,1	164 ± 15	17,3 ± 0,3	28,2 ± 2,4
	E	459 ± 16	16,8 ± 0,5	77,0 ± 3,8	191 ± 12	17,4 ± 0,5	33,1 ± 1,4
	T	457 ± 13	16,8 ± 0,3	76,9 ± 3,1	193 ± 5,3	17,2 ± 0,5	33,4 ± 1,8
	N	477 ± 10*	17,1 ± 0,6	81,2 ± 1,7*	188 ± 14	17,2 ± 0,6	32,1 ± 1,9

Groups of animals	Investigated compounds	Treatment			
		The ventral prostate			The ventral prostate
		Weight (mg)	Protein content		Weight (mg)
			%	mg	
Adult mature rats	C	332 ± 21	10,2 ± 0,5	33,7 ± 2,3	30,2 ± 1,9
	E	375 ± 23	10,5 ± 0,5	39,1 ± 1,9	30,9 ± 2,6
	T	389 ± 7,3*	10,2 ± 0,5	39,8 ± 1,4*	33,2 ± 1,9
	N	416 ± 26*	9,7 ± 0,7	39,2 ± 2,3	28,2 ± 1,9
Immature intact rats	C	50,0 ± 6,1	11,5 ± 1,2	5,4 ± 0,4	16,8 ± 1,0
	E	63,1 ± 7,9	11,7 ± 1,4	7,2 ± 1,2	15,4 ± 1,7
	T	65,2 ± 2,6*	11,6 ± 0,3	7,5 ± 0,3*	18,4 ± 0,7
	N	115,3 ± 12,1*	11,8 ± 0,5	13,8 ± 1,9*	18,8 ± 0,8
Immature gonadectomized rats	C	25,7 ± 2,3	11,2 ± 0,5	2,9 ± 0,3	–
	E	28,6 ± 3,6	11,1 ± 0,4	3,2 ± 0,5	–
	T	31,0 ± 1,7	11,2 ± 0,4	3,5 ± 0,2	–
	N	92,0 ± 6,9*	11,8 ± 0,6	10,7 ± 0,8*	–

Note: C—control, E—ecdysterone, T—turkesterone, N—Nerobol, *the significance of differences compared with the control ($p < 0,05$)

ecdysterone, sileneoside A, and turkesterone, liver mass increased by 23.0%, 25.3%, and 31.2%, respectively. Corresponding increases were observed in heart mass (19.8%, 21.8%, and 34.0%), kidney mass (21.5%, 25.1%, and 27.9%), and anterior tibial muscle mass (19.2%, 23.4%, and 30.0%). In the Nerobol-treated group, increases of 17.8%, 19.3%, 32.5%, and 33.7% were recorded for the liver, heart, kidneys, and anterior tibial muscle, respectively.

The obtained data, however, are sufficient to indicate certain differences between the effects of phytoecdysteroids and steroidal anabolics on protein synthesis, which underlie their anabolic activity in animals. In particular, experiments demonstrated a pronounced selective effect of Nerobol on muscle and kidney tissues in intact rats of both adult and immature groups. The differences in effects between the compounds in these two groups were most clearly revealed in comparison experiments on gonadectomized animals. As shown in Tables 4.1 and 4.3 and Fig. 4.1, castration of male rats led to a substantial reduction in the overall anabolic effect of phytoecdysteroids. In contrast, administration of Nerobol to castrated animals produced an even more pronounced anabolic response than that observed in intact animals of the corresponding age group. From both cases, it follows that, although there was a change in the extent of the anabolic response in rats injected with phytoecdysteroids and with Nerobol, there was no change in their type. In both cases, it manifested itself in a rather marked, about equally pronounced proneness to stimulate protein synthesis in the liver, heart, skeletal muscle, and kidney of animals injected with phytoecdysteroids, and as selectively myotropic and renotropic activity in those injected with Nerobol.

It should be noted that when assessing anabolic effects, responses across organs to the administered compounds vary markedly. Here, primarily, are the ventral prostate and seminal vesicles, with their weights in immature castrated rats tested in this system and used to index individual androgenic responses to injected compounds (Pivnitskiy 1974). As shown in Table 4.2 and Fig. 4.1, phytoecdysteroids stimulated protein synthesis in these organs in both adult and immature rats, leading to increased organ mass. Similar changes in the weights of other organs and tissues were also observed in animals of the corresponding age groups. However, in castrated rats, phytoecdysteroids failed to considerably affect the weight of the ventral prostate and seminal vesicles. In contrast, in adult, sexually mature rats, Nerobol caused a reduced intensity of protein synthesis in the prostate gland and a statistically significantly reduced mass of the prostate gland and a concomitantly evident reduced mass of the vesicles of semen. The mentioned effects of the synthetic derivatives of the male sex hormones administered to sexually mature male rats with high blood levels of endogenous male sex hormones are well known and, naturally, are due to their antigonadotropic action (Yudaev and Pokrovsky 1966; Grokett et al. 1992; Mitchell and Wilson 1987; Kuoppasalmi et al. 1980; Ryan and Hoffman 1978).

In young growing animals, regardless of hormonal status, Nerobol induced a marked increase in the mass of the ventral prostate and the seminiferous vesicles. The one-way effect induced by the phytoecdysteroids and Nerobol in the accessory male sex organs of young castrated rats could have suggested the existence of specific androgenic properties in the plant compounds studied, to some extent characteristic of the aforementioned steroidal compounds, which act as derivatives of male sex hormones (Pivnitskiy 1974). However, the most compelling argument from our data is that these compounds lack classical androgen-like specificity. Both the present experiments and our previous studies conducted using the method of L. Hershberger et al. demonstrated that the tested compounds exerted no detectable effect on the intensity of protein synthesis in the prostate gland, nor on the mass of the ventral prostate or seminal vesicles in castrated animals. Apparently, the effects of phytoecdysteroids on the male reproductive organs of castrated rats do not differ from the general stimulation of protein synthesis observed throughout the organism. Since these compounds lack the properties characteristic of male sex hormones and their synthetic analogs, they cannot exert a replacement effect following gonadectomy. Consequently, their anabolic activity cannot be attributed to androgenic mechanisms. As is well known, the lack of endogenous synthesis of male sex hormones results in the reduced amount of the mentioned polyribosomes in the organs and tissues and, naturally, in the reduced intensity of protein synthesis. Clearly, these effects can be completely normalized only by administration of androgens or their analogs with appropriate hormonal activity (Kochakian et al. 1965; Johnsonbaugh et al. 1976). Because phytoecdysteroids lack androgen-like hormonal activity, they are unable to fully prevent postcastration changes. Consequently, they cannot stimulate protein synthesis to the same extent as in intact animals, particularly in androgen-sensitive tissues. This explains our previous findings regarding the relatively low anabolic properties of phytoecdysteroids (Aizikov et al. 1978). This conclusion was based on the assessment of anabolic activity using changes in the weight of the levator ani muscle, a perineal muscle group, in immature castrated rats. Measurement of levator ani muscle mass represents one of the most widely accepted bioassays for evaluating the anabolic effects of steroid compounds (Pivnitskiy 1974).

All these findings suggest that an anabolic effect of phytoecdysteroids can be seen only in individuals with an intact natural hormonal background. In summarizing the comparative study of the anabolic effects of phytoecdysteroids and Nerobol in male rats, it should be noted that an additional important difference was identified. None of the phytoecdysteroids tested possessed thymolytic properties. On the contrary, an increase in their weight was noted in accordance with the growth of weight of other organs (increase by 7–12% in fully grown rats, $p < 0.05$; by 35–45% in immature intact rats). In contrast to this finding with phytoecdysteroids, Nerobol, like all other steroids (Sergeev et al. 1991; Chernykh et al. 1990; Semeĭkin et al. 1991; Sergeev et al. 1994), showed a marked inhibition in thymic tissue development with an increase in the dose of compounds tested. This was most easily demonstrated when these compounds (using ecdysterone, turkesterone, or Nerobol as examples) were injected in immature castrated rats. When these compounds (ecdysterone, turkesterone, or Nerobol) were tested in immature intact rats, the mean weight of the thymus glands in intact controls was 282 ± 26.3 mg. Administration of phytoecdysteroids at doses of 1, 5, 10, 25, and 50 mg/kg resulted in dose--dependent changes in thymus mass. In rats treated with ecdysterone, thymus mass was 307 ± 3.5, 376 ± 28.0, 333 ± 24.4, 326 ± 26.0, and 336 ± 38.0 mg, respectively, whereas turkesterone treatment produced values of 308 ± 19.9, 368 ± 28.4, 379 ± 28.9, 377 ± 24.1, and 386 ± 24.5 mg. In contrast, Nerobol treatment led to a progressive reduction in thymus mass, with values decreasing to 226 ± 9.6, 210 ± 10.2, 177 ± 22.1, 167 ± 20.1, and 140 ± 18.2 mg, respectively.

These findings indicate that phytoecdysteroids produce a considerable anabolic effect on castrated male rats by influencing the synthesis of protein in all investigated organs and tissues, the intensity of which depends on their age and hormone status. In intact animals, the anabolic effects of certain phytoecdysteroids—particularly turkesterone—were found to be fully comparable to those of Nerobol. At the same time, phytoecdysteroids completely lacked the specific hormone-like (androgenic) activity characteristic of Nerobol, especially with respect to their action on androgen-sensitive target organs in male animals. The administration of phytoecdysteroids to female rats clearly demonstrated an anabolic effect also in these animals, as was observed for Nerobol. As observed in the previous series, the anabolic effects of the tested substances were strongly age-dependent. Sexually mature female rats (160–180 g) in the control group gained 8.1 ± 0.83 mg/g of initial body mass per day, whereas age-matched females treated with ecdysterone, turkesterone, and Nerobol gained 11.7 ± 1.8, 13.4 ± 0.8, and 13.2 ± 1.6 mg/g/day, respectively, corresponding to increases of 44%, 65%, and 63%. In contrast, sexually immature females (60–70 g) in the control group gained 16.8 ± 1.6 mg/g/day, while those receiving ecdysterone, turkesterone, and Nerobol gained 31.9 ± 4.9, 33.8 ± 1.6, and 40.4 ± 3.6 mg/g/day, respectively. These increases corresponded to stimulation levels of 89%, 101%, and 140% compared with controls ($p < 0.05$).

This was because, as a result of ecdysterone, turkesterone, and Nerobol treatment, protein-synthesizing processes occurring in all organs of immature female rats, as was the case also in males, were more readily activated as compared to mature females (Fig. 4.2). As a consequence, increased mass of liver, heart, kidneys, and skeletal muscles due to pronounced accumulation of protein was also observed to a greater extent in these animals (Table 4.3). In ovariectomized female rats, as also observed in males, the anabolic effects of both phytoecdysteroids and Nerobol were markedly reduced. This became evident not only through decreased ^{14}C-amino acid incorporation into protein molecules across different organs and tissues (Fig. 4.2) and reduced mass (Table 4.3), but also through a reduced rate of increase in body mass. Thus, the experimental animals used as controls gained 25.9 ± 3.1 mg/g/day, whereas those that received ecdysterone, turkesterone, and Nerobol gained 35.2 ± 5.1, 37.5 ± 2.7, and 33.9 ± 0.0 mg/g/day, respectively (simulating 35.9, 44.7, and 30%) ($p < 0.05$). It is important to note here that a similar reduction also occurred when ecdysteroids were administered to males. On the other hand, Nerobol exhibited an opposite action. It is important to note the specific basic distinction between the effect of phytoecdysteroids and Nerobol on protein synthesis in the uterus. There was no selective uterotropic effect of ecdysterone and turkesterone. Protein synthesis in the uterus induced by the two compounds was the same as in other tissues of the organisms. As a result, the weights of the uterus were increased by 12.9–17.2% in intact adult rats, 26.2–30.4% in intact immature rats, and 11.2–20.6% in ovariectomized immature rats.

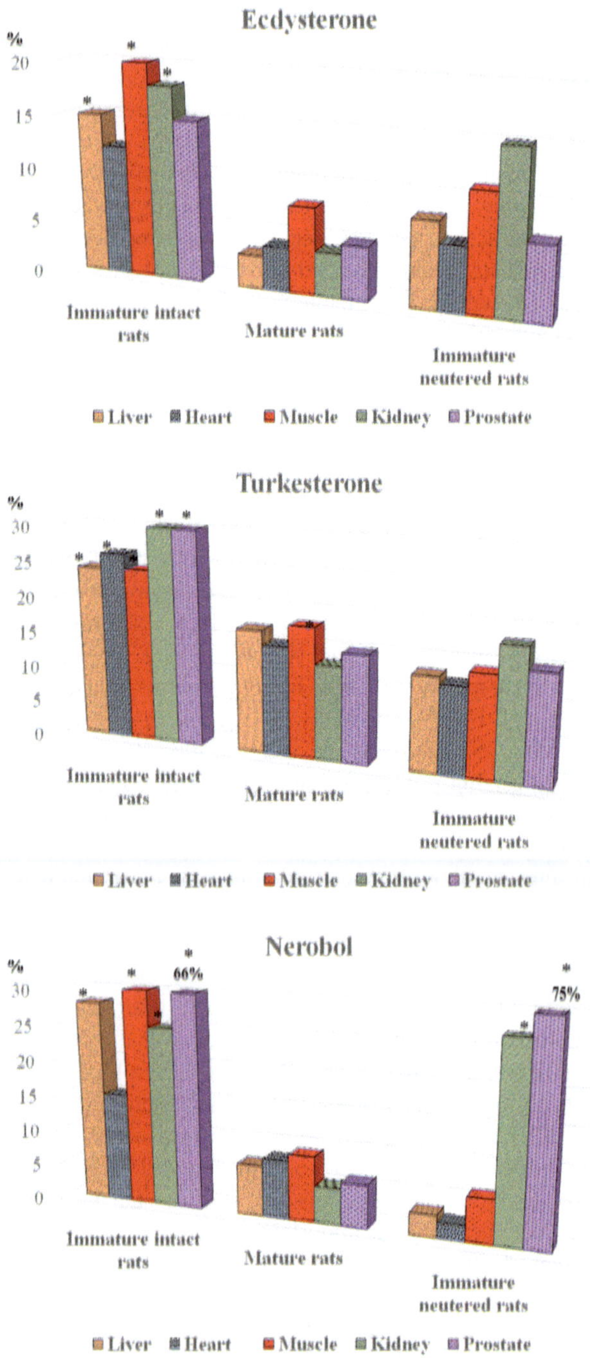

Fig. 4.2 The effect of ecdysterone, turkesterone, and Nerobol on the intensity of incorporation of ^{14}C amino acids into the proteins of internal organs and skeletal muscle (imp/min/mg) in female rats, depending on age and hormonal status of the body (in % of the control), n = 6–8. The asterisk indicates the reliability of the differences compared to the control ($p < 0,05$)

Nerobol exhibits androgen-like activity comparable to that of male sex hormones and their synthetic analogs, as previously reported by Semeĭkin et al. (1991) and Kochakian et al. (1965). In rats, this activity is reflected in a selective effect on the uterus. Specifically, uterine mass increased by 25%, 130%, and up to 257% in the experimental groups (Table 4.3). In intact female rats, none of the tested compounds produced significant changes in ovarian mass. A slight tendency toward a decrease in ovarian weight was observed only in adult females treated with Nerobol. In female rats, both phytoecdysteroids and Nerobol exerted effects on thymic tissue similar to those observed in males. Accordingly, phytoecdysteroids demonstrated anabolic activity in females that was comparable in pattern and magnitude to that seen in male animals. Importantly, unlike Nerobol and other classical steroid hormones, the anabolic effects of phytoecdysteroids were not sex-dependent. This sex--independent profile is consistent with earlier observations reported by Dessaulles (1960), Matsuoka (1964), Egginton (1987), and Sillence and Rodway (1990) and represents a distinctive pharmacological feature of phytoecdysteroids.

4.2.2 Changes in the Serum Protein Profile and in Hemoglobin and Erythrocyte Levels Under the Action of Phytoecdysteroids

Administration of phytoecdysteroids (ecdysterone and turkesterone) as well as Nerobol to intact male rats resulted in an increase in total serum protein and albumin levels. This effect was more pronounced in younger animals, consistent with the age-dependent enhancement of protein biosynthesis observed in other tissues. In adult rats, phytoecdysteroid treatment increased total serum protein by 3–7% and albumin levels by 1.7–2.4% compared with controls ($p < 0.05$). In immature rats, the response was markedly greater, with total protein levels rising by 13–16% and albumin by 12–24%. In both age groups, phytoecdysteroids produced slightly stronger effects than Nerobol. After 10 days of treatment, substantial alterations in serum globulin fractions were observed, again more pronounced in immature animals. Under phytoecdysteroid treatment, α_1-globulin levels showed a slight increase, whereas α_2-globulins decreased. In contrast, Nerobol administration resulted in increases in both α_1- and α_2-globulin fractions. Notably, phytoecdysteroids induced a significant increase in β- and especially γ-globulin fractions, whereas Nerobol led to a reduction in these components. These differences in the serum protein spectrum, illustrated using turkesterone (the most active phytoecdysteroid) and Nerobol as representative compounds, are presented in Fig. 4.3.

In addition to their effects on serum protein composition, these anabolic compounds exert a pronounced stimulatory effect on erythropoiesis. Therefore, in evaluating the anabolic properties of phytoecdysteroids, particular attention was paid to their influence on erythrocyte counts and hemoglobin levels in the blood (Syrov et al. 1997). The results of these experiments are summarized in Table 4.4. In

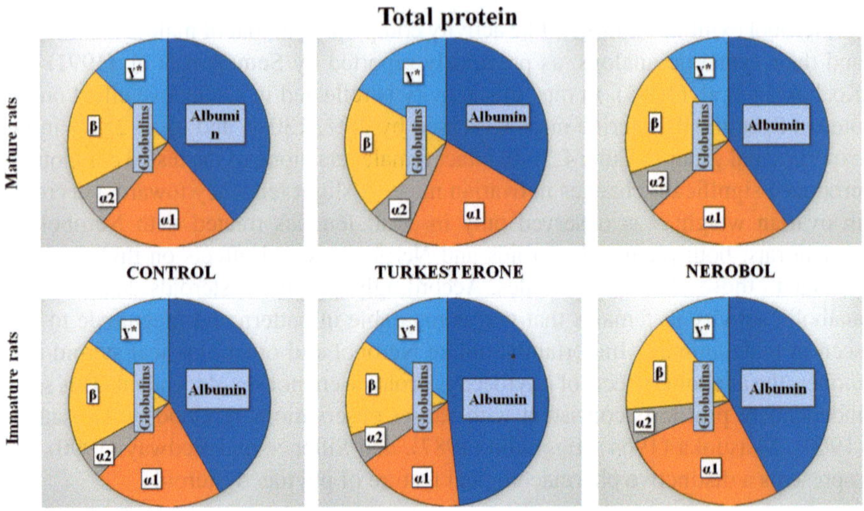

Fig. 4.3 The effect of turkesterone and Nerobol on the protein spectrum of rat blood serum (in % of control), n = 6. The vertical diameter of the circle corresponds to the value reflecting the total protein content. The asterisk marks the defined indicators, the changes of which are reliable in relation to the control (p < 0,05)

Table 4.4 The effect of phytoecdysteroids and Nerobol on red blood cell counts in rats (M ± m, n = 8)

Experimental groups	Parameter	Days of observation		
		Initial	Day 7	Day 14
Control	Hemoglobin	12,7 ± 0,6	12,8 ± 0,5	13,0 ± 0,5
	Red blood cells	5,89 ± 0,3	5,82 ± 0,1	5,93 ± 0,3
α-Ecdysone	Hemoglobin	13,2 ± 0,4	13,9 ± 0,3	14,7 ± 0,7
	Red blood cells	6,01 ± 0,4	6,43 ± 0,4	6,73 ± 0,4
2-deoxy-ecdysterone	Hemoglobin	12,1 ± 0,6	13,0 ± 0,4	13,7 ± 0,5
	Red blood cells	6,28 ± 0,4	6,79 ± 0,3	7,15 ± 0,2
Ecdysterone	Hemoglobin	12,6 ± 0,6	14,1 ± 0,5	15,2 ± 0,5*,**
	Red blood cells	5,44 ± 0,2	6,22 ± 0,2*,**	6,57 ± 0,3*,**
Cyasterone	Hemoglobin	13,0 ± 0,5	14,6 ± 0,7	15,8 ± 0,4*,**
	Red blood cells	5,88 ± 0,3	6,78 ± 0,2*,**	7,20 ± 0,3*,**
Sileneoside A	Hemoglobin	12,9 ± 0,5	14,4 ± 0,4*,**	15,8 ± 0,3*,**
	Red blood cells	5,86 ± 0,2	6,81 ± 0,3*,**	7,22 ± 0,3*,**
Turkesterone	Hemoglobin	12,7 ± 0,3	15,1 ± 0,5*,**	16,4 ± 0,4*,**
	Red blood cells	5,75 ± 0,3	6,78 ± 0,2*,**	7,33 ± 0,2*,**
Nerobol	Hemoglobin	12,4 ± 0,6	14,5 ± 0,3*,**	16,2 ± 0,4*,**
	Red blood cells	6,15 ± 0,3	7,32 ± 0,3*,**	7,87 ± 0,1*,**

Note: *The reliability of the differences compared to the baseline values, **compared to the control (p < 0,05) control rats maintained on a standard diet for 2 weeks, no significant changes in erythrocyte count or hemoglobin concentration were observed, and both parameters remained close to baseline values throughout the observation period. In contrast, administration of the tested compounds resulted in an increase in both erythrocyte number and hemoglobin content. The magnitude of this effect depended on the specific phytoecdysteroid administered. The weakest stimulatory effects were observed for α-ecdysone and 2-deoxyecdysone. In these groups, erythrocyte counts increased by 6.9% and 8.1% on day 7, and by 11.9% and 13.8% on day 14, respectively. Hemoglobin levels showed a similar trend, increasing by 5.3–7.4% on day 7 and by 11.4–13.2% on day 14; however, these changes did not reach statistical significance relative to baseline values. More pronounced effects were observed following treatment with ecdysterone, cyasterone, and sileneoside A. In these groups, erythrocyte counts increased by 14.3%, 15.3%, and 16.2% on day 7, and by 20.7%, 22.4%, and 23.2% on day 14, respectively. Corresponding increases in hemoglobin concentration reached 11.6–12.3% on day 7 and 20.6–22.5% on day 14, with most of these changes being statistically significant. Turkesterone exhibited the strongest erythropoiesis-stimulating activity. As early as day 7, erythrocyte counts increased by 17.9%, and by day 14 reached 7.33×10^6 cells/mm^3, corresponding to a 27.5% increase over baseline. Hemoglobin levels rose by 18.9% on day 7 and by 29.1% on day 14. These effects were comparable to those observed with Nerobol, which increased erythrocyte counts by 19.0% and hemoglobin by 16.9% on day 7, and by 27.9% and 30.6%, respectively, by day 14.

The experimental results in this section show that in the blood and organs of rats treated with phytoecdysteroids, changes occur that resemble those observed after administering anabolic agents. These changes include an increase in total serum protein, albumin, and α_1-globulin levels. At the same time, phytoecdysteroids, unlike Nerobol, caused a slight decrease in α_2-globulins and raised the levels of β- and γ-globulins. Furthermore, phytoecdysteroids, like steroidal anabolics, significantly stimulated erythropoiesis.

4.2.3 Investigation of the Mechanisms Behind the Effects of Phytoecdysteroids on Protein-Synthesizing Processes

Data on how phytoecdysteroids regulate protein biosynthesis in mammals were gathered from studies comparing their effects with those of steroidal anabolics on the functional activity of the hepatic protein-synthesizing system in mice, using both in vivo and in vitro experimental models.

Figure 4.4 shows that the activation of the protein-synthesizing system in hepatocytes, following a single intraperitoneal administration of phytoecdysteroids and the anabolic steroids, was increased. The level of incorporation of ^{14}C-amino acids into proteins of the liver homogenates of the experimental mice was 14–91% for phytoecdysteroids and 94–105% for the anabolic steroids, higher than that in the

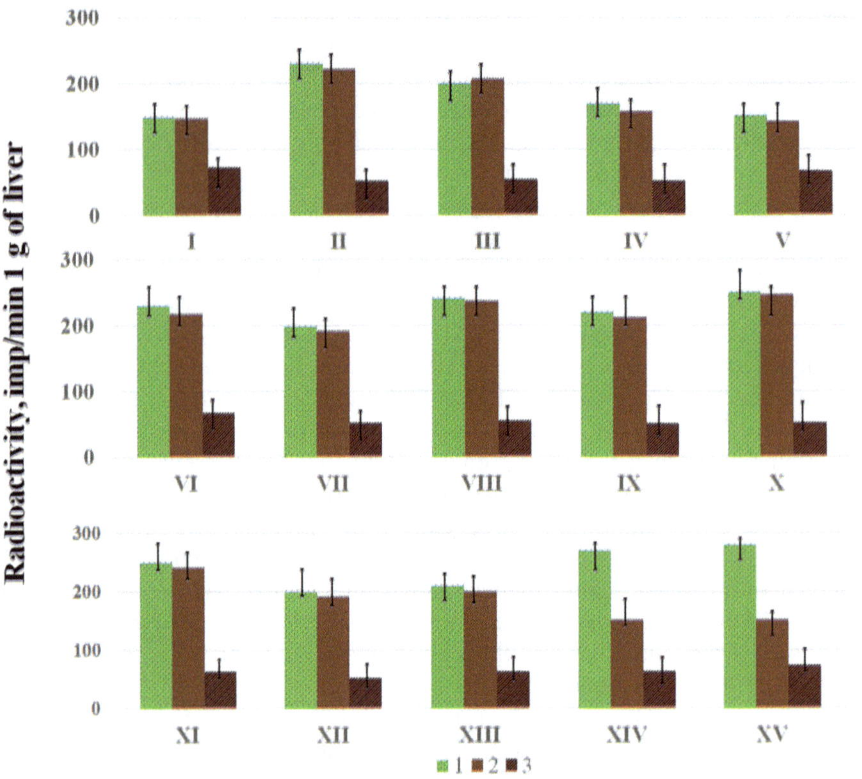

Fig. 4.4 Effect of phytoecdysteroids and steranaboles on the incorporation of labeled amino acids into the total proteins of mouse liver homogenate (1, 3) and acid-soluble fraction (2) (n = 6–8). I—control; II—ecdysterone; III—viticosterone E; IV—2-deoxyecdysterone; V—ecdysone; VI—sileneoside A; VII—sileneoside E; VIII—cyasterone; IX—22-acetylcyasterone; X—turkesterone; XI—tetraacetate turkesterone; XII—integristerone A; XIII—24(28)-dehydromakisterone A; XIV—methylandrostenediol; XV—Nerobol. 1—Without pretreatment; 3—After treatment with actinomycin D

controls ($p < 0.001$). The values reported above were obtained at the time point corresponding to the maximal activity of the compounds, namely 4 h after administration. Comparison of the protein-anabolic effects of phytoecdysteroids observed in rats during long-term administration (Table 4.5), expressed as body weight gain, with those recorded in mice (Fig. 4.4), expressed as the intensity of incorporation of labeled precursors into liver cell proteins, indicates that the overall pattern of action of these compounds remained consistent under both experimental conditions. However, the magnitude of the anabolic response was greater in mice, particularly for anabolic steroids, suggesting a higher sensitivity of this model to protein-biosynthetic stimulation.

However, when actinomycin D-pretreated animals were used, the protein-anabolic effect of steroidal anabolics became almost completely inhibited (Fig. 4.4).

Table 4.5 Incorporation of ^{14}C-labeled amino acids into total liver polypeptides of mice under the action of turkesterone and Nerobol (radioactivity expressed as imp/min/g of liver) (M ± m, n = 9)

Treatment	Total polypeptide radioactivity	Completed polypeptide radioactivity	Growing polypeptide radioactivity	Fraction of growing polypeptide radioactivity	Polypeptide synthesis time (s)
Control	98,325 ± 1449	92,383 ± 1251	5942	0,0604	84
Turkesterone	162,523 ± 1777*	158,533 ± 1383*	3990	0,0245	33
Nerobol	168,135 ± 1460*	165,544 ± 2284*	2591	0,0154	21

Note: The time from the moment the label was introduced to the immersion of the liver in liquid nitrogen, necessary for calculating the time of synthesis of the polypeptide according to formula (110), was determined in each case by a stopwatch. In the control, it was 663 s, in animals treated with turkesterone and Nerobol—643 and 647, respectively; *significantly at p < 0,001

The differences in mechanisms of action of phytoecdysteroids and steroidal anabolics, observed in the mentioned experiments, were further verified while considering the site of action of test agents on individual components of protein-synthesizing systems in in vitro crossover experiments with polyribosomes and cytosolic fractions isolated from control and test animals. In vitro cell-free protein synthesis systems exhibited increased protein-synthetic activity when polyribosomes isolated from hepatocytes of treated animals were combined with cytosolic fractions from either control or treated hepatocytes. These findings indicate that the protein synthesis–stimulating effects of both phytoecdysteroids and steroidal anabolics in the cell-free system are primarily associated with alterations in the polyribosomal machinery of hepatocytes (Table 4.6).

In contrast to phytoecdysteroids, whose effects at the level of cytosolic factors were negligible, methylandrostenediol and Nerobol exhibited clear and, in most cases, statistically significant cytosol-dependent effects. This was demonstrated in experiments combining control polyribosomes with cytosolic fractions from treated animals, where protein-synthesizing activity increased by 10.7% ($p < 0.05$) and 9.7% ($p > 0.05$), respectively. Additional evidence was obtained by comparing experimental polyribosomes combined with control cytosol versus experimental polyribosomes combined with cytosol from treated animals, in which case protein synthesis in the cell-free system increased by a further 13–14% ($p = 0.05$). Analysis of sucrose density gradient sedimentation profiles of ribosomal structures isolated from the livers of treated animals revealed substantial differences between phytoecdysteroids and steroidal anabolics. Treatment with ecdysterone did not increase the proportion of polyribosomes within the total hepatic ribosome population. Accordingly, the ratio of translating to non-translating ribosomal material remained comparable to that observed in control animals (Table 4.5). In contrast, treatment with methylandrostenediol led to a marked increase in the ratio of translating to non-translating ribosomal material. This effect was associated with an increased proportion of polyribosomes and a concomitant decrease in ribosomal dimers and monomers. These observations indicate that the protein-anabolic effects of

Table 4.6 Results of cross-experiments using fractions from control and experimental protein synthesis systems (M ± m, n = 9)

Polyribosomal fraction	Cytosolic fraction	^{14}C amino acids incorporation (imp/min/g)
Control	Control	12,212 ± 374
	Ecdysterone	12,916 ± 362
	2-Deoxyexydysterone	12,720 ± 315
	Turkesterone	13,044 ± 481
	Methylandrostenediol	13,522 ± 412*
	Nerobol	13,407 ± 597
Ecdysterone	Control	22,470 ± 159**
	Ecdysterone	23,728 ± 294**
2-Deoxyecdysterone	Control	15,152 ± 287**
	2-Deoxyecdysterone	15,879 ± 302**
Turkesterone	Control	23,133 ± 301**
	Turkesterone	24,823 ± 224**
Methylandrostenediol	Control	23,927 ± 325**
	Methylandrostenediol	27,115 ± 502**
Nerobol	Control	25,561 ± 131**
	Nerobol	29,181 ± 430**

Note: Asterisks indicate the significance of differences from the indicator reflecting the test result of the reconstructed cell-free protein synthesis system with polyribosomes and cell juice from the liver of control animals: *significant at $p < 0.05$; **significant at $p < 0,001$

phytoecdysteroids are primarily mediated by activation of protein biosynthesis at the translational level. Most likely, this involves coordinated stimulation of translation initiation and elongation. As a result, the absolute rate of protein synthesis increases without detectable changes in polyribosome profiles. In contrast, regulation of protein biosynthesis by steroidal anabolics in mammals appears to depend mainly on activation of transcriptional processes, leading to enhanced synthesis of ribonucleic acids, particularly mRNA. This interpretation is supported by the short time interval between drug administration and the onset of the anabolic response. Consequently, the proportion of actively translating ribosomes increases, and their functional activity in cell-free protein synthesis systems is enhanced. However, if the anabolic effects of steroidal anabolics were determined solely by increased mRNA synthesis, an elevation in the absolute rate of protein biosynthesis would not be expected. Since the experimental results demonstrate such an increase, it can be concluded that anabolic steroids, in addition to their actinomycin D-dependent stimulation of transcription, also substantially accelerate translational processes. This translational activation may result either from increased efficiency of newly synthesized mRNAs as templates or from stimulation of translation through cytosolic factors acting on both newly synthesized and pre-existing mRNAs. The involvement of tRNA in this process cannot be excluded, given its established regulatory role in translation (Kirchner and Ignatova 2015). Taken together, these findings demonstrate that phytoecdysteroids differ fundamentally from steroidal anabolic agents, synthetic analogues of male sex hormones, in the mechanisms by which they stimulate protein synthesis in mammalian organisms.

4.2.4 Effects of Phytoecdysteroids on Carbohydrate Metabolism

Compounds with anabolic modes of action generally exert a pronounced influence on carbohydrate metabolism. Numerous studies have reported a characteristic hypoglycemic effect associated with such agents (Vijayakumar et al. 2010; Efimov and Karabun 1972; Kliaczko et al. 1970). Our experiments with phytoecdysteroids likewise highlighted this important aspect of their biological activity.

All phytoecdysteroids administered to animals for 7 days produced a reduction in blood glucose levels. A relatively weak and statistically insignificant hypoglycemic effect was observed after treatment with α-ecdysone and 2-deoxy-α-ecdysone. In contrast, ecdysterone, sileneoside A, cyasterone, turkesterone, turkesterone tetraacetate, and polypodine B exhibited a pronounced hypoglycemic effect comparable to that of Nerobol, with glucose-lowering activity ranging from 15 to 20% (Table 4.7). These compounds demonstrated even greater hypoglycemic efficacy in experiments using rats with alloxan-induced hyperglycemia.

After 1 week of treatment, ecdysterone, sileneoside A, and turkesterone significantly reduced blood glucose levels in alloxan-diabetic rats. In animals with an

Table 4.7 Comparative hypoglycemic activity of phytoecdysteroids and Nerobol (M ± m, n = 10)

Investigated drugs	Blood serum glucose, mg%		The effect in %
	Initial	After 7 days of drug administration	
Control	96,5 ± 2,6	95,8 ± 2,0	–
Ecdysterone	101,9 ± 2,1	86,6 ± 2,1*	−15,1
Viticosterone E	101,8 ± 1,9	87,4 ± 2,5*	−14,2
Ecdysterone tetraacetate	101,5 ± 3,2	87,2 ± 2,3*	−14,1
2-Deoxyecdysterone	99,9 ± 3,3	88,6 ± 2,1*	−11,3
α-Ecdysone	104,6 ± 3,7	96,2 ± 1,8	−8,1
2-deoxy-α-ecdysone	104,5 ± 4,8	97,3 ± 1,8	−6,9
Sileneoside A	104,2 ± 2,0	85,2 ± 1,6*	−18,2
Sileneoside E	100,9 ± 3,1	89,5 ± 2,2*	−11,3
Cyasterone	96,4 ± 4,3	80,6 ± 3,1*	−16,4
22-Cyasterone acetate	97,8 ± 3,2	84,0 ± 3,1*	−14,2
Turkesterone	99,1 ± 2,8	79,3 ± 3,1*	−19,9
Turkesterone tetraacetate	97,4 ± 4,4	79,8 ± 3,8*	−18,1
Polypodine B	101,0 ± 3,0	84,9 ± 3,2*	−15,9
Integristerone A	102,2 ± 3,3	90,3 ± 2,0*	−11,7
24(28)-Dehydromakisterone	103,6 ± 3,2	88,6 ± 2,3*	−14,5
Nerobol	96,5 ± 2,6	81,9 ± 3,3*	−16,2

Note: *The reliability of the differences compared to the initial values ($p < 0,05$)

initial glycemia of approximately 150 mg%, glucose levels decreased by 29–38%. A comparable hypoglycemic effect was also observed in rats with baseline glycemia of up to 250 mg%. In animals with more severe hyperglycemia (approximately 400 mg%), the glucose-lowering effect was attenuated, although it remained statistically significant. Overall, the hypoglycemic activity of these phytoecdysteroids was essentially equivalent to that of Nerobol (Table 4.8). The observed reduction in blood glucose does not appear to result from stimulation of insulin secretion. In intact animals, no significant differences in plasma insulin levels were detected between control rats and those treated with phytoecdysteroids or Nerobol for 1 week. Similarly, during glucose tolerance tests in diabetic rats with moderate hyperglycemia (up to 150 mg%), treatment with ecdysterone or Nerobol did not increase insulin levels. On the contrary, at 30 and 60 minutes after glucose administration, insulin concentrations were slightly lower than in controls, while at other time points, no significant differences were observed (Fig. 4.5, I). These findings argue against the possibility that ecdysterone or Nerobol suppresses insulin synthesis, as glucose tolerance was not impaired. Instead, both fasting and postload blood glucose levels were consistently lower in treated animals compared with controls (Fig. 4.5, I).

This pattern suggests that the hypoglycemic effect is more likely due to enhanced tissue sensitivity to insulin rather than increased hormone secretion. To test this hypothesis, we examined the effects of the studied anabolically active compounds

Table 4.8 Effect of phytoecdysteroids and Nerobol on blood sugar levels in rats with alloxan diabetes (M ± m, n = 6–8)

Treatment	Group I		The effect in %	Group II		The effect in %	Group III		The effect in %
	Initial value	After 7 days		Initial value	After 7 days		Initial value	After 7 days	
Control	154 ± 8,3	157 ± 7,3	+1,9	239 ± 12,3	248 ± 27,3	+3,7	388 ± 16,3	368 ± 10,4	−5,2
Ecdysterone	158 ± 9,4	109 ± 5,4*	−30,8	245 ± 14,5	176 ± 10,2*	−28,2	413 ± 19,6	318 ± 25,9*	−23,1
Sileneoside A	152 ± 8,7	107 ± 5,7*	−29,4	255 ± 11,8	174 ± 14,4*	−31,8	357 ± 17,0	282 ± 6,4*	−21,1
Turkesterone	159 ± 6,8	98,0 ± 6,6*	−38,4	226 ± 11,1	152 ± 8,1*	−32,8	352 ± 20,6	266 ± 22,7*	−24,5
Nerobol	150 ± 8,1	98,0 ± 5,6*	−34,7	240 ± 14,6	168 ± 16,3*	−30,0	389 ± 28,7	279 ± 20,1*	−28,3

Note: In Group I, animals with blood sugar levels in the range of 150 mg% were selected, in Group II—up to 250 mg%, in Group III—up to 400 mg%. *The reliability of the differences compared to the initial values ($p < 0,05$)

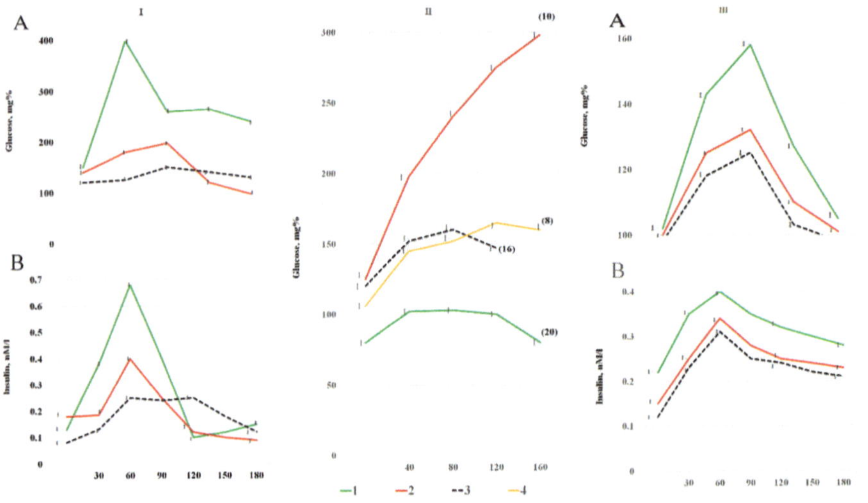

Fig. 4.5 The effect of ecdysterone and Nerobol on insulin-dependent processes in experimental insulin resistance. I—Glycemic (**a**) and insulinemic (**b**) reactions to oral glucose (5 g/kg) in rats with mild alloxan diabetes (1) treated with ecdysterone (2) and Nerobol (3). Each point is the arithmetic mean of 6 rats. II—Glycemic reaction to intravenous infusion of insulin with glucose in control group rats (1) treated with hydrocortisone (2), hydrocortisone and ecdysterone (3), hydrocortisone and Nerobol (4). The number of animals in the experiment is indicated in parentheses. III—Glycemic (A) and insulinemic (B) reactions to oral glucose administration in rats with hydrocortisone-induced insulin resistance (1) treated with hydrocortisone on the background of ecdysterone (2) or Nerobol (3). Each point is the arithmetic mean of 5 rats. In all cases, the ordinate axis shows the glucose level (mg%) or insulin (nM/L), and the abscissa axis shows the time after glucose administration (min)

on insulin sensitivity in epididymal adipose tissue, a tissue known for its high responsiveness to insulin. In the absence of insulin, ecdysterone and Nerobol (Kosovskiĭ et al. 1989), as well as turkesterone and methylandrostenediol (Table 4.9), did not affect basal glucose uptake or adrenaline-induced lipolysis. However, in the presence of insulin, these compounds significantly increased glucose uptake and suppressed lipolysis. Because the physiological concentration of insulin used in these experiments already induced a maximal tissue response, the observed effects indicate an increase in the maximal insulin-mediated response. This finding is consistent with an enhancement of tissue insulin sensitivity. To obtain further confirmation at the whole-organism level, we subsequently employed a model of hypercorticism, which is characterized by marked insulin resistance despite near-normal blood glucose levels (Dobre et al. 2025).

Treatment with ecdysterone or Nerobol significantly reduced blood glucose levels in rats with experimentally induced hypercorticism caused by hydrocortisone administration. In this model, blood glucose served as an integrated indicator of insulin-dependent glucose utilization (Fig. 4.5, II). Baseline differences in

Table 4.9 Effect of turkesterone and methylandrostenediol on insulin-stimulated glucose uptake and inhibition of adrenaline-induced lipolysis in isolated epididymal adipose tissue of alloxan--diabetic rats (M ± m, n = 7–9)

Treatment	Insulin (0.8 nM)	Glucose uptake (mmol/g wet tissue)	FFA release (mEq/g wet tissue)
Control	–	13,6 ± 0,6	29,3 ± 2,77
	+	25,7 ± 1,4	7,8 ± 0,85
Turkesterone	–	12,8 ± 0,5	31,8 ± 3,20
	+	54,5 ± 3,3*	0,98 ± 0,08*
Methylandrostenediol	–	13,1 ± 0,5	31,2 ± 2,10
	+	49,1 ± 2,7*	1,21 ± 0,17*

Note: *differences from the control were significant ($p < 0.05$); (–)—absence, (+)—presence of insulin

endogenous insulin secretion were not considered relevant, as pronounced hyperinsulinemia was achieved during the infusion protocol (194 ± 30 μIU/mL). Under these conditions, pancreatic insulin release and hepatic glucose output are known to be effectively suppressed. Therefore, the development of hypoglycemia during combined insulin and glucose infusion in animals treated with the studied compounds provides clear evidence of an enhanced systemic sensitivity to insulin. Given the presence of sustained hyperinsulinemia, this effect can be attributed primarily to increased glucose uptake by peripheral tissues rather than changes in insulin availability. Additional support for this interpretation comes from glucose tolerance tests. In rats with hypercorticism treated with ecdysterone or Nerobol, as well as in alloxan-diabetic animals, insulin levels measured after glucose loading were lower than those in control animals (Fig. 4.5, III). Despite this reduced insulin response, blood glucose levels remained significantly lower in the treated groups, indicating preserved or even improved glucose tolerance. A plausible explanation for the diminished insulin response observed in animals receiving anabolically active steroids is a reduced physiological demand for insulin. Lower blood glucose levels would require less insulin secretion, which is consistent with enhanced insulin sensitivity in peripheral tissues. Having established the pronounced hypoglycemic activity of phytoecdysteroids and clarified key aspects of their mechanism of action, additional findings further support their role in regulating carbohydrate metabolism. In all experimental series, repeated administration of phytoecdysteroids, similarly to Nerobol, consistently increased glycogen content in all examined organs and tissues (Table 4.10). Moreover, treatment with these compounds led to marked changes in tissue levels of lactic and pyruvic acids, indicating a substantial influence on major pathways of carbohydrate metabolism.

The content of lactic acid (lactate) in the liver, heart, and anterior tibial muscle of rats treated with all the studied compounds was reduced, whereas the content of pyruvic acid in the organs and tissues of experimental animals was noticeably higher than in controls (Table 4.11).

Particular attention was paid to changes in the ratio of lactic acid (LA) to pyruvic acid (PA) in different organs. Together with evidence of increased glucose

Table 4.10 Effect of phytoecdysteroids and Nerobol on glycogen content in the liver, heart, and skeletal muscle of rats (M ± m, n = 8)

Experimental group	Treatment duration	Glycogen, mg%		
		Liver	Heart	Skeletal muscle
Control (intact animals)	–	2018 ± 196	310 ± 11,8	342 ± 32,8
Ecdysterone	7th	2680 ± 134*	456 ± 43,9*	471 ± 40,2*
Cyasterone	7th	2734 ± 113*	448 ± 29,1*	476 ± 27,5*
Sileneoside A	7th	2695 ± 205*	439 ± 20,4*	479 ± 16,7*
Turkesteron	7th	2710 ± 157*	458 ± 40,6*	506 ± 46,4*
Nerobol	7th	2521 ± 162	425 ± 28,8*	464 ± 39,8*
Ecdysterone	14th	2710 ± 117*	426 ± 38,0*	486 ± 36,9*
Cyasterone	14th	2762 ± 127*	436 ± 29,5*	450 ± 38,2*
Sileneoside A	14th	2779 ± 188*	436 ± 46,9*	489 ± 46,0*
Turkesteron	14th	2850 ± 157*	474 ± 39,1*	512 ± 40,8*
Nerobol	14th	2793 ± 100*	441 ± 35,3*	452 ± 39,4*

Note: *The reliability of the differences compared to the control ($p < 0{,}05$)

utilization in treated animals, these parameters allowed calculation of lactate excess and assessment of the redox state of the LA/PA system. This approach made it possible to determine whether phytoecdysteroids preferentially influence aerobic or anaerobic pathways of carbohydrate metabolism. Administration of phytoecdysteroids and Nerobol resulted in a clearly negative lactate excess in rats, indicating a shift toward enhanced aerobic metabolism (Table 4.11). Such a shift indicates a predominance of aerobic metabolic processes in the organs and tissues of treated animals. This conclusion is further supported by an increase in the LA/PA redox potential, reflecting enhanced oxidative metabolism. In the liver and heart, the magnitude of these effects was comparable for phytoecdysteroids and Nerobol. In contrast, skeletal muscle exhibited a more pronounced response to Nerobol treatment. In this tissue, lactate excess became strongly negative, and the LA/PA redox potential increased by 20.7 mV. By comparison, treatment with ecdysterone, sileneoside A, cyasterone, and turkesterone increased the LA/PA redox potential in skeletal muscle by 7.2, 8.4, 8.6, and 10.3 mV, respectively.

When evaluating the overall effects of phytoecdysteroids on carbohydrate metabolism, it is important to consider the previously observed increase in glycogen content in the liver, heart, and skeletal muscle following their administration. At first glance, this finding may appear unexpected, given the enhanced glucose utilization by tissues, which could be expected to reduce carbohydrate reserves. Indeed, earlier studies reported glycogen depletion under conditions of substrate deficiency, such as prolonged fasting (Aizikov et al. 1978). In contrast, under standard vivarium conditions, repeated administration of phytoecdysteroids did not result in glycogen depletion, despite increased glucose consumption by tissues.

This phenomenon can be explained by the fact that activation of glycogen breakdown under physiological conditions is known to stimulate glycogen-synthesizing enzymes (Rosenfeld and Popova 1979). Enhanced activity of these enzymes—apparently further promoted by phytoecdysteroids and Nerobol—allows glycogen stores to be maintained and, in some cases, increased. Taken together, these results demonstrate

Table 4.11 Effect of phytoecdysteroids and Nerobol on lactate and pyruvate levels, excess lactate, and the lactate/pyruvate ratio in rat tissues (M ± m, n = 6)

Experimental study	Treatment					
	Control	Ecdysterone	Cyasterone	Sileneoside A	Turkesterone	Nerobol
Liver	85,7 ± 4,2	64,2 ± 4,6*	59,8 ± 6,0*	59,0 ± 5,5*	54,5 ± 5,3*	64,4 ± 2,0*
MK, mg%	2,09 ± 0,10	2,47 ± 0,12*	2,44 ± 0,13	2,49 ± 0,14*	2,55 ± 0,16*	2,25 ± 0,15
PVC, mg%	-	−37,1	−40,2	−43,1	−50,1	−27,8
Excess lactate						
ORP = MK/PVC mV	−253,6	−247,7	−246,8	−246,6	−245,4	−249,0
Heart	60,6 ± 2,7	46,6 ± 2,9*	46,8 ± 2,4*	43,6 ± 2,8*	39,7 ± 3,1*	38,8 ± 2,8*
MK, mg%	1,57 ± 0,06	1,95 ± 0,09*	1,91 ± 0,11*	2,02 ± 0,13*	2,01 ± 0,12*	1,92 ± 0,09*
PVC, mg%	-	−28,6	−26,9	−34,4	−37,8	−35,3
Excess lactate						
ORP = MK/PVC mV	−252,9	−246,4	−246,7	−245,3	−243,7	−244,0
Skeletal muscle	89,7 ± 4,4	68,9 ± 4,6*	64,0 ± 2,8*	61,9 ± 1,9*	57,7 ± 3,1*	56,5 ± 2,6*
MK, mg%	2,12 ± 0,12	2,78 ± 0,16*	2,81 ± 0,14*	2,74 ± 0,13*	2,92 ± 0,13*	2,99 ± 0,17*
PVC, mg%	-	−48,7	−54,8	−54,0	−65,0	–
Excess lactate						
ORP = MK/PVC mV	−254,3	−247,1	−245,9	−245,7	−244,0	−233,6

Note: *The reliability of the differences compared to the control ($p < 0,05$)

that phytoecdysteroids exert a pronounced regulatory effect on carbohydrate metabolism in mammals. In many respects, this effect resembles that of steroidal anabolic agents. Importantly, the observed metabolic changes appear to be directed toward ensuring adequate energy supply for the anabolic processes stimulated by these compounds.

4.2.5 *Effects of Ecdysterone and Turkesterone on Respiratory Rate, Oxidative Phosphorylation, and the Activity of Multienzyme Systems of Liver Mitochondrial Membranes*

To clarify how energy supply supports the enhanced anabolic processes observed in animals treated with phytoecdysteroids, we examined the functional state of liver mitochondria in rats after administration of ecdysterone and turkesterone (Baev et al. 2022). As in previous experiments, a group of animals treated with Nerobol was included for comparison. After 7 days of treatment, none of the studied compounds significantly altered the mitochondrial respiratory rate of rat liver when glutamate or succinate was used as the oxidation substrate. However, the coupling between respiration and phosphorylation was clearly improved, particularly during succinate oxidation. When treatment was extended to 14 days, both phytoecdysteroids and Nerobol induced a pronounced activation of substrate oxidation across different metabolic states. This stimulation of mitochondrial respiration was especially evident in the succinate-dependent pathway. Specifically, parameters reflecting glutamate oxidation (V_2, V_3, V_4, and V_dnph) increased by only 14–28% compared with controls ($p < 0.05$). In contrast, the corresponding indices of succinate oxidation increased by 23–50% ($p < 0.05$). Under these conditions, the respiratory control ratio remained essentially unchanged, whereas the ADP/O ratio showed a decrease.

An increase in oxidative activity accompanied by a relative reduction in the ADP/O ratio is a well-known feature of intensified energy mobilization in living cells. Such conditions are characterized by a sharp rise in mitochondrial respiration and a somewhat smaller increase in phosphorylation efficiency, resulting in lower oxygen utilization efficiency (Kondrashova, 1966). Importantly, a decrease in the ADP/O ratio does not indicate impaired ATP production. On the contrary, it may reflect an increase in the overall rate of high-energy phosphate synthesis per unit time (Kondrashova 1966; Skulachev 1962, 1969). Thus, activation of mitochondrial oxidation in the liver of rats treated with phytoecdysteroids or Nerobol for 14 days, despite the reduced ADP/O ratio, is likely associated with increased production of macroergic phosphate compounds. This response appears to meet the elevated energy demands imposed by intensified anabolic processes. Through this mechanism, the rate of ATP synthesis is adjusted to the metabolic requirements of the tissues. To better understand the basis of the enhanced respiratory activity of liver mitochondria, we analyzed the activity of key multienzyme systems. After 7 days of treatment, glutamate dehydrogenase activity increased by 45%, 58%, and 18% in rats treated with ecdysterone, turkesterone, and Nerobol, respectively, compared with controls. Activities of succinate dehydrogenase and succinate oxidase in the mitochondrial respiratory chain were also elevated. These activities increased by 33% and 42% with ecdysterone, by 38% and 25% with turkesterone, and by 71% and 38% with Nerobol, respectively. A similar pattern of activation was observed for the NADH dehydrogenase and NADH oxidase systems of the mitochondrial membranes (Table 4.12).

When NADH oxidase activity was assessed in the presence and absence of rotenone, an inhibitor of the internal oxidation pathway (Skulachev 1969), distinct differences between the compounds became evident. Ecdysterone predominantly enhanced NADH oxidation through a rotenone-insensitive pathway, whereas turkesterone showed the opposite tendency. In contrast to phytoecdysteroids, Nerobol increased electron transfer through both the internal and external pathways of NADH oxidation. The enhanced transfer of electrons from oxidation substrates to molecular oxygen observed in animals treated with phytoecdysteroids or Nerobol appears to be linked to increased catalytic activity of mitochondrial enzymes. Taken

Table 4.12 Effect of phytoecdysteroids and Nerobol on the activity of multienzyme systems in rat liver mitochondrial membranes (M ± m, n = 8)

	Experimental group			
Enzyme	Control	Ecdysterone	Turkesterone	Nerobol
Glutamate dehydrogenase	7,6 ± 1,0	11 ± 1,0	12 ± 1,0	9 ± 0,5
p		<0,05	<0,02	<0,25
Succinate dehydrogenase	21 ± 1,0	28 ± 2,0	29 ± 2,0	36 ± 4,0
p		<0,02	<0,01	<0,01
Succinate oxidase	222 ± 14	316 ± 21	278 ± 15	307 ± 22
p		<0,01	<0,05	<0,01
NADH dehydrogenase	8,7 ± 0,8	10 ± 0,4	10 ± 0,5	11 ± 0,6
p		<0,25	<0,25	<0,05
NADH oxidase	53 ± 4,0	60 ± 3,0	64 ± 2,0	70 ± 3,0
p		<0,25	<0,05	<0,01
Rotenone-sensitive NADH oxidase	39 ± 3,0	42 ± 2,0	52 ± 3,0	54 ± 3,0
p		<0,5	<0,02	<0,01
Rotenone-sensitive NADH oxidase	14 ± 1,0	19 ± 0,6	15 ± 2,0	17 ± 0,5
p		<0,002	<0,5	<0,055

Note: Dehydrogenase activity is expressed as μmol of 2,6-dichlorophenolindophenol (DCPIP) reduced per minute per mg of protein. Oxidase activity is expressed as ng-atoms of oxygen consumed per minute per mg of mitochondrial protein

together, these findings indicate that administration of anabolic phytoecdysteroids (ecdysterone and turkesterone) and the synthetic androgen derivative Nerobol induces a functional reorganization of liver mitochondria. This reorganization is directed toward stimulating oxidative metabolism and increasing the formation of high-energy phosphate compounds, thereby supporting the elevated energetic demands associated with anabolic processes.

4.2.6 *Effects of Phytoecdysteroids on Lipid Metabolism*

One of the least explored aspects of the biological activity of phytoecdysteroids in higher animals is their effect on lipid metabolism, and the available data remain inconsistent. For example, studies investigating the influence of α-ecdysone on cholesterol levels in rat serum and liver did not reveal any significant hypocholesterolemic effect (Lupien et al. 1969). In contrast, another study reported that prolonged administration of ecdysterone to rabbits fed a cholesterol-enriched diet produced measurable changes in lipid metabolism (Matsuda et al. 1974).

Diets containing 1% cholesterol were previously shown to produce markedly lower levels of total and free cholesterol in the serum compared with control animals. At the same time, other studies have reported that the phytoecdysteroid ponasterone A increases plasma cholesterol levels in chickens (Matsuoka et al. 1969). Given these inconsistent findings, we considered it necessary to examine the effects of several phytoecdysteroids available to us, primarily those with pronounced protein-synthesizing activity, on lipid metabolism. For comparison, we included Nerobol, a well-characterized steroidal anabolic whose effects on lipid metabolism are relatively predictable (Semenov et al. 1985; Uzbekova 1974; Doyle et al. 1974). Repeated administration of phytoecdysteroids resulted in a modest reduction in serum cholesterol and triglyceride levels. In contrast, serum phospholipid levels were generally higher than in control animals. A particularly notable finding was the marked increase in serum free fatty acids (FFAs) in treated rats. In some cases, such as after 21 days of treatment with diasterone or turkesterone, FFA levels increased by 48.7% and 55%, respectively (Table 4.13). Similar trends were observed in the liver. According to the data presented in Table 4.15, following 14 days of treatment with phytoecdysteroids, total hepatic lipid content decreased by 20.4–28.8%, while hepatic cholesterol levels were reduced by 15.3–21.4%. At the same time, triglyceride and phospholipid fractions showed opposite changes. Hepatic triglyceride levels

Table 4.13 Effect of phytoecdysteroids and Nerobol on selected lipid components in the serum of intact rats (M ± m, n = 8)

Experimental group	Duration of treatment (days)	Measured parameters (mg/100 mL serum)			
		Triglycerides	Phospholipids	Cholesterol	Free fatty acids mEq/ml
Control	7	41,3 ± 3,6	98,3 ± 5,1	75,1 ± 3,9	0,88 ± 0,043
	14	37,3 ± 2,2	97,6 ± 6,3	70,2 ± 1,4	0,92 ± 0,036
	21	38,0 ± 2,9	104,8 ± 6,3	73,5 ± 3,9	0,80 ± 0,040
Ecdysterone	7	34,9 ± 2,1	111,9 ± 6,8	68,9 ± 4,2	1,06 ± 0,10
	14	35,2 ± 3,2	124,2 ± 7,5*	59,4 ± 4,3*	1,25 ± 0,13*
	21	31,5 ± 1,3	114,9 ± 7,1	62,3 ± 3,9	1,06 ± 0,077*
Cyasterone	7	36,4 ± 1,7	116,1 ± 6,4	67,3 ± 5,0	1,02 ± 0,095
	14	33,9 ± 1,9	117,8 ± 3,5*	57,9 ± 4,4*	1,27 ± 0,15*
	21	29,8 ± 1,4*	129,3 ± 5,5*	61,6 ± 3,8	1,14 ± 0,087*
Sileneoside A	7	32,7 ± 1,2*	119,5 ± 3,0*	62,8 ± 3,1*	1,07 ± 0,051*
	14	30,4 ± 1,3*	124,8 ± 4,8*	53,9 ± 3,6*	1,15 ± 0,047*
	21	30,8 ± 1,5	119,9 ± 4,6	50,8 ± 3,7*	1,19 ± 0,062*
Turkesteron	7	33,3 ± 2,0	107,0 ± 8,1	66,9 ± 4,2	1,07 ± 0,11
	14	30,2 ± 2,0*	127,8 ± 6,2*	56,5 ± 5,0*	1,28 ± 0,14*
	21	29,6 ± 1,2*	126,9 ± 6,6*	57,8 ± 4,4*	1,24 ± 0,084*
Nerobol	7	35,1 ± 1,4	95,7 ± 3,4	59,9 ± 3,8*	1,29 ± 0,095*
	14	28,6 ± 0,98*	127,6 ± 11*	57,6 ± 4,4*	1,27 ± 0,14*
	21	31,9 ± 2,2	124,2 ± 8,6	60,6 ± 4,2*	1,04 ± 0,073*

Note: Here and in Tables 4.15 and 4.16* there are significant differences compared to the control ($p < 0,05$)

decreased by 24.8–31.6%, whereas phospholipid content increased by 16.6–23.5%. The beneficial effects of phytoecdysteroids on lipid metabolism were also evident under pathological conditions associated with disrupted lipid homeostasis. In particular, all tested phytoecdysteroids prevented the elevation of serum cholesterol and triglycerides induced by Triton WR-1339 administration (Table 4.14). As a result, serum cholesterol levels in treated animals were 15–47% lower, and triglyceride levels were 21–39% lower than those observed in control rats (Table 4.15).

When ecdysterone and turkesterone were administered to rabbits with experimental atherosclerosis, the following results were obtained. In control animals, the serum cholesterol level was 1259 ± 213 mg%, whereas in animals treated with ecdysterone and turkesterone, it was 748 ± 75 mg% and 564 ± 64 mg%, respectively—representing a reduction of 40–55% ($p < 0.05$). The decrease in triglyceride levels was less pronounced, amounting to 27–33% (control: 690 ± 48.7 mg%; treated groups: 506 ± 35 mg% and 464 ± 40.9 mg%, $p < 0.05$). Similar changes were observed in the examined organs and tissues. There was a clear reduction in total cholesterol and triglyceride content in the aorta and liver of the treated animals (Fig. 4.6).

In the series of comparative experiments, it was established that the changes induced by phytoecdysteroids in the levels of individual lipid components in blood serum, as well as in organs and tissues, were analogous to those observed following administration of Nerobol to animals.

Table 4.14 Effect of phytoecdysteroids and Nerobol on Triton WR-1339-induced hyperlipidemia in rats (M ± m, n = 8)

Experimental group	The content of lipids in blood serum (mg/100 mL)	
	Total lipids	Triglycerides
Control (Triton WR-1339)	249 ± 13,1	674 ± 45,0
Ecdysterone+triton	167 ± 28,9*	507 ± 33,0*
Sileneoside A+triton	158 ± 10,7*	484 ± 45,9*
Cyasterone+triton	172 ± 28,7*	461 ± 78,6*
Turkesterone+triton	132 ± 32,5*	409 ± 27,7*
Nerobol+triton	144 ± 6,0*	476 ± 36,4*

Note: *The reliability of the differences compared to the control ($p < 0,05$)

Table 4.15 The effect of phytoecdysteroids and Nerobol on the content of total lipids, triglycerides, phospholipids, and cholesterol in the liver of rats (M ± m, n = 8)

Experimental group	The content of lipids in blood serum (mg/100 mL)			
	Total lipids	Triglycerides	Total lipids	Triglycerides
Control	5265 ± 380	849 ± 44,8	2656 ± 187	294 ± 12,0
Ecdysterone	4194 ± 250*	623 ± 49,4*	3098 ± 231	238 ± 13,4*
Cyasterone	4105 ± 294*	638 ± 43,0*	3241 ± 142*	248 ± 14,2*
Sileneoside A	4016 ± 267*	609 ± 52,4*	3198 ± 181	231 ± 11,3*
Turkesterone	3748 ± 323*	581 ± 32,5*	3281 ± 163*	249 ± 16,9*
Nerobol	4373 ± 284	684 ± 50,9*	3125 ± 219	275 ± 16,6

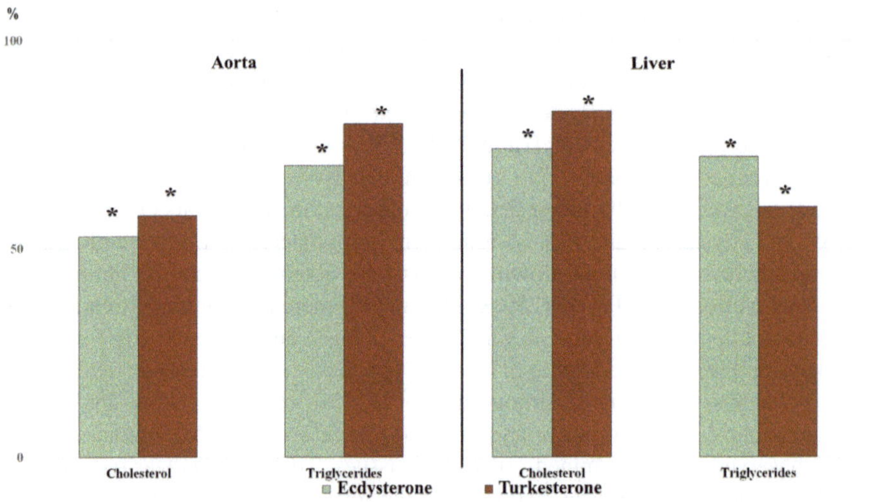

Fig. 4.6 The effect of ecdysterone and turkesterone on cholesterol and triglycerides in the aorta and liver of rabbits with experimental atherosclerosis (% of control), n = 6. *- reliability compared to the control ($p < 0,05$)

Thus, phytoecdysteroids exert a hypolipidemic effect in experiments on intact rats, inhibit the development of hyperlipidemia in rats treated with Triton WR-1339, and alleviate the course of experimental atherosclerosis in rabbits. Administration of phytoecdysteroids also leads to an increase in phospholipid content in both serum and liver and produces a lipid-mobilizing (lipolytic) effect.

4.2.7 *Antioxidant Activity of Phytoecdysteroids*

The biological effects of steroids are largely driven by primary changes in the physicochemical properties of cellular and subcellular membranes (Sergeev et al. 1971). These changes are mediated, at least in part, by their influence on lipid peroxidation (LPO) processes within the membrane lipid fraction (Gukasov and Fedorov 1977). On this basis, we examined the effects of phytoecdysteroids on LPO. Our results show that ecdysterone, turkesterone, and cyasterone significantly affect LPO in vivo, although their actions differ across specific parameters. The effects of various concentrations of these compounds on iron-induced chemiluminescence (CL) in liver homogenates, ecdysterone and turkesterone generally suppressed lipid peroxidation. Analysis of CL kinetics, however, revealed that ecdysterone most closely resembled estrone, a classical natural antioxidant (Sergeev et al. 1974). Within the tested concentration range, ecdysterone delayed the onset of CL, reduced the intensity of both the "fast" and "slow" flashes, and decreased steady-state luminescence (Table 4.16). In contrast, the inhibitory effect of turkesterone was limited to a narrower set of parameters, including prolongation of the latent period, a reduced rate of CL development, and a lower integrated light output of the "slow" flash. At the same time, turkesterone produced a clear dose-dependent increase in the intensity of the "fast" CL flash. Cyasterone showed a distinctly different profile. Rather than inhibiting lipid peroxidation, it exhibited a pronounced pro-oxidant effect. Under its action, the intensity of the "fast" CL flash increased more than tenfold, and both the rate of CL rise and the integrated light output of the "slow" flash increased markedly. The antioxidant activity of the tested phytoecdysteroids was subsequently quantified using the following formula (Vladimirov et al. 1973):

$$\mathrm{A} = (\delta_0 / \delta_1) / \ln \mathrm{H}.$$

For this purpose, the concentration of the compound that reduced the slope of the semilogarithmic apomorphs of the "slow" CL flash by twofold was determined graphically. The comparative data obtained illustrate changes in the δ_0/δ_{-1} ratio as a function of the molar concentrations of the steroids (Fig. 4.7).

Analysis of the presented data indicates that ecdysterone and turkesterone, with antioxidant activities of 3.3×10^2 M^{-1} and 2.8×10^2 M^{-1}, respectively, are markedly less potent than estrone, whereas no antioxidant activity was detected for cyasterone (Table 4.16).

The antioxidant activity previously determined by us for Nerobol in analogous experiments amounted to 2.5×10^3 M^{-1}. Because ecdysterone, similar to Nerobol, inhibited all stages of iron-induced chemiluminescence in these experiments, both compounds were subsequently evaluated following direct administration to animals.

In these experiments, LPO in the examined rat organs was also less pronounced than in control animals. This effect was especially evident after 2 h of ischemia at 37 °C in excised organs, a condition known to activate LPO processes (Bilenko 1978) strongly. LPO indices in the liver, heart, and kidneys of rats pretreated with ecdysterone or Nerobol were not only significantly lower than those of the control group but were also close to the values observed in organs from animals that were not subjected to ischemia (Table 4.17).

Overall, the chemiluminescence data clearly indicate that ecdysterone and turkesterone possess direct antioxidant activity. Given the wide range of biological effects exhibited by these compounds, it can be assumed that, as with other steroid hormones and anabolic steroids, their action is initially mediated by the suppression of lipid peroxidation in organ cells, particularly within biological membranes. This primary effect likely leads to changes in membrane permeability and enzyme activity (Vladimirov and Archakov 1972; Gukasov et al. 1974), which, in turn, indirectly influences the functional state of the protein-synthesizing system.

The pro-oxidant effect observed for cyasterone, despite its belonging to the phytoecdysteroid group and sharing its typical biological activity, suggests that

Table 4.16 Effect of the studied steroids on parameters of iron-induced chemiluminescence in rat liver homogenates (M ± m)

Compounds	Concentration (M)	Parameters of CL kinetics				AOA (M^{-1})
		I (relative units)	τ (minutes)	S (relative units)	δ	
Ec	0	5,0 ± 0,40	2,4 ± 0,26	143 ± 12,8	1,73 ± 0,072	$3,3 \cdot 10^{2}$
	$2,6 \cdot 10^{-3}$	4,0 ± 0,33	2,8 ± 0,30	133 ± 13,1	1,23 ± 0,068*	
	$5,2 \cdot 10^{-3}$	4,0 ± 0,36	3,1 ± 0,24	122 ± 14,1	0,81 ± 0,032*	
	$7,8 \cdot 10^{-3}$	3,0 ± 0,48*	7,4 ± 0,49*	73,5 ± 7,8*	0,36 ± 0,011*	
T	0	5,0 ± 0,36	2,3 ± 0,24	160 ± 15,4	1,8 ± 0,033	$2,8 \cdot 10^{2}$
	$3,0 \cdot 10^{-3}$	6,0 ± 0,38	2,6 ± 0,21	142 ± 12,1	1,54 ± 0,021*	
	$6,0 \cdot 10^{-3}$	9,0 ± 0,42*	2,9 ± 0,18	132 ± 14,2	1,23 ± 0,018*	
	$9,0 \cdot 10^{-3}$	9,0 ± 044*	10,9 ± 0,52*	112 ± 9,7*	0,38 ± 0,012*	
C	0	5,0 ± 0,46	2,3 ± 0,28	140 ± 11,2	1,49 ± 0,031	–
	$5,0 \cdot 10^{-3}$	64,0 ± 2,7*	3,0 ± 0,27	237 ± 14,3*	2,05 ± 0,98*	
	$1,0 \cdot 10^{-2}$	72,0 ± 3,3*	2,9 ± 0,19	235 ± 18,4*	2,25 ± 0,044*	
	$5,0 \cdot 10^{-2}$	52,0 ± 1,8*	3,0 ± 0,18	224 ± 10,2*	1,73 ± –0,019*	
Es	0	4,5 ± 0,41	2,2 ± 0,22	137 ± 10,4	1,8 ± 0,071	$2,5 \cdot 10^{4}$
	$3,7 \cdot 10^{-5}$	3,0 ± 0,44*	3,1 ± 0,19*	110 ± 16,1	1,15 ± 0,020*	
	$7,4 \cdot 10^{-5}$	2,0 ± 0,50*	3,8 ± 0,25*	86 ± 7,6*	0,70 ± 0,026*	
	$1,1 \cdot 10^{-4}$	2,5 ± 0,38*	3,4 ± 0,18*	50 ± 9,2*	0,46 ± 0,023*	

Note: Ec-ecdysterone, T-turkesterone, C-cyasterone, Es-estrone; δ-tg of the angle of inclination of a straight line expressing the dependence of the logarithm of the luminescence intensity on time at the initial exponential stage of a slow CL flash. The tabular numerical values of the determined parameters are the average values from the five observations; an asterisk marks the reliable values ($p < 0.05$) relative to the corresponding control

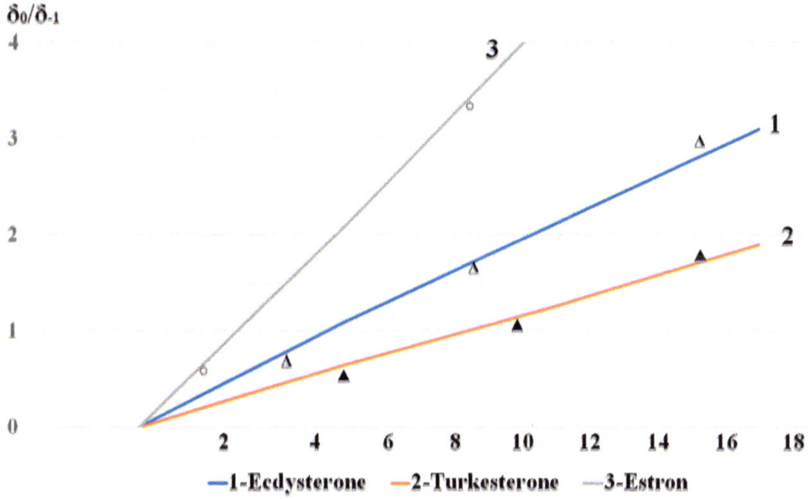

Fig. 4.7 The dependence of δ_0/δ_{-1} on the concentration of ecdysterone (1), turkesterone (2), and estrone (3). The abscissa axis shows the concentration of steroids (10^{-5} M)

Table 4.17 Effect of pretreatment with ecdysterone and Nerobol on chemiluminescence parameters in homogenates of excised and 2-h ischemic rat organs (M ± m, n = 8)

Experimental condition		ChL speed (δ)*		
		Liver	Heart	Kidney
Organs of intact animals		1,03 ± 0,06	0,62 ± 0,01	0,68 ± 0,02
Animal organs that were in an ischemic state after excision for 2 h	Control points	1,35 ± 0,05	0,76 ± 0,07	0,88 ± 0,07
	p_1	<0,02	<0,05	<0,02
	Previously treated with ecdysterone at a dose of 5 mg/kg	0,98 ± 0,06	0,53 ± 0,06	0,60 ± 0,04
	p_1	<0,05	<0,25	<0,05
	p_2	<0,001	<0,05	<0,01
	Previously treated with ecdysterone at a dose of 50 mg/kg	0,70 ± 0,04	0,40 ± 0,0,1	0,50 ± 0,02
	p_1	<0,001	<0,001	<0,001
	p_2	<0,001	<0,001	<0,001
	Previously treated with Nerobol at a dose of 10 mg/kg	0,86 ± 0,04	0,44 ± 0,07	0,44 ± 0,03
	p_1	<0,05	<0,05	<0,001
	p_2	<0,001	<0,01	<0,001

Note: *δ-tg of the angle of inclination of the straight line expressing the dependence of the logarithm of the luminescence intensity on time at the initial exponential stage of the slow flash of CL. p1 is the reliability of differences from the indicators of intact animals; p2 is the reliability of differences from the control

Table 4.18 Effect of phytoecdysteroids and Nerobol on malondialdehyde content in rat liver (M ± m, n = 6)

Experimental group	The content of MDA, nmol/mg of protein	
	Day 7	Day 14
Control	0,479 ± 0,04	0,493 ± 0,03
Ecdysterone	0,369 ± 0,02	0,334 ± 0,03
p	<0,05	<0,002
Turkesteron	0,372 ± 0,02	0,337 ± 0,02
p	<0,05	<0,002
Cyasterone	0,383 ± 0,02	0,375 ± 0,02
p	<0,05	<0,01
Nerobol	0,384 ± 0,02	0,380 ± 0,02
p	<0,05	<0,01

inhibition of LPO may occur through an indirect mechanism. This effect is likely mediated by activation of the endogenous antioxidant defense system via feedback regulation. Consistent with this assumption, measurement of malondialdehyde (MDA), a terminal product of lipid peroxidation, in the livers of rats treated with the studied phytoecdysteroids for 2 weeks showed a reduction in MDA levels. Notably, cyasterone decreased hepatic MDA content in the same direction as ecdysterone and turkesterone (Table 4.18). A comparable decrease in MDA levels was also observed following treatment with the anabolic agent Nerobol.

4.2.8 Effects of Phytoecdysteroids on Electrolyte Metabolism

The effects of ecdysteroids on electrolyte metabolism in arthropods have been extensively studied (Kroeger 1963, 1966). In contrast, much less information is available for organisms in which ecdysteroids do not serve specific metamorphic functions. Among the limited data, particular attention should be given to the study by Kholodova (1979), which examined ionic regulation in neurons of the snail *Helix pomatia*. In that work, ecdysterone was shown to cause a marked increase in intracellular K^+ concentration and a modest increase in Ca^{2+} levels, while Na^+ concentrations remained unchanged. Additional experiments using ion-selective microelectrodes demonstrated that ecdysterone increased the resting membrane potential, K^+ activity, and intracellular pH in snail neurons (Sorokina and Kholodova 1974). Based on these findings, the authors suggested that ecdysterone enhances the transport of K^+, and possibly H^+, against their electrochemical gradients. A slight elevation of blood K^+ levels under the action of phytoecdysteroids has also been reported in rachitic chicks, whereas Na^+ levels remained largely unaffected. Similar changes in K^+ and Ca^{2+} content were observed in avian muscle tissue (Kholodova 1979). To our knowledge, no studies have previously examined the effects of phytoecdysteroids on electrolyte content in the organs and tissues of mammals. To address

Table 4.19 Effect of phytoecdysteroids and Nerobol, alone and in combination, on myocardial potassium and sodium content in rats (M ± m, n = 6)

Experimental group	Total K^+	Total N^+	K^+/N^+ ratio
	mEq/kg dry tissue		
Intact animals	268,6 ± 7,6	148,6 ± 6,8	1,81 ± 0,06
Ecdysterone	296,0 ± 7,1*	154,3 ± 13,8	1,99 ± 0,19
Cyasterone	319,3 ± 19,1*	149,8 ± 5,3	2,16 ± 0,14*
Turkesterone	322,8 ± 15,2*	155,3 ± 12,2	2,11 ± 0,12*
Nerobol	294,3 ± 6,3*	145,2 ± 2,1	2,02 ± 0,03*
Strophanthin K	224,6 ± 7,3*	166,3 ± 2,5*	1,35 ± 0,05*
Strophanthin K+Ecdysterone	287,0 ± 6,7**	156,3 ± 5,3	1,85 ± 0,10**
Strophanthin K+Cyasterone	318,3 ± 20,1*,**	143,6 ± 4,3**	2,22 ± 0,17*,**
Strophanthin K+Turkesteron	310,6 ± 8,9*,**	152,2 ± 7,4	2,05 ± 0,06*,**
Strophanthin K+Nerobol	300,3 ± 10,0*,**	144,3 ± 4,5**	2,09 ± 0,09*,**

Note: *$p < 0.05$ vs. intact animals; **$p < 0.05$ vs. animals treated with strophanthin K

this gap, we conducted a series of experiments in rats. Thirty-day administration of phytoecdysteroids resulted in a pronounced increase in total potassium content in cardiac muscle (Table 4.19).

Treatment with ecdysterone increased myocardial potassium levels by 10.2%, while cyasterone and turkesterone produced even greater increases of 18.8% and 20.2%, respectively. A similar trend was observed after prolonged administration of the anabolic steroid Nerobol, although the increase in this case was less pronounced (9.6%). In contrast, total sodium content in the heart did not change significantly under any experimental condition. Minor variations—slight increases following phytoecdysteroid treatment or slight decreases after Nerobol administration—did not exceed 1–4% and were not statistically significant. As a result, the total potassium-to-sodium ratio increased by 9.9%, 19.3%, and 16.5% following treatment with ecdysterone, cyasterone, and turkesterone, respectively, and by 11.6% after Nerobol administration. It is well established that prolonged administration of cardiac glycosides can disrupt myocardial electrolyte balance by inhibiting Na^+/K^+-ATPase activity, thereby promoting potassium–sodium imbalance (Azizova et al. 1992).

For this reason, we examined the effects of combined administration of cardiac glycosides with phytoecdysteroids. Under identical experimental conditions, one group of rats received subcutaneous injections of strophanthin K for 1 month at a dose of 0.36 mg/kg (equivalent to 3 cardiac units per kilogram, 3 CU/kg). Four additional groups received the same dose of strophanthin K in combination with one of the tested phytoecdysteroids or with Nerobol. Thirty-day administration of strophanthin K alone caused a pronounced disturbance of electrolyte balance in cardiac muscle (Table 4.19). Total potassium content decreased by 16.4%, while total sodium content increased by 11.9%. As a consequence, the K_total/Na_total ratio declined by 25.4% compared with intact animals. A markedly different pattern was observed in animals receiving strophanthin K together with anabolically active steroids. In all combined-treatment groups, myocardial potassium content was significantly higher than in animals treated with the cardiac glycoside alone. Co-administration with ecdysterone increased myocardial potassium levels by 27.7%, whereas combinations with cyasterone, turkesterone, or Nerobol produced even greater increases of 41.7%, 38.2%, and 33.7%, respectively.

Importantly, the combined use of phytoecdysteroids or Nerobol with strophanthin K not only prevented the potassium depletion induced by prolonged glycoside treatment but also promoted potassium accumulation in the myocardium to levels exceeding those in intact control animals. In addition, phytoecdysteroids and Nerobol effectively counteracted the increase in myocardial sodium content caused by strophanthin K. In all combined-treatment groups, sodium levels were 6–13% lower than in animals receiving strophanthin K alone and were close to values observed in intact rats. As a result, the K_total/Na_total ratio in the hearts of rats treated with strophanthin K together with phytoecdysteroids or Nerobol increased substantially compared with animals treated with the cardiac glycoside alone (Table 4.19). Similar observations have been reported previously (Khushbaktova et al. 1982).

Additional evidence was obtained from studies on the combined administration of turkesterone with 3,5-dirhamnoside strophanthidin, a cardiac glycoside recently introduced into clinical practice (Dadabaeva et al. 1993). In accordance with the methodological approach of Benson et al. (1956), these studies demonstrated that turkesterone—similar to Nerobol—not only increased intracellular potassium concentration in myocardial cells without affecting sodium levels, but also exerted a pronounced normalizing effect on the reduced intracellular/extracellular potassium ratio and the elevated intracellular/extracellular sodium ratio induced by prolonged administration of 3,5-dirhamnoside strophanthidin (Khushbaktova et al. 1982). Taken together, these findings indicate that phytoecdysteroids, like steroidal anabolic agents, possess potassium-retaining properties in higher animals. The ability of phytoecdysteroids to restore electrolyte balance during long-term treatment with cardiac glycosides may therefore represent a promising potassium-sparing strategy, contributing both to a reduction in glycoside toxicity and to enhancement of their therapeutic efficacy.

4.2.9 Effects of Phytoecdysteroids on the Endogenous Hormonal Background of Animals

Investigating the effects of phytoecdysteroids on the endogenous hormonal background of experimental animals was considered particularly important, as such data can help clarify the mechanisms underlying their biological activity. Some observations in this area have already been discussed, especially the effects of ecdysterone on plasma insulin levels in relation to carbohydrate metabolism. In this section, we focus on the influence of ecdysterone on plasma concentrations of corticosterone, somatotropic hormone (growth hormone), and testosterone, key regulators of the balance between anabolic and catabolic processes. Because the biological effects of phytoecdysteroids were consistently evaluated in comparison with those of steroidal anabolic agents, the experiments included not only animals treated with ecdysterone but also a group of rats receiving the anabolic steroid methandrostenolone under identical conditions. Single administration of ecdysterone at doses of 1 and 5 mg/kg resulted in an increase in plasma corticosterone levels at 2 and 4 h after treatment. In contrast, somatotropic hormone levels remained unchanged at 1 h and decreased only at 4 h following administration of the higher ecdysterone dose (5 mg/kg) (Fig. 4.8).

Methandrostenolone produced a markedly different hormonal response. This steroid caused a much stronger increase in plasma corticosterone at both tested doses, particularly 4 h after administration (Fig. 4.8). At the same time, it induced a statistically significant reduction in somatotropic hormone levels under all experimental conditions. By 24 h after treatment, plasma concentrations of both corticosterone and somatotropic hormone in animals treated with ecdysterone or methandrostenolone had returned to levels not significantly different from those of control animals.

With regard to testosterone, the phytoecdysteroid used in this study—ecdysterone—did not produce any meaningful changes in plasma levels of this hormone 1 or 4 h after a single administration. After 24 h, rats receiving ecdysterone at a dose

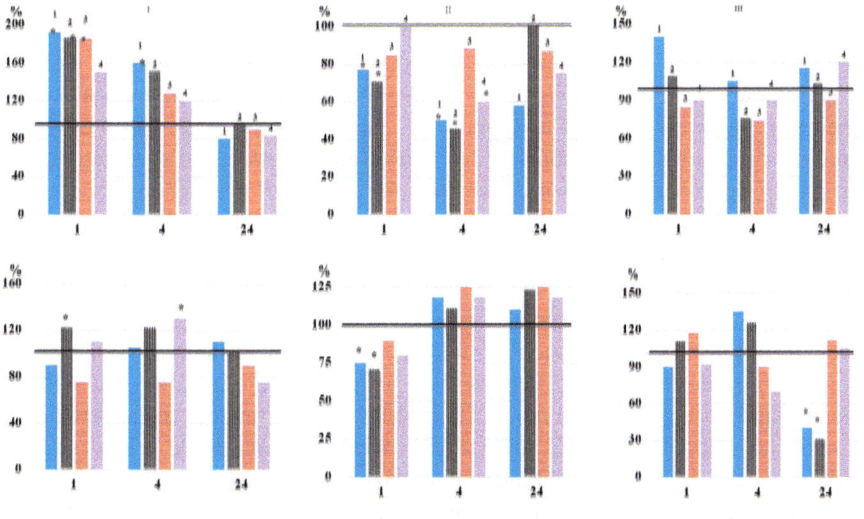

Fig. 4.8 The effect of a single injection of methandrostenolone and ecdysterone (below) on changes in the concentration of 11-deoxycorticosterone (I), corticosterone (II), and aldosterone (III) in the blood plasma of rats. The ordinate axis shows the concentration of steroids in % relative to the control. 1, 2—methandrostenolone (1 and 10 mg/kg; 3, 4—ecdysterone (1 and 5 mg/kg. The asterisk indicates the significance of the differences compared to the control ($p < 0.05$)

of 5 mg/kg showed only a slight and statistically insignificant increase in testosterone concentration. Administration of methandrostenolone resulted in a modest elevation of plasma testosterone at 1 and 24 h; however, these changes were also not statistically significant. After 8 days of repeated administration, increased plasma corticosterone levels were detected in rats treated with ecdysterone only at the higher dose (5 mg/kg) and only 4 h after the final injection. Somatotropic hormone (STH) levels remained unchanged at 1 h, but showed a clear tendency toward elevation at 4 and 24 h. In contrast, methandrostenolone produced elevated corticosterone concentrations at all observation time points, with the most pronounced effect observed at the higher dose of 10 mg/kg. At both doses, methandrostenolone induced a transient reduction in STH levels, which was evident 1 h after the final administration.

Comparison of plasma testosterone levels after 1 week of treatment revealed distinct differences between the tested compounds. In rats treated with ecdysterone, testosterone concentrations remained relatively stable throughout the observation period, with only minor increases detected at 1 h following the lower dose (1 mg/kg) and at 24 h after both doses. In contrast, methandrostenolone produced a markedly different pattern: testosterone levels increased at 4 h, but subsequently declined sharply at later time points, reaching reductions of 50–60%. The hormone-related effects of phytoecdysteroids described here were identified by us for the first time. In contrast, the hormonal changes induced by anabolic steroid drugs are consistent with previously published experimental and clinical data, which demonstrate their ability to elevate circulating glucocorticoid levels (Hervey et al. 1976; Sharpe et al. 1984) while simultaneously reducing STH (Donaldson et al. 1981) and testosterone concentrations (Ryan and Hoffman 1978).

Analysis of the temporal patterns of hormonal responses suggests that both ecdysterone and methandrostenolone share the ability to induce a transient increase in plasma corticosterone. This response may contribute to the activation of biochemical pathways required to support the enhanced protein synthesis observed after administration of anabolic agents. However, the effects of ecdysterone and methandrostenolone on STH and testosterone clearly differ, reflecting the specific endocrine properties of methandrostenolone as a synthetic steroid hormone analog. These differences were consistently observed not only at the systemic level but also at the level of individual organs and tissues. Overall, our findings indicate that administration of both phytoecdysteroids and steroidal anabolic agents leads to substantial changes in the endogenous hormonal profile of healthy animals. Importantly, however, treatment with phytoecdysteroids such as ecdysterone does not shift the plasma corticosterone:testosterone:STH ratio beyond the range associated with disruption of anabolic–catabolic balance. In contrast, steroanabolic drugs typically induce more specific hormonal alterations linked to their genetically programmed control of metabolic processes in various tissues. Finally, it should be noted that for several parameters describing the hormonal effects of ecdysterone, subsequent studies confirmed similar responses in humans.

Clinical evaluations demonstrated that course administration of ecdysterone, as in experimental animals, was associated with increased serum levels of somatotropic hormone and testosterone, and in some cases estradiol (Skosyreva et al. 1988; Skosyreva 1993). Taken together, the results presented in this chapter demonstrate that phytoecdysteroids exert a pronounced protein-anabolic effect in mammals, which in many respects resembles that of steroidal anabolic drugs. However, only certain phytoecdysteroids—most notably turkesterone—approach the anabolic potency of Nerobol. Fundamental differences in the regulation of protein biosynthesis distinguish these two classes of compounds. In particular, phytoecdysteroids lack the specific hormonal side effects characteristic of steroidal anabolics, and their anabolic activity is markedly reduced following gonadectomy. At the same time, both phytoecdysteroids and steroidal anabolics induce similar alterations in carbohydrate, lipid, and related metabolic pathways, ultimately shifting cellular metabolism toward enhanced energy production required to sustain increased protein synthesis.

4.3 Stress-Protective Activity of Phytoecdysteroids

One of the earliest issues that attracted our attention was the potential of phytoecdysteroids to exert corrective effects on the stress response. Stress represents a nonspecific, standardized adaptive reaction that plays a fundamental role in the survival of living organisms and was first described by Selye (1936). Our interest in this problem was also motivated by the views expressed by Selye himself and by other researchers, who emphasized that in many pathological conditions, the damaging factor is often of secondary importance, whereas the primary role belongs to an insufficiency or imbalance of adaptive mechanisms (Davydovsky 1962; Selye 1936). From this standpoint, experimental attempts to correct or support the organism's own adaptive responses to stress through phytoecdysteroid-induced metabolic reprogramming in organs and tissues may be of considerable practical value. Experiments conducted in this direction showed that, under our experimental conditions—where stress in rats was induced by immobilization in the supine position for 16 h—a typical pattern of the general adaptation (stress) response developed (Brekhman 1968; Dardymov 1976). Stress exposure resulted in a 28.6% increase in the relative mass of the adrenal glands compared with intact animals, accompanied by a marked depletion of their metabolic reserves (Table 4.20). Specifically, the content of ascorbic acid and cholesterol in the adrenal glands decreased by 47.4% and 42.6%, respectively. At the same time, the mass of the thymus decreased by 36.6%, and that of the spleen by 45.3%. Pronounced ulcerative lesions were observed in the gastric mucosa. Stress also caused a significant reduction in liver mass, which was accompanied by a 45.6% decrease in hepatic glycogen content. In parallel, a pronounced activation of lipid peroxidation was detected in the liver, as indicated by a 67.2% increase in MDA levels. In the blood serum of stressed animals, substantial elevations in urea and glucose concentrations were observed, amounting to increases of 75.7% and 54.5%, respectively. The total body mass of the animals decreased by 16.5%. All of the changes described above were

Table 4.20 Effects of phytoecdysteroids and Eleutherococcus extract on stress responses in rats subjected to 5-h forced swimming (M ± m, n = 10)

Measured parameters	Experimental conditions					
		Stress				
	Intact animals	Control	Ecdysterone	Turkesterone	Eleutherococcus extract	Nerobol
Adrenal glands, mg/100 g	22.4 ± 1.1	28.8 ± 1.2*	23.3 ± 1.4**	24.5 ± 1.5**	25.6 ± 1.5	27.8 ± 1.1*
Ascorbic acid of the adrenal glands, mg%	380 ± 19.1	200 ± 16.8*	306 ± 12.0*,**	292 ± 11.6*,**	248 ± 11.0*,**	218 ± 18.8*
Adrenal cholesterol, mg%	2773 ± 199	1592 ± 165*	2273 ± 83*,**	2209 ± 77*,**	1858 ± 84*	1603 ± 87.6*
Thymus, mg/100 g	273 ± 17.1	173 ± 10.6*	218 ± 4.5*,**	228 ± 11.8*,**	213 ± 6.5*,**	165 ± 6.3*
Spleen mg/100 g	662 ± 33.6	362 ± 11.8*	464 ± 17.8*,**	498 ± 17.0*,**	418 ± 13.9*,**	394 ± 20.4*
Liver, mg/100 g	5365 ± 184	3877 ± 194*	4818 ± 155*,**	5093 ± 192**	4335 ± 155*	4712 ± 168*,**
Liver glycogen, mg%	2109 ± 60.0	1147 ± 49.7*	1859 ± 66.1*,**	1926 ± 59.1*,**	1483 ± 80.4*,**	1597 ± 83.9*,**
MDA of liver, nmol/mg of protein	0.451 ± 0.02	0.754 ± 0.06*	0.515 ± 0.03**	0.531 ± 0.02*,**	0.657 ± 0.04*	0.684 ± 0.03*
The number of ulcers in the stomach	–	2.1 ± 0.61	0.6 ± 0.26**	0.5 ± 0.26**	0.9 ± 0.29	0.5 ± 0.27**

Note: Here and in Table 4.21: $^{*}p < 0.05$ vs. intact animals; $^{**}p < 0.05$ vs. with control group statistically significant. The phytoecdysteroids examined in this study (Table 4.20) demonstrated a comparable ability to counteract stress-induced adrenal hypertrophy, as well as to prevent the depletion of ascorbic acid and cholesterol reserves in the adrenal glands.

Ecdysterone and turkesterone also provided marked protection of the thymus and spleen against stress-induced involution, prevented the pronounced decrease in liver mass, and showed a clear tendency toward normalization of hepatic glycogen content and MDA levels. In addition, phytoecdysteroids significantly reduced trophic damage to the gastric mucosa, leading to a substantial decrease in the number of bleeding ulcerative lesions. In rats treated with ecdysterone or turkesterone, serum urea levels were 25.9–32.0% lower and glucose levels were 22.4–26.1% lower than in control animals ($p < 0.05$). Body weight loss under stress conditions was also less pronounced: it reached only 8.9% in rats receiving ecdysterone and 5.6% in those treated with turkesterone. Under these experimental conditions, both phytoecdysteroids clearly surpassed the well-known adaptogenic agent Eleutherococcus extract in their overall efficacy (Brekhman 1968). In contrast, Nerobol did not show a positive effect for most of the evaluated parameters. Its activity was limited to partial prevention of liver mass reduction, less pronounced than that observed with phytoecdysteroids, and to a decrease in the number of gastric ulcerations, comparable to the effects of phytoecdysteroids. Both Eleutherococcus extract and Nerobol were also less effective than phytoecdysteroids in preventing stress-induced body weight loss, which amounted to 12.0% and 10.4%, respectively. In animals treated with Eleutherococcus extract, serum urea and glucose levels decreased relative to control values by 10.5% ($p > 0.05$) and 15.6%, respectively, whereas in rats receiving Nerobol, the corresponding reductions were 31.2% and 17.9%.

In an additional series of experiments, we demonstrated that the changes observed during the alarm stage of stress after administration of phytoecdysteroids reflect a favorable reorganization of adaptive processes. The experimental design allowed simultaneous evaluation of both the ability of these compounds to modify the course of the alarm reaction and their intrinsic adaptogenic activity. Rats received phytoecdysteroids 1 h before the start of the experiment and were then placed, together with control animals, in a water bath. After 5 h, some animals were removed and sacrificed, while the remaining rats continued swimming until death. Thus, this model combined a standardized stress load, enabling assessment of stress-induced alterations, with a lethal load, allowing evaluation of adaptogenic efficacy. Five hours of swimming was sufficient to induce pronounced stress-related changes. Similar to prolonged immobilization, swimming stress caused marked adrenal hypertrophy accompanied by depletion of adrenal ascorbic acid and cholesterol compared with animals maintained under standard vivarium conditions (Table 4.21).

Phytoecdysteroids, like Eleutherococcus extract, effectively counteracted these alterations. The tested compounds also significantly prevented the reduction in thymus mass observed in stressed control animals. In contrast, spleen mass did not differ substantially between intact animals and those subjected to swimming stress. Animals that continued swimming until death survived significantly longer when

Table 4.21 Effects of phytoecdysteroids and Eleutherococcus extract on stress responses in rats subjected to 5-h forced swimming (M ± m, n = 10)

Measured parameters	Experimental condition				
	Intact animals	Stress			
		Control	Ecdysterone	Turkesterone	Eleutherococcus extract
Adrenal glands, mg/100 g	21.2 ± 0.9	26.2 ± 0.8*	22.5 ± 0.9**	21.6 ± 0.8**	23.5 ± 0.7**
Ascorbic acid of the adrenal glands, mg%	360 ± 14.3	223 ± 12.6*	316 ± 8.8*,**	326 ± 7.8**	271 ± 6.9*,**
Adrenal cholesterol, mg%	2700 ± 144	1692 ± 128*	2372 ± 99.3**	2431 ± 81.8**	2061 ± 88.2*,**
Thymus, mg/100 g	259 ± 13.2	182 ± 8.5*	222 ± 10.2*,**	250 ± 13.0**	213 ± 9.5*,**
Spleen mg/100 g	654 ± 24.5	582 ± 17.4*	644 ± 24.2	640 ± 22.0	610 ± 23.4

treated with phytoecdysteroids than did control animals. The increase in swimming time to death ranged from 13.7% to 16.6% ($p < 0.05$). By comparison, Eleutherococcus extract prolonged survival by only 8.4%. Taken together, these results indicate that the favorable changes observed during the alarm stage of stress develop in parallel with the adaptogenic effects of phytoecdysteroids.

The obtained results indicate that the reorganization of adaptive activity induced by the tested compounds is beneficial for the organism (Shakhmurova et al. 2010). Thus, phytoecdysteroids attenuate the consequences of stress exposure in animals by normalizing metabolic processes, which under stress conditions typically acquire a predominantly catabolic character. In terms of their ability to enhance the organism's adaptation to adverse conditions, ecdysterone and turkesterone proved to be more effective than Eleutherococcus extract and exhibited clear qualitative advantages over Nerobol.

4.3.1 *Potential Applications of Phytoecdysteroids and Their Analogs in Enhancing Physical Performance and Reducing Fatigue*

In the previous section, we already noted the ability of phytoecdysteroids to enhance physical performance in animals, as reflected by the increased swimming time to exhaustion observed after their administration. To further clarify this effect, a more detailed investigation was undertaken. First, a comparative study was performed to evaluate the effects of several ecdysteroids, Eleutherococcus extract, the plant-derived stimulant Saparal (Sokolov 1972), and the steroidal anabolic agents methylandrostenediol and Nerobol on the duration of work performed by mice until complete fatigue. The results showed that as early as 1 h after administration, all tested compounds produced a pronounced increase in running performance in the "endless climbing rope" test (Table 4.22). Among the phytoecdysteroids, α-ecdysone increased physical performance by 11.5%, whereas 2-deoxyecdysterone produced a stronger effect of 29.3%. Ecdysterone and sileneoside A showed nearly identical activity, enhancing performance by 34.9% and 35.7%, respectively, while turkesterone exerted the most pronounced effect, increasing performance by 38.3%. The reference adaptogenic agents Eleutherococcus extract and Saparal increased physical performance by 22.5% and 25.1%, respectively. The steroidal anabolic agents, methylandrostenediol and Nerobol stimulated physical performance to a degree comparable with that of ecdysterone and sileneoside A, although their effects were somewhat weaker than that observed with turkesterone. A similar pattern was observed in experiments assessing swimming performance after a single administration of the tested compounds (Table 4.22). Notably, an even greater enhancement of physical performance was detected when phytoecdysteroids were administered preliminarily for 7 days.

In these experiments, administration of the most active compounds, ecdysterone, sileneoside A, and turkesterone, produced a marked increase in physical endurance in mice. Specifically, running time to complete exhaustion increased by 65.2%, 67.5%, and 72.3%, respectively, while swimming time to exhaustion increased by 50.8%, 52.9%, and 69.5%. Under these conditions, the superiority of phytoecdysteroids over Eleutherococcus extract and Saparal became even more evident. At the same time, the effectiveness of phytoecdysteroids in enhancing physical performance after repeated administration was somewhat lower than that of anabolic steroid drugs. The latter showed a much stronger stimulatory effect under these conditions than after a single administration (Table 4.22).

This observation, together with the fact that the use of steroanabolics to stimulate muscular activity has in recent years been regarded as incompatible with antidoping regulations (Rogozkin and Tchaikovsky 1981), indicated that further comparison of phytoecdysteroids with methylandrostenediol or Nerobol in this context was inappropriate. Accordingly, in subsequent experiments, plant-derived stimulants, which have attracted considerable and well-deserved interest in recent years (Bobkov et al. 1984; Kiplinger 1967), were generally used as reference compounds. In these experiments, the effects of phytoecdysteroids on physical performance were also investigated, but their stimulatory activity was evaluated either against the background of developing fatigue or under conditions of already enhanced performance resulting from prolonged training. In the first experimental series, rats carrying an additional load equal to 6% of their body mass were subjected to forced swimming to complete exhaustion twice: before and 1 h after administration of the tested compounds.

These experiments showed that, in control animals, the duration of the second swimming bout was reduced by 25.2% ($p < 0.05$). In contrast, rats treated with ecdysterone, sileneoside A, or turkesterone demonstrated a statistically significant increase in the duration of the second swim relative to the first, amounting to 27.8%,

Table 4.22 Effects of the tested compounds on endurance in mice during forced muscular activity (M ± m, n = 10)

Experimental group	Duration of running until exhaustion (min)				Duration of swimming until complete exhaustion (min)			
	Running time (single)	% to control	Running time (7 days)	% to control	Swimming time (single)	% to control	Swimming time (7 days)	% to control
Control	35,5 ± 3,3	–	35,7 ± 3,4	–	344 ± 32,5	–	342 ± 31,5	–
α-Ecdysone, 5 mg/kg	39,6 ± 2,8	11,5	43,0 ± 2,8	20,4	387 ± 29,8	12,5	410 ± 31,2	19,8
2-Deoxyecdysterone, 5 mg/kg	45,9 ± 3,3*	29,3	48,2 ± 2,3*	35,0	393 ± 29,5	14,2	413 ± 31,6	20,7
Ecdysterone, 5 mg/kg	47,9 ± 4,1*	34,9	59,0 ± 4,5*	65,2	427 ± 19,4*	24,1	516 ± 38,2*	50,8
Sileneoside A, 5 mg/kg	48,2 ± 2,7*	35,7	59,8 ± 2,8*	67,5	421 ± 27,1	22,4	523 ± 38,6*	52,9
Turkesterone, 5 mg/kg	49,1 ± 2,5*	38,3	61,5 ± 2,9*	72,3	445 ± 19,3*	29,4	580 ± 37,3*	69,5
Eleutherococcus extract, 0.2 mg/100 g	43,5 ± 2,8	22,5	50,2 ± 2,9*	40,6	389 ± 30,2	13,1	411 ± 32,0	20,2
Saparal, 20 mg/kg	44,4 ± 3,6	25,1	52,4 ± 3,4*	46,8	396 ± 28,7	15,1	430 ± 15,4*	22,7
Methylandrostenediol, 10 mg/kg	46,6 ± 2,6*	31,3	64,0 ± 2,9*	79,3	398 ± 29,2	15,7	607 ± 28,7*	77,5
Nerobol, 10 mg/kg	47,2 ± 2,8*	32,9	66,7 ± 3,2*	86,8	400 ± 28,9	16,3	629 ± 17,4*	83,9

Note: *The reliability of the differences compared to the control ($p < 0,05$)

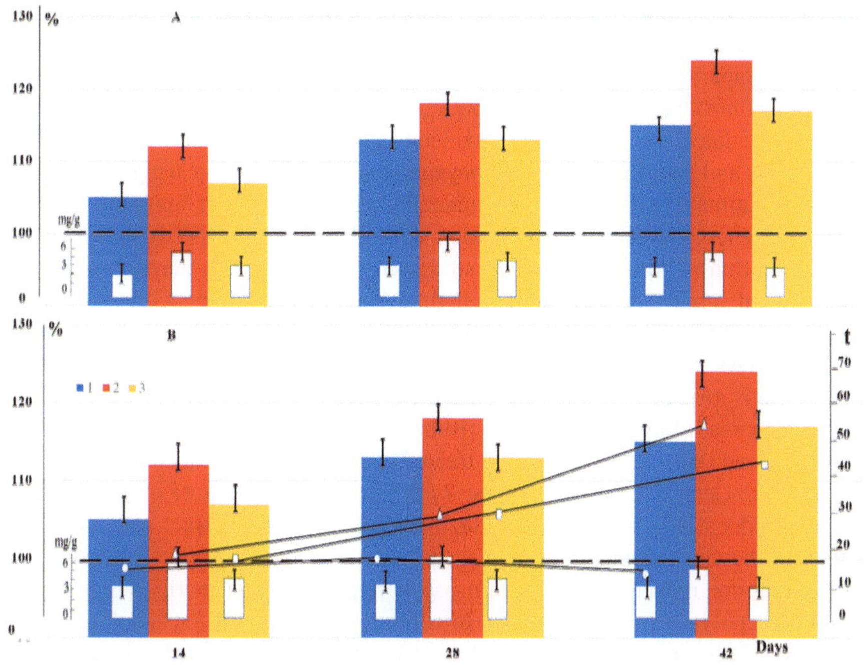

Fig. 4.9 The effect of ecdysterone and saparal on the weight of rats and the maximum swimming time under maximum load conditions (n = 10). Along the abscissa axis: on the left—an increase in body weight (in % of the initial) and weight gain in terms of the initial (in mg/g/day); on the right—swimming time (min). A—intact animals, B—subjected to systematic swimming training. 1—control; 2—ecdysterone; 3—saparal

29.1%, and 34.3%, respectively ($p < 0.05$). Once again, the performance-enhancing effects of phytoecdysteroids exceeded those of Eleutherococcus extract and Saparal, whose effects were limited to 10.9% and 17.0%, respectively ($p < 0.05$).

In the second series of experiments, it was demonstrated that long-term administration of the tested compounds to rats maintained under a training regimen significantly increased overall physical endurance, consistent with the results of the previous experiments. In this series, only ecdysterone was studied in comparison with Saparal, and the animals again swam with an additional load equal to 6% of body mass. After 14 days of administration, the maximum swimming duration in rats receiving ecdysterone was 22% higher than in controls, whereas in rats receiving Saparal, the increase was only 6% (Fig. 4.9). At later time points, the enhancement of physical performance became even more pronounced. After 28 days, it reached 27% and 32%, and after 42 days, 100% and 77% for ecdysterone and Saparal, respectively ($p < 0.05$). It should also be noted that body mass increased more rapidly in animals receiving the tested compounds, particularly ecdysterone, than in the control group subjected to training alone (Fig. 4.9). As a result, the amount of mechanical work performed during locomotion was also greater in these animals. When the performed work was calculated taking into account both body mass gain and swimming duration, the increase in work capacity by the end of the experiment amounted to 3.1-fold in rats receiving ecdysterone and 1.8-fold in those receiving Saparal. In contrast, in intact trained rats, the corresponding increase exceeded control values by only 42–66% ($p < 0.05$).

The calculated energy charge of skeletal muscle in rats treated with the tested compounds and subjected to regular physical loading was 21–28% higher than that in control animals ($p < 0.05$). In rats maintained under standard, nontraining conditions, the energy charge still exceeded control values by approximately 14%. Administration of the studied compounds—most notably ecdysterone—to physically trained animals helped maintain high levels of reserve carbohydrates in working muscles and reduced the accumulation of lactic acid, which is a major factor limiting physical performance (Table 4.23) (Hollman et al. 1980). These changes suggest activation of gluconeogenesis and enhanced glycogen resynthesis, particularly via increased lactate utilization (Bobkov et al. 1984). Ultrastructural analysis of cardiac and skeletal muscle in both trained and intact rats, conducted by researchers from the Central Research Laboratory of the Tashkent Institute of Postgraduate Medical Education (Vaisbrot and Tadzhikbaev), revealed distinct morphological effects of the tested compounds.

Ecdysterone induced pronounced myocyte hypertrophy, primarily due to expansion of the polyribosomal apparatus. In contrast, the action of Saparal was mainly reflected in the abundance of mitochondria with densely packed cristae and an electron-dense matrix. In addition, treatment with both compounds, especially ecdysterone, was associated with increased capillary activity in the myocardium and skeletal muscle; both capillary number and morphofunctional state indicated an elevated level of tissue metabolism. Notably, in experiments in which animals were sacrificed after the final maximal swimming bout under peak load, rats receiving long-term ecdysterone treatment exhibited complete preservation of skeletal muscle ultrastructure. By contrast, rats treated with Saparal exhibited multiple pathological changes in myocytes, including severe mitochondrial swelling, disruption of cristae, fragmentation of intracellular contractile elements, and interstitial edema. These observations indicate that, based on morphological criteria, ecdysterone promotes intracellular and cellular hypertrophy throughout the heart and skeletal muscle, whereas the effects of Saparal are mediated mainly through mitochondrial adaptation. Taken together, these findings demonstrate that phytoecdysteroids enhance physical performance through favorable biochemical and functional adaptations that support metabolic homeostasis and energy production, both during training loads and during exercise to complete exhaustion. Comparative studies further showed that the most metabolically active phytoecdysteroids, characterized by strong protein-anabolic activity (ecdysterone, sileneoside A, and turkesterone), exhibit greater actoprotective effects than plant-derived stimulants such as Eleutherococcus extract and Saparal, although they are less effective than methylandrostenediol and Nerobol when administered for prolonged periods.

This chapter summarizes experimental evidence indicating that PEs, particularly ecdysterone, turkesterone, and cyasterone, represent potent and safe adaptogens with broad stress-protective activity. Unlike androgenic anabolic steroids,

Table 4.23 Effects of ecdysterone and Saparal on carbohydrate metabolism in skeletal and cardiac muscles of intact and trained rats (M ± m, n = 6)

Experimental condition	Analyzed parameters: Glycogen (mg/100 g tissue)	Lactic acid, LA (mg/100 g)	Pyruvic acid, PA (mg/100 g)	LA/PA ratio	Lactate excess	Redox potential (ORP, LA/PA), mV
Skeletal muscle						
Intact animals						
Control	348 ± 15.0	83.3 ± 3.9	2.09 ± 0.12	39.8 ± 2.9	–	−253.1
Ecdysterone	394 ± 18.0	79.5 ± 7.3	2.94 ± 0.21**	27.04 ± 1.8**	−37.6	−247.9
Saparal	403 ± 16.0**	74.0 ± 6.8	2.83 ± 0.22**	26.1 ± 1.8**	−38.7	−247.4
Intact animals						
Control	376 ± 49.0	89.0 ± 5.2	2.52 ± 0.18	35.3 ± 4.2	−11.4	−251.5
Ecdysterone	454 ± 34.0*	52.7 ± 3.9*,**	3.62 ± 0.22*,**	14.5 ± 16*,**	−75.1	−239.6
Saparal	434 ± 28.0*	71.1 ± 3.6*,**	2.70 ± 0.18*,**	26.3 ± 2.6*	−36.5	−247.6
Myocardium						
Intact animals						
Control	268 ± 21.0	67.9 ± 6.4	1.59 ± 0.06	43.2 ± 3.8	–	−254.2
Ecdysterone	330 ± 16.0**	61.7 ± 4.8	2.72 ± 0.19**	22.7 ± 3.8**	−55.8	−245.6
Saparal	271 ± 13.0	62.6 ± 5.4	2.50 ± 0.07**	25.0 ± 3.6**	−45.5	−246.9
Intact animals						
Control	313 ± 17.0	86.2 ± 4.5*	2.52 ± 0.14*	34.2 ± 3.6	−22.7	−251.7
Ecdysterone	340 ± 22.0**	58.7 ± 6.9**	2.98 ± 0.22*	19.7 ± 1.8*,**	−43.2	−243.7
Saparal	309 ± 18.0	73.6 ± 3.8**	2.96 ± 0.18*	24.8 ± 2.1*,**	−54.3	−246.8

Note: *The reliability of differences ($p < 0.05$) with the indicators of intact animals; **with the appropriate control phytoecdysteroids do not act via classical steroid hormone receptors. Nevertheless, they exert a strong anabolic–metabolic effect by stabilizing protein, carbohydrate, lipid, and electrolyte homeostasis across a wide range of stress conditions, including adrenaline overload, carbon tetrachloride–induced hepatotoxicity, oxidative stress, hypoxia, and fatigue. Quantitative analysis showed that during toxic liver injury, the decrease in total serum protein—approximately 16% in untreated controls—was limited to 2.9–7.3% in animals receiving phytoecdysteroids (compared with 8.7% in animals treated with Nerobol). Albumin levels increased markedly during chronic injury, reaching +59.7% with ecdysterone, +55.1% with turkesterone, and +51.5% with cyasterone. Nitrogen metabolism was also normalized: serum urea levels increased only 1.4–1.6-fold with phytoecdysteroids, compared with a 2.5-fold rise in untreated animals and a 1.8-fold rise with Nerobol.

Redox balance and mitochondrial function improved rapidly under phytoecdysteroid treatment. The lactate–pyruvate redox potential increased by approximately 8.9 mV by day 3 and returned to near-baseline levels by day 14, while hepatic malondialdehyde concentrations remained close to initial values throughout the treatment period. Lipid metabolism also recovered, with restoration of the bile acid–to–cholesterol ratio by day 14, indicating normalization of hepatobiliary function. At the systemic level, phytoecdysteroids preserved myocardial structure and function under catecholaminergic stress and fatigue, enhanced physical performance, and accelerated tissue recovery. In fracture models, they promoted mineral deposition in callus tissue with coordinated redistribution of calcium and phosphorus. Neuroendocrine and central nervous system data further indicate wake-promoting and stress-attenuating effects. Specifically, sedation-induced sleep duration was reduced by approximately 27% with ecdysterone and 35.6% with turkesterone, whereas Nerobol prolonged sleep by 31–62%. Across multiple experimental models, phytoecdysteroids matched or exceeded the efficacy of Nerobol and classical plant adaptogens such as Eleutherococcus and Saparal, while demonstrating superior safety. These compounds exhibited very low toxicity (LD_{50} > 6–9 g/kg) and showed no androgenic, teratogenic, or mutagenic effects. Overall, phytoecdysteroids can be regarded as non-hormonal anabolic adaptogens that integrate mitochondrial protection, antioxidant defense, and metabolic regulation into a unified and clinically promising strategy for stress protection.

References

Arif Y, Singh P, Bajguz A, Hayat S (2022) Phytoecdysteroids: Distribution, Structural Diversity, Biosynthesis, Activity, and Crosstalk with Phytohormones. Inter J Mol Sci 23:8664. https://doi.org/10.3390/ijms23158664

Aizikov MI, Kurmukhov AG, Syrov VN (1978) Physiological activity and correlative changes in protein, carbohydrate and fat metabolism under the influence of ecdysones and nerobol. Farm Prirod Vesch 107:107–125.

Azizova SS, Markov IV, Mikhailov IB, Khushbaktov ZA (1992) Cardiac glycosides (cardiac and extracardial effects), Abstract p 104 (in Russ)

Baev AY, Charishnikova OS, Khasanov FA, Nebesnaya KS, Makhmudov AR, Rakhmedova MT, Khushbaktova ZA, Syrov VN, Levitskaya YV (2022) Ecdysterone prevents negative effect of acute immobilization stress on energy metabolism of rat liver mitochondria. J Ster Bioch Mol Biol 219:106066. https://doi.org/10.1016/j.jsbmb.2022.106066

Bajguz A, Bąkała I, Talarek M (2015) Ecdysteroids in plants and their pharmacological effects in vertebrates and humans. Studies in natural products chemistry 45:121–145. https://doi.org/10.1016/B978-0-444-63473-3.00005-8

Baltaev UA and Abubakirov NK (1987) Phytoecdysteroids of *Rhaponticum carthamoides*. Khim Prirod Soed 23:565–568. https://doi.org/10.1007/BF00598674

Bazulko AS and Rogozkin VA (1973) Effect of nerobol on the aspartate aminotransferase activity of rat skeletal muscles under conditions of experimental training. Ukrain Biochem J 1:35 (in Russ)

Benson ES, Freier E, Hallaway BE, Jonson M (1956) Distribution of fluid and electrolytes and concentration of actomyosin and other proteins in the myocardium of dogs with chronic congestive heart failure. Am J Physiol 187:483–492. https://doi.org/10.1152/ajplegacy.1956.187.3.483

Bilenko MV (1978) Biological aspects of kidney allotransplantation. Moscow: Medicine. p 384 (in Russ)

Bobkov YuG, Vinogradov VM, Katkov VF et al. (1984) Pharmacological correction of fatigue. Medicine p 208 (in Russ)

Brekhman I I (1968) Eleutherococcus. Nauka p 188 (in Russ)

Chernykh NS, Dukhanin AS, Andreev YuA, Karymshakova ChT (1990) Mechanisms of the thymolytic effect of methandrostenolone. Bull Exp Biol Med 109:631–634. https://doi.org/10.1007/BF00839887

Dadabaeva NA, Zavodovskaya LI, Khushbaktov ZA et al. (1993) Clinical assessment of the cardiotonic activity of a new cardiac glycoside, rafantoside. Med J Uzb 1:69–71 (in Russ)

Dardymov NV (1976) Ginseng, eleutherococcus (on the mechanism of biological action). Moscow: Nauka Publ p 184 (in Russ)

Davydovsky IV (1962) Problems of causality in medicine (etiology). Moscow: Meditsina Publ p 176 (in Russ)

Dessaulles PA (1960) Les hormones anabolisantes du point de vue expérimental. Helv Med Acta 27:479–482 (in French)

Dobre MZ, Virgolici B, Cioarcă-Nedelcu R (2025) Lipid Hormones at the Intersection of Metabolic Imbalances and Endocrine Disorders. Curr Iss Mol Bio 47:565. https://doi.org/10.3390/cimb47070565

Donaldson JA, Hart JC, Heitzman RJ (1981) Growth hormone, insulin, prolactin and total thyroxine in the plasma of sheep implanted with anabolic steroid trenbolone acetate alone or with estradiol. Res Vet Sci 30:7–13. https://doi.org/10.1016/S0034-5288(18)32598-0

Doyle AE, Pinkus NB, Green J (1974) The use of oxandrolone in hyperlipidemia. Med J Aust 1:127–129. https://doi.org/10.5694/j.1326-5377.1974.tb47682.x

Dzhukarova MK, Saatov Z, Abdullaev ND (1995) Phytoecdysteroids of plants of the genus *Silene* XVI. 5α-sileneoside E from *Silene brahuica*. Chem Nat Compd 31:207–210. https://doi.org/10.1007/BF01170207

Efimov AS and Karabun PM (1972) Mechanism of action of anabolic steroids in the treatment of diabetes mellitus. Terapevt arkhiv 44:90–94 (in Russ)

Egginton S (1987) Effect of an anabolic hormone on striated muscle growth and performance. Pflugers Arch 410:349–355. https://doi.org/10.1007/BF00586510

Fedorov VN, Punegova NV, Zainullin VG, Punegov VV, Sychev RL (2009) Extraction of Ecdysterone-80 from *Serratula coronate* L. and evaluation of its pharmacological actions. II. Cardioprotective properties. Effects on hormone-transmitter balance in chronic cardiac failure. Pharm Chem J 43:36–40. https://doi.org/10.1007/s11094-009-0232-0

Fedorova AV, Yerimbetov KT, Obvinceva OV (2020) The effect of the nanoscale form of 20-hydroxyecdysone on protein metabolism, growth, and development in wistar rats.

Scientific Notes of the Kazan State Academy of Veterinary Medicine named after NE Bauman 243(3):267–272 https://doi.org/10.25687/1996-6733.prodanimbiol.2020.2.57-65 (in Russ)
Grokett BH, Ahmad N, Warren DW (1992) The effects of an anabolic steroid (oxandrolone) on reproductive development in the male rat. Acta endocrinol 126:173–178. https://doi.org/10.1530/acta.0.1260173
Gukasov VM, Sergeev PV, Seifulla RD, Vladimirov YuA (1974) On the antioxidant effect of steroid hormones on lipid peroxidation of mitochondrial membranes in vivo and in vitro. Buell exper biol and med 11:54–57 (in Russ)
Gukasov VM, Fedorov VK (1977) The effect of hormones on the process of lipid peroxidation of biological membranes. The role of membrane structure changes in cellular pathology. p 8–52 (in Russ)
Gusol LO, Guzovskaya EV, Serebrennikova SN, Seminskiy IJ (2022) Stress (The General Adaptation Syndrome): A Lecture. Baykal Med J 1:70–80. https://doi.org/10.57256/2949-0715-2022-1-1-70-80 (in Russ)
Hervey GR, Hutchinson J, Knibbs AV et al (1976) Anabolic effects of methandienone in men undergoing athletic training. The Lancet 2:699–702. https://doi.org/10.1016/S0140-6736(76)90001-5
Hollman W, Heck H, Mader A, Liesen H (1980) Kinetics of lactate formation during work training. J Clin Chem and Clin Biochem 18:652–656
Johnsonbaugh RE, Dalldorf FG, French FS, Nayfeh SN (1976) Androgen action in the rat ventral prostate: effect of castration and testosterone treatment on polyribosomes. J Steroid Biochem 7:73–79. https://doi.org/10.1016/0022-4731(76)90138-2
Kholodova YuD (1979) Phytoecdysones—biologically active polyoxyhydroxylated sterides. Ukr biochem J 51:560–575
Khushbaktova ZA, Syrov VN, Sultanov MV (1982) The effect of cardiac glycosides, anabolic active steroids and their combined use on electrolyte metabolism in the myocardium. Dokl Acad Sci Uzbek SSR 8:35–36 (in Russ)
Kiplinger GL (1967) The effects of drugs on the rate of development on fatigue in mice. Texas Rep Biol Med 25:531–540
Kirchner S, Ignatova Z (2015) Emerging roles of tRNA in adaptive translation, signalling dynamics and disease. Nat Rev Genet 16:98–112. https://doi.org/10.1038/nrg3861
Kliaczko WR, Tirkina TN, Michajlowa IN (1970) Diabetic retinopathy and its complex treatment with particular reference to therapy with anabolic steroids. Endocr Pol 21:383–392 (in Polish)
Kochakian CD, Tillotson C, Endahl G (1965) Castration and the growth of muscles in the rat. Endocrinology 58:226–231. https://doi.org/10.1210/endo-58-2-226
Kondrashova MN (1966) The mild decoupling of oxidative and energetic processes is the basis of the action of cardiac glycosides. Bull Acad Med Sci 4:44–49 (in Russ)
Kosovskiĭ MI, Katkova SP, Mirakhmedov MM, Rakhimdzhanov RT (1989) Insulin resistance in experimental hypo- and hyperthyroidism. Probl endocrinol 35:50–54 (in Russ)
Krasnov EA, Saratikov AS, Yakunina GD (1976) Inokosterone and ecdysterone from *Rhaponticum carthamoides*. Chem Nat Compd 12:494–495. https://doi.org/10.1007/BF00564836
Kroeger H (1963) Chemical nature of the system controlling gene activities in insect cells. Nature 200:1234–1235. https://doi.org/10.1038/2001234a0
Kroeger H (1966) Potentialdifferenz und Puffmuster. Elektrophysiologische und cytologische Untersuchungen an den Speicheldrüsen von *Chironomus tentans.* Exptl Cell Res 41:64–80. https://doi.org/10.1016/0014-4827(66)90547-7
Kuoppasalmi R, Naveri H, Harkonen M, Adlercreutz H (1980) Plasma cortisol, androstenedione, testosterone and luteinizing hormone in running exercise of different intensities. Scand J Clin Lab Invest 40:403–409. https://doi.org/10.3109/00365518009101862
Lupien PJ, Hinse C, Chaudhary KD (1969) Ecdysone as a hypocholesterolemic agent. Arch Int Physio Biochem 77:206–212. https://doi.org/10.3109/13813456909109700
Matsuda H, Kawaba T, Yamamoto Y (1974) Effect of ecdysterone on experimental atherosclerosis in rabbits. Nihon Yakurigaku Zasshi, Folia. Pharmacol Jap 70:325–328. https://doi.org/10.1254/fpj.70.325 (in Jpn)

Matsuoka T, Imai S, Sakai M, Kamada M (1969) Studies on phytoecdysones: a review of our works. Ann Rep Takeda Res Lab 28:221–271

Matsuoka Y, Tanabe S, Moriyama T, Iton N (1964) The effect of an anabolic hormone on tryptophan metabolism. J Nara Med Assoc 15:285–287

Mitchell J, Wilson MC (1987) The behavioral and toxicological effect of testosterone pellet implantation. Physiol and Behav 41:427–432. https://doi.org/10.1016/0031-9384(87)90076-x

Pivnitskiy KK (1974) Anabolic steroid hormones. Problems of testing and the relationship of activity with structure. Probl Endocrinol 4:111–122 (in Russ)

Politova NK, Beshley IV, Volodin VV (2005) Institute of Biology, Komi Scientific Center, Russian Academy Sciences, Syktyvkar, Russia. Functional Foods for Cardiovascular Diseases: Functional Foods Can Help Reduce the Risks of Cardiovascular Diseases p 240 (in Russ)

Ramazanov NSh, Bobayev ID, Yusupova UY, Aliyeva NK, Egamova FR, Yuldasheva NKh, Syrov VN (2016) Phytoecdysteroids-containing extract from *Stachys hissarica* plant and its wound-healing activity. Nat Prod Res 31:593–597. https://doi.org/10.1080/14786419.2016.1205058

Ramazonov N, Yusupova U, Egamova F, Syrov V (2020) Anti stress activity of phytoecdysteroids isolated from aerial part of *Silene claviformis*. Chem Sci Int J 29:9-17. https://doi.org/10.9734/CSJI/2020/v29i730190

Rogozkin VA, Tchaikovsky VS (1981) Radioimmunodetermination - a method of screening analysis of anabolic steroids. Medical and doping control of athletes p 27–41 (in Russ)

Rosenfeld EL, Popova IA (1979) Glycogen disease, Moscow: Medicine p 288 (in Russ)

Ryan J, Hoffman B (1978) Trenbolone acetate: experiences with bound residues in cattle tissues. J Assoc Anall Chem 61:1274–1279. https://doi.org/10.1093/jaoac/61.5.1274

Saatov Z, Gorovits MB, Abubakirov NK (1993) Phytoecdysteroids of plants of the genus *Silene.* Chem Nat Compd 29:551–557. https://doi.org/10.1007/BF00630195

Selye H (1936) A syndrome produced by diverse nocuous agents. Nature 3479:32. https://doi.org/10.1038/138032a0

Semeĭkin AV, Stanevskaia TIu, Chermnykh NS, Sergeev PV (1991) Mekhanizm timoliticheskogo deĭstviia anabolicheskikh steroidov [The mechanism of the thymolytic action of anabolic steroids]. Farmakol Toksikol 54:37–38

Semenov VV, Glumova VA, Trusov VV (1985) Anabolic steroids in experimental and clinical medicine Kazan Medical Journal 1:48–50. https://doi.org/10.17816/kazmj60547 (in Russ)

Sergeev PV, Seifulla RD, Maisky AI (1971) Molecular aspects of the action of steroid hormones. Moscow: Nauka p 221 (in Russ)

Sergeev PV, Semeikin AV, Dukhanin AS, Solovyova EV (1991) The effect of anabolic steroids on proliferative activity of thymocytes. Biull 'Eksper Bio Med 112:393–395 (in Russ)

Sergeev PV, Semeykin AV, Dukhanin AS, Stonas EY (1994) A comparative study of the interaction of steroid preparations with the lymphocytes. Eksper Klin Farm 57:20–22 (in Russ)

Sergeev PV, Vladimirov YuA, Seifulla RD et al (1974) The role of the chemical structure of steroid hormones in inhibiting lipid peroxidation in mitochondrial membranes Vopr med chemistry 4:359–362 (in Russ)

Shakhmurova GA, Batybekov AA, Egamova FR, Khushbaktova ZA, Syrov VN (2013) Experimental evaluation of immunotropic actions of total ecdysteroid-containing drugs from *Silene brahuica* and *Ajuga turkestanica*. Immunology 34:24–26 (in Russ)

Shakhmurova GA, Syrov VN, Khushbaktova ZA (2010) Immunomudulating and antistress activity of ecdysterone and turkesterone under immobilization-induced stress conditions in mice. Pharm Chem J 44:7–9 https://doi.org/10.1007/s11094-010-0387-8

Sharpe PM, Buttery PJ, Haynes NB (1984) Glucocorticoid status and growth manipulation in sheep. Can J Anim Sci 64:310–311. https://doi.org/10.4141/cjas84-275

Sidorova YS, Shipelin VA, Petrov NA, Zorin SN, Mazo VK (2021) Adaptogenic Properties of a Phytoecdysteroid-Rich Extract from the Leaves of *Spinacia oleracea* L. Plants 10:2555. https://doi.org/10.3390/plants10122555

Sillence MN, Rodway RG (1990) Effects of trenbolone acetate and testosterone on growth and on plasma concentrations of corticosterone and ACTH in rats. J Endocrinol 126:461–466. https://doi.org/10.1677/joe.0.1260461

Skosyreva OV (1993) Ecdysten in the complex therapy of duodenal ulcer: Abstract of the dissertation of the Candidate of Medical Sciences. p 17 (in Russ)

Skosyreva OV, Askarov UA, Sahibov AD (1988) The use of ecdystene in the complex treatment of patients with peptic ulcer disease. Proceedings of the Plenum of the Board of Directors. An INTRO dedicated to the memory of the Academy. Vasilenko VKh, Moscow: Smolensk. p 542–544 (in Russ)

Skulachev VP (1962) The ratio of oxidation and phosphorylation in the respiratory chain. Moscow: Publishing House of the USSR Academy of Sciences p 156 (in Russ)

Skulachev VP (1969) Energy accumulation in the cell. Moscow: Nauka Publ p 440 (in Russ)

Sokolov SYa (1972) Comparative pharmacological study of some triterpene saponins of neutropic action: Abstract of the dissertation of Doctor of Medical Sciences. p 36 (in Russ)

Solovyova AG, Yerimbetov KT, Obvinceva OV, Fedorova AV, Mikhailov VV (2021) Physiological mechanisms of action and prospects for the use of phytoecdysteroids in biomedical technologies. Problems of biology of productive animals 1:26–40 https://doi.org/10.25687/1996-6733.prodanimbiol.2021.1.26-40 (in Russ)

Sorokina ZA, Kholodova YuD (1974) On the effect of ecdysterone on ion transport processes in neurons of the grape snail. Structural lability of membranes and its role in regulating the functional activity of cells: Abstract p 84–85 (in Russ)

Syrov VN (1996) On adaptogenic properties of phytoecdysteroids. Reports AS RUz 11:61–64 (in Russ)

Syrov VN, Islamova ZhI, Egamova FR, Yuldasheva NKh, Khushbaktova ZA (2014) Stress-protective properties of phytoecdysteroids. Exp and Clin Pharmacol 77(7):35–38 (in Russ)

Syrov VN, Nasyrova SS, Khushbaktova ZA (1997) The results of experimental study of phytoecdysteroids as erythropoiesis stimulators in laboratory animals. Eksp Klin Farm 60:41–44 (in Russ)

Usmanov D, Yusupova U, Syrov V, Casanola-Martin GM and Rasulev B (2025) A Quantitative Structure–Activity Relationship Study of the Anabolic Activity of Ecdysteroids. Computation 13(8):195. https://doi.org/10.3390/computation13080195.

Usmanov DA, Ramazanov NS, Yusupova UY, Kurilov DV, Kachala VV, Krukovskaya NV, Kolotyrkina NG and Zavarzin IV (2019) Iridoids from Phlomis ostrowskiana Regel plant. Russ Chem Bull 68(12):2275–2278. https://doi.org/10.1007/s11172-019-2699-5.

Usmanov VZ, Gorovits MB, Abubakirov NK (1971) Phytoecdysones of *Ajuga turkestanica*. Chem Nat Compd 7:520. https://doi.org/10.1007/BF00564772

Uzbekova DG (1974) Anabolic steroid drugs (literature review). Pharmacol and toxicol 4:484–490 (in Russ)

Vijayakumar A, Novosyadlyy R, Wu Y, Yakar S, LeRoith D (2010) Biological effects of growth hormone on carbohydrate and lipid metabolism. Growth Hormone & IGF Research 20:1–7. https://doi.org/10.1016/j.ghir.2009.09.002

Vladimirov YuA, Archakov AI (1972) Lipid peroxidation in biological membranes, Moscow: Nauka Publ p 252 (in Russ)

Vladimirov YuA, Sergeev PV, Seifulla RD, Rudnev YuN (1973) The effect of steroids on lipid peroxidation of liver mitochondrial membranes. Mol biol 7:247–253 (in Russ)

Yudaev NA, Pokrovsky BV (1966) The anabolic and androgenic effect of methandrostenolone in experiments on male rats. Vopr Med Chem 5:527–597 (in Russ)

Yusupova UY, Ramazonov NSh, Syrov VN, Sagdullaev SS (2022a) Biological Activities of Phytoecdisteroids. In: Phytoecdysteroids. Springer, Singapore. https://doi.org/10.1007/978-981-16-6711-4_6

Yusupova U, Sasmakov S, Usmanov D, Ramazonov N, Azimova S and Sagdullaev S (2020) Phytoecdysteroids from Silene claviformis and their antibacterial and antifungal activities. Pol. J. Natur. Sc. 35(3):313–321.

Yusupova UY, Bobakulov KM, Girault JP, Usmanov DA, Ramazonov NS and Abdullaev ND (2022b) New Phytoecdysteroid from Silene popovii. Chem Nat Compd 58(6):1082–1084. https://doi.org/10.1007/s10600-022-03874-9.

Yusupova UY, Bobakulov KM, Khurramov AR, Syrov VN, Egamova FR, Karuth A, Usmanov DA, Quadir M and Rasulev B (2025) Phytochemical constituents isolated from Silene popovii Schischk. Med Chem Res 34(2):423–431. https://doi.org/10.1007/s00044-024-03345-6.

Yusupova UY, Ramazonov NS, Bobakulov KM, Egamova FR, Syrov VN and Usmanov DA (2021) Biological active components of silene tomentella and their pharmacological properties. Khim. Rastit. Syrya 1: 197–202. https://doi.org/10.14258/jcprm.2021018323

Yusupova UY, Usmanov DA and Ramazonov NS (2019) Phytoecdysteroids from the Plant Dianthus helenae. Chem Nat Compd 55(2):393–394. https://doi.org/10.1007/s10600-019-02701-y.

Chapter 5
Cardioprotective and Hepatoprotective Actions of Phytoecdysteroids: Molecular and Clinical Perspectives

Vladimir Nikolayevich Syrov and Egamova Feruza

Abstract Phytoecdysteroids (PEs) are a distinctive class of polyhydroxylated plant-derived steroids that exhibit pronounced cardioprotective and hepatoprotective activities, clearly differentiating them from synthetic anabolic steroids and conventional hepatoprotective agents. Experimental and clinical studies show that ecdysterone, turkesterone, cyasterone, and related analogs act through multiple interconnected mechanisms. These include inhibition of lipid peroxidation, stabilization of mitochondrial bioenergetics, and preservation of membrane integrity. In addition, they normalize carbohydrate and lipid metabolism and modulate intracellular signaling pathways. In cardiac models, phytoecdysteroids reduce myocardial necrosis, preserve glycogen and ATP stores, prevent arrhythmias by maintaining Na^+/K^+ homeostasis, and protect the ultrastructural integrity of sarcomeres and mitochondria. In models of liver injury, they significantly lower serum transaminase and bilirubin levels, attenuate centrilobular necrosis, reduce hepatic cholesterol and triglyceride accumulation, enhance bile secretion, and preserve mitochondrial respiration. Clinical use of phytoecdysteroid-based preparations, such as Ecdysten and Eksumid, has confirmed these beneficial effects in patients with ischemic heart disease, myocardial dystrophy, and toxic or viral hepatitis. Treatment was associated with improved biochemical parameters, functional recovery, and overall well-being. Notably, phytoecdysteroids exhibit an excellent safety profile, lacking androgenic, hepatotoxic, or endocrine-disrupting effects even during prolonged administration. Collectively, these findings indicate that phytoecdysteroids act as multifunctional adaptogenic agents. They show substantial translational potential in cardiovascular and hepatic medicine by safely protecting and restoring vital organs under metabolic, toxic, and ischemic stress.

Keywords Phytoecdysteroids · Cardioprotective activity · Hepatoprotective activity · Nephroprotective activity · Antioxidant defense · Ecdysten · Eksumid

V. N. Syrov (✉) · E. Feruza
Department of Pharmocology and Toxicology, Institute of the Chemistry of Plant Substances, Academy of Science of the Republic of Uzbekistan, Tashkent, Uzbekistan

V. N. Syrov et al. (eds.), *Phytoecdysteroids II*,
https://doi.org/10.1007/978-981-95-8423-9_5

PEs are naturally occurring polyhydroxylated steroids that were initially identified as plant analogs of insect molting hormones. Subsequent studies, however, demonstrated that these compounds exhibit a wide range of biological activities in mammals, including anabolic, adaptogenic, antioxidant, cardioprotective, hepatoprotective, and nephroprotective effects(Yusupova et al. 2019; Yusupova et al. 2020; Yusupova et al. 2022; Yusupova et al. 2025; Yusupova et al. 2021). Their broad distribution in the plant kingdom and their relatively high abundance compared with animal-derived sources stimulated extensive pharmacological research, particularly in the former Soviet Union. In this context, species belonging to the genera *Rhaponticum*, *Silene*, and *Ajuga* were intensively investigated as potential adaptogens and regulators of metabolic processes (Solovyova et al. 2021; Yusupova et al. 2022a). Early experimental studies showed that phytoecdysteroids enhance protein biosynthesis, stabilize cellular membranes, and increase resistance to various stress factors in mammals (Gorelick-Feldman et al. 2008; Báthori et al. 2008). Later, more targeted investigations revealed pronounced protective effects on vital organs, particularly the heart and liver, which are highly vulnerable to metabolic overload and oxidative stress. These findings provide a strong rationale for considering phytoecdysteroids as promising agents in metabolic pharmacology. Their potential applications are particularly relevant to conditions associated with ischemic injury, myocardial dystrophy, arrhythmias, hepatitis, and toxin-induced liver damage (Mamarasulov et al. 2020).

A defining feature of phytoecdysteroids is their ability to correct cardiometabolic and hepatic dysfunction caused by diverse stressors. Across experimental models of myocardial injury, hepatotoxicity, ischemia-reperfusion, and toxic nephropathy, pretreatment or cotreatment with ecdysterone, turkesterone, cyasterone, or silenoside A markedly reduced tissue damage (Cahlíková et al. 2011). In several experimental models, the protective effects of phytoecdysteroids were comparable to those of synthetic anabolic steroids or well-established hepatoprotective drugs, while exhibiting a more favorable safety profile. From a mechanistic perspective, the cardioprotective and hepatoprotective actions of phytoecdysteroids are mediated through multiple interconnected biological pathways. A central role is played by their antioxidant activity, as reflected by a marked reduction in malondialdehyde accumulation and modulation of chemiluminescence parameters in cardiac and hepatic tissues, indicating effective suppression of lipid peroxidation and attenuation of oxidative stress (Syrov et al. 2014). Equally important is the preservation of mitochondrial function. By improving oxidative phosphorylation efficiency, respiratory control, and the activity of mitochondrial multienzyme complexes, phytoecdysteroids stabilize energy metabolism in hepatocytes and cardiomyocytes exposed to toxic injury or excessive functional load (Bin Sayeed et al. 2016). Maintenance of electrolyte homeostasis further contributes to their protective effects. Modulation of myocardial potassium and sodium balance underlies the membrane-stabilizing and antiarrhythmic actions of phytoecdysteroids (Yusupova et al. 2022b). Importantly, unlike classical anabolic steroids, phytoecdysteroids regulate metabolic and functional processes without significant interference with the hypothalamic–pituitary–gonadal axis. Enhancement of protein synthesis, normalization of lipid and carbohydrate metabolism, and stimulation of bile secretion occur without inducing marked hormonal imbalance. Finally, the membrane-stabilizing and cytoprotective properties of phytoecdysteroids provide an additional level of protection. By limiting oxidative and ischemic damage to hepatocyte and cardiomyocyte membranes, these compounds preserve cellular architecture and maintain functional integrity under stress conditions (Grebenchikov et al. 2016).

According to data reported by international researchers, administration of 20-hydroxyecdysone (20E) to rats at doses of 5–20 mg/kg effectively alleviates cardiometabolic disturbances induced by a high-fat, high-fructose diet and by deprivation of female sex hormones (Buniam et al. 2020). Rats treated with 20-hydroxyecdysone showed reduced body weight and abdominal fat accumulation, maintenance of normal blood pressure, and significantly lower levels of low-density lipoproteins compared with untreated control animals. These beneficial effects are believed to be mediated, at least in part, through activation of the AMPK and FGF21 signaling pathways. In addition, 20-hydroxyecdysone positively affects systemic energy metabolism by enhancing the synthesis and activity of key metabolic enzymes, including glutamate dehydrogenase, succinate dehydrogenase, NADH dehydrogenase, succinate oxidase, and NADH oxidase. Through these mechanisms, phytoecdysteroids stimulate aerobic glucose oxidation and improve overall metabolic efficiency (Cevostyanov et al. 2006). Experimentally observed outcomes with clear clinical relevance include a reduction in myocardial necrosis, improvement of electrocardiographic (ECG) parameters, normalization of serum transaminase levels in toxic hepatitis, stimulation of bile secretion, and acceleration of regenerative processes in both cardiac and hepatic tissues. Collectively, these findings indicate that phytoecdysteroids possess a distinctive pharmacological profile as dual cardioprotective and hepatoprotective agents. Their broad range of molecular targets, together with adaptogenic and anabolic properties, clearly distinguishes them from conventional steroidal and nonsteroidal drugs. Importantly, chronic toxicity studies demonstrate a favorable safety profile for phytoecdysteroids. Long-term administration to rats, rabbits, and dogs for up to 6 months did not produce significant adverse effects, supporting their potential clinical application as metabolic regulators and organ-protective agents (Lafont and Dinan 2003). This chapter provides a comprehensive analysis of the cardioprotective and hepatoprotective properties of phytoecdysteroids, with particular emphasis on molecular mechanisms, preclinical evidence, and clinical perspectives. Subsequent sections examine experimental data illustrating the therapeutic potential of ecdysterone, turkesterone, silenoside A, and cyasterone in preserving the structural and functional integrity of the heart, liver, and kidneys under various pathological conditions.

5.1 Prevention of Adrenaline-Induced Myocardial Damage by Phytoecdysteroids

Normalization of pathologically altered myocardial metabolism is an important strategy for improving contractile function in various cardiac disorders. In this context, anabolically active compounds from different chemical classes have shown efficacy in both experimental and clinical settings. A substantial body of literature has documented their beneficial effects in conditions associated with myocardial dystrophy (Bardin 1978; Palladino et al. 2019; Meyers and Townsend 2019; Belenky et al. 1973). Among phytoecdysteroids investigated in our studies, ecdysterone, known for its pronounced protein-synthesizing activity, previously demonstrated the ability to normalize key components of carbohydrate-phosphate metabolism in myocardial dystrophy. Notably, its effects exceeded those observed with the anabolic steroid Retabolil (Bakhteev 1987). Based on these observations, we conducted a comparative evaluation of the pharmacocorrective effects of phytoecdysteroids, including comparison with Nerobol, using a widely applied experimental model of adrenaline-induced myocardial dystrophy. Three days after administration of a necrotizing dose of adrenaline, mortality in the control group reached 21.4%. In the myocardium of control rats, creatine phosphokinase (CPK) and lactate dehydrogenase (LDH) activities decreased by 38–42% ($p < 0.05$), while their activities in serum increased by 25–33% ($p < 0.05$). Myocardial glycogen content declined by 51.2%, and creatine phosphate levels decreased by 30.6%. A pronounced reduction was also observed in the calculated redox potential of the lactate-pyruvate (L/P) system compared with intact animals.

Significant disturbances were detected in myocardial levels of adenine nucleotides, potassium, and sodium (Table 5.1). Electrocardiographic analysis revealed marked alterations in R-wave voltage. As early as 2 h after adrenaline administration, R-wave amplitude decreased by 39.6%. After 24 h, pathological changes persisted, and although partial normalization occurred on days 2–3, complete recovery was not achieved. In contrast, prophylactic-therapeutic administration of ecdysterone, silenoside A, or turkesterone completely prevented mortality following adrenaline injection, whereas mortality reached 14.2% in the Nerobol-treated group. In rats receiving phytoecdysteroids, myocardial creatine phosphokinase (CPK) activity was 24–30% higher and lactate dehydrogenase (LDH) activity was 33–35% higher than in controls ($p < 0.05$). At the same time, serum enzyme activities were reduced toward intact values: by 16–19% with ecdysterone, 18–21% with silenoside A, and 20–25% with turkesterone ($p < 0.05$). In contrast, Nerobol produced only minor, statistically insignificant reductions in serum CPK (9%) and LDH (14%). All treatment groups showed increased myocardial glycogen content (51–60% above control), accompanied by decreased lactate levels (13–24%) and increased pyruvate concentrations (16–25%). These changes resulted in an elevation of the lactate/pyruvate redox potential by 4.3–6.7 mV and a shift toward negative lactate excess, indicating improved aerobic metabolism.

Table 5.1 Effects of phytoecdysteroids and Nerobol on biochemical and functional parameters of cardiac muscle in rats with experimental adrenaline-induced myocardiodystrophy (M ± m, n = 10)

Measured parameters		Experimental study					
		Intact animal	Control (adrenaline)	Adrenaline + ecdysterone	Adrenaline + sileneoside A	Adrenaline + turkesterone	Adrenaline + Nerobol
Glycogen, mg%		313 ± 24	153 ± 17*	245 ± 18*,**	253 ± 14*,**	278 ± 13**	232 ± 18*,**
PVC, mg%		1,51 ± 0,06	1,06 ± 0,03*	1,24 ± 0,03*,**	1,25 ± 0,04*,**	1,33 ± 0,08**	1,26 ± 0,02*,**
MK, mg%		60,7 ± 2,7	86,1 ± 4,2*	69,6 ± 2,4*,**	71,7 ± 2,5*,**	65,0 ± 3,4**	74,2 ± 1,3*,**
ORP MK/PVC, mV		−253,2	−262,6	−257,7	−257,9	−255,8	−258,3
Excess lactate		–	+43,5	−31,1	−29,8	−43,0	−28,1
Creatine phosphate, mmol/g of tissue		4,9 ± 0,6	3,4 ± 0,2*	4,2 ± 0,2**	4,1 ± 0,2**	4,4 ± 0,3**	4,0 ± 0,2**
ATP, mmol/g of tissue		2,64 ± 0,11	2,18 ± 0,05*	2,49 ± 0,12**	2,50 ± 0,11**	2,57 ± 0,10**	2,42 ± 0,13
ADP, mmol/g of tissue		1,95 ± 0,14	1,36 ± 0,07*	1,56 ± 0,08*	1,56 ± 0,07*	1,88 ± 0,07**	1,55 ± 0,07*
AMP, mmol/g of tissue		1,21 ± 0,05	1,45 ± 0,06*	1,26 ± 0,03**	1,24 ± 0,02**	1,19 ± 0,05**	1,25 ± 0,03**
The sum of adenine nucleotides, mmol/g of tissue		5,80 ± 0,18	4,97 ± 0,12*	5,32 ± 0,09*,**	5,30 ± 0,09*,**	5,65 ± 0,15**	5,22 ± 0,13*
Inorganic phosphorus, mmol/g of tissue		5,70 ± 0,53	7,2 ± 0,36*	6,1 ± 0,18**	6,2 ± 0,24**	5,8 ± 0,51**	6,4 ± 0,22
K_{total}, mkmol/g of tissue		265 ± 18	204 ± 5,7*	250 ± 15**	258 ± 16**	264 ± 17**	240 ± 13**
Na_{total}, mkmol/g of tissue		154 ± 5,6	185 ± 4,6*	162 ± 4,9**	158 ± 6,1**	153 ± 8,8**	167 ± 8,6
Voltage of the R-wave on the mV ECG	After 2 h	0,53 ± 0,04	0,32 ± 0,03*	0,42 ± 0,03*,**	0,45 ± 0,02**	0,47 ± 0,03**	0,38 ± 0,02*
	After 3 days	0,53 ± 0,04	0,41 ± 0,03*	0,52 ± 0,04**	0,53 ± 0,03**	0,59 ± 0,06**	0,49 ± 0,02**

Note: *The reliability of differences ($p < 0.05$) with the indicators of intact animals; **with the appropriate control

Phytoecdysteroids were also more effective than Nerobol in restoring myocardial creatine phosphate, adenine nucleotide pools, and potassium and sodium balance (Table 5.1). Electrocardiographic analysis further confirmed the cardioprotective action of phytoecdysteroids. Reductions in R-wave amplitude were markedly attenuated as early as 2 h after adrenaline injection, and by day 3, R-wave voltage in treated animals was comparable to—or exceeded—that of intact controls, particularly in the turkesterone group. Overall, phytoecdysteroids demonstrated substantially greater cardioprotective efficacy than Nerobol in this experimental model.

Thus, prophylactic-therapeutic administration of phytoecdysteroids markedly enhances resistance to adrenaline-induced necrotic myocardial injury and does so more effectively than Nerobol. This protection is evidenced by improved survival in animals treated with ecdysterone, silenoside A, and turkesterone, rapid normalization of myocardial and serum enzyme activities, restoration of cardiac carbohydrate-phosphate and electrolyte metabolism, and a pronounced cardiostimulatory effect under these experimental conditions.

5.2 Investigation of the Hepatoprotective Properties of Phytoecdysteroids

The effects of phytoecdysteroids on carbon tetrachloride (CCl_4)-induced liver injury were of particular interest, as this hepatotoxin disrupts major metabolic pathways (Hilchuk 2013), many of which are known targets of phytoecdysteroid action. Comparative evaluation with Nerobol is also relevant when considering the hepatoprotective potential of anabolically active compounds, given the ongoing controversy surrounding their use in hepatitis therapy (Syrov et al. 1986). Although some studies report therapeutic benefits, significant concerns remain regarding adverse effects associated with anabolic agents (Albano et al. 2021; Nemeth and Bernath 2008). In this context, ecdysterone, turkesterone, and cyasterone were examined in experimental models of both acute and chronic toxic liver injury.

5.2.1 Results Obtained in Animals with Acute Toxic Liver Injury

Concomitant administration of phytoecdysteroids with carbon tetrachloride markedly attenuated hepatotoxic injury already at the early stages of exposure. In contrast, the therapeutic effect of Nerobol became evident only at later time points, coinciding with spontaneous recovery of liver function observed in control animals after cessation of CCl_4 administration. Immediately after CCl_4 intoxication, control rats exhibited a 16% reduction in total serum protein compared with intact animals, accompanied by a 2.5-fold increase in serum urea levels. In animals that had received three doses of ecdysterone, turkesterone, or cyasterone by this time point, serum protein levels decreased by only 5.8%, 2.9%, and 7.3%, respectively. In contrast, Nerobol-treated animals showed a greater reduction in serum protein (8.7%). Serum urea concentrations increased by only 1.5-, 1.4-, and 1.6-fold in animals treated with phytoecdysteroids, whereas a 1.8-fold increase was observed in the Nerobol group (Table 5.2).

The superiority of phytoecdysteroids over Nerobol during the early phase of intoxication was also evident from analysis of serum alanine aminotransferase (ALT) and aspartate aminotransferase (AST) activities. On day 3 of treatment with ecdysterone, turkesterone, and cyasterone, ALT activity was reduced by 32%, 35%, and 37.7%, respectively, compared with untreated controls, while AST activity decreased by 25.1%, 23.6%, and 26.3%. In Nerobol-treated animals, reductions in serum transaminase activity were modest and not statistically significant, amounting to 15.6% for ALT and 15.0% for AST ($p > 0.1$). By days 7 and 14, the effects of phytoecdysteroids and Nerobol on restoration of total serum protein and urea levels became more comparable, although turkesterone consistently produced a more pronounced response. The normalizing effect of phytoecdysteroids on serum enzyme activity persisted at later time points. By day 14, ALT and AST activities in phytoecdysteroid-treated animals were almost completely normalized. In contrast, Nerobol-treated rats still exhibited elevated ALT and AST levels, exceeding intact values by 31% and 7%, respectively. In control animals, transaminase activities remained markedly increased at this time point (80% for ALT and 97% for AST).

An additional marker of therapeutic efficacy in toxic hepatitis was the consistently lower hepatic malondialdehyde (MDA) content observed in treated animals compared with controls throughout the study period. Under treatment with ecdysterone and turkesterone, hepatic MDA levels differed only minimally from those of intact animals, whereas the effect of cyasterone, similar to that of nerobol, was noticeably weaker. Concomitant administration of phytoecdysteroids with CCl_4 also substantially prevented depletion of hepatic glycogen stores. In control rats, hepatic glycogen content decreased by 51.7% relative to intact animals following CCl_4 exposure. In contrast, glycogen reductions in phytoecdysteroid-treated rats were limited to 36.6%, 35.2%, and 37.8%, respectively (Table 5.3), whereas Nerobol produced only a weak and statistically insignificant effect. At later observation periods, both phytoecdysteroids and nerobol accelerated the recovery of hepatic glycogen levels compared with controls. By day 14, glycogen content had returned to values comparable with those of intact animals in all experimental groups.

Treatment with phytoecdysteroids, and to a lesser extent with Nerobol, produced a more pronounced favorable shift in lactic and pyruvic acid levels compared with controls. Accordingly, the redox potential (RP) of the lactate–pyruvate (L/P) system in animals treated with ecdysterone, turkesterone, and cyasterone exceeded control values by 8.9, 8.9, and 8.4 mV on day 3; by 8.0, 8.9, and 7.2 mV on day 7; and by 4.9, 6.3, and 4.3 mV on day 14, respectively. In contrast, Nerobol treatment increased RP by only 3.8, 3.0, and 2.4 mV at the corresponding time points. The investigated compounds also attenuated impairment of bile secretion induced by CCl_4, with phytoecdysteroids exerting a stronger protective effect during the early stages of intoxi-

Table 5.2 Effects of phytoecdysteroids and Nerobol on serum total protein, urea, and alanine, and aspartate aminotransferase activities in rats with acute toxic hepatitis ($M \pm m$, $n = 8$)

Experimental group	Administration (days)	Investigated indicators			
				AlAT	AsAT
		Protein (g%)	Urea (mg%)	mmol of PVC per 1 ml of serum for 1 h of incubation	
Intact animals	–	6,9 ± 0,07	29,9 ± 1,8	0,96 ± 0,08	1,54 ± 0,12
CCl_4, end of seeding (control)		5,8 ± 0,23*	74,9 ± 4,7*	3,98 ± 0,36*	2,67 ± 0,20*
CCl_4 + ecdysterone	3	6,5 ± 0,13*,**	46,8 ± 3,7*,**	2,71 ± 0,18*,**	2,00 ± 0,14*,**
CCl_4 + turkesterone	3	6,7 ± 0,13**	43,5 ± 3,5*,**	2,59 ± 0,15*,**	2,04 ± 0,14*,**
CCl_4 + cyasterone	3	6,4 ± 0,15*,**	47,0 ± 1,3*,**	2,48 ± 0,10*,**	1,97 ± 0,16*,**
CCl_4 + Nerobol	3	6,3 ± 0,13*	54,5 ± 5,8*,**	3,36 ± 0,23*	2,27 ± 0,17*
CCl_4 (control)		6,2 ± 0,10*	51,2 ± 3,7*	2,93 ± 0,15*	2,05 ± 0,11*
CCl_4 + ecdysterone	7	6,7 ± 0,14**	34,8 ± 2,9**	1,58 ± 0,14*,**	1,63 ± 0,09**
CCl_4 + turkesterone	7	6,9 ± 0,07**	30,0 ± 2,8**	1,47 ± 0,17*,**	1,67 ± 0,12**
CCl_4 + cyasterone	7	6,7 ± 0,07**	40,7 ± 2,8*,**	1,56 ± 0,17*,**	1,69 ± 0,08**
CCl_4 + Nerobol	7	6,6 ± 0,09*,**	31,6 ± 3,4**	2,24 ± 0,26*,**	1,98 ± 0,08*
CCl_4 (control)		6,6 ± 0,12*	39,2 ± 3,6*	1,73 ± 0,19*	1,68 ± 0,07
CCl_4 + ecdysterone	14	6,9 ± 0,08	30,2 ± 1,9**	0,97 ± 0,10**	1,50 ± 0,06
CCl_4 + turkesterone	14	7,1 ± 0,04*,**	27,8 ± 2,4**	0,98 ± 0,11**	1,52 ± 0,07
CCl_4 + cyasterone	14	6,9 ± 0,11	32,3 ± 1,5	1,08 ± 0,12**	1,52 ± 0,07
CCl_4 + nerobol	14	6,8 ± 0,08	29,4 ± 2,4**	1,26 ± 0,10*,**	1,65 ± 0,04

Note: Here and in Table 5.27. *The reliability of differences ($p < 0.05$) with the indicators of intact animals; **with the appropriate control

Table 5.3 Effects of phytoecdysteroids and Nerobol on hepatic malondialdehyde levels and selected indicators of carbohydrate metabolism in rats with acute toxic hepatitis (M ± m, n = 8)

Experimental group	Administration (days)	Investigated indicators				ORP MK/PVC, mV
		MDA, nmol per 1 mg of protein	Glycogen (mg%)	Lactic acid (mg%)	PVC (mg%)	
Intact animals	–	0,500 ± 0,03	2054 ± 152	88,4 ± 3,6	2,32 ± 0,11	−252,5
CCl_4, end of seeding (control)		1125 ± 0,09*	992 ± 94*	158,0 ± 6,4*	1,48 ± 0,10*	−266,3
CCl_4 + ecdysterone	3	0,776 ± 0,04*,**	1302 ± 97*,**	103,4 ± 4,1*,**	1,89 ± 0,10*,**	−257,4
CCl_4 + turkesterone	3	0,790 ± 0,03*,**	1331 ± 125*,**	98,0 ± 4,8**	1,79 ± 0,06*,**	−257,4
CCl_4 + cyasterone	3	0,785 ± 0,04*,**	1279 ± 74*,**	104,8 ± 3,5*,**	1,83 ± 0,03*,**	−257,9
CCl_4 + Nerobol	3	0,844 ± 0,04*,**	1038 ± 95*	138,8 ± 6,5*	1,72 ± 0,09*	−262,5
CCl_4 (control)		0,882 ± 0,05*	1380 ± 95*	143,2 ± 10,4*	1,70 ± 0,08*	−263,1
CCl_4 + ecdysterone	7	0,548 ± 0,05**	1708 ± 79**	94,7 ± 4,2**	2,05 ± 0,09**	−255,1
CCl_4 + turkesterone	7	0,533 ± 0,04**	1822 ± 67**	91,7 ± 3,6**	2,12 ± 0,09**	−254,2
CCl_4 + cyasterone	7	0,652 ± 0,06*,**	1670 ± 64*,**	98,2 ± 3,8**	2,00 ± 0,04*,**	−255,9
CCl_4 + Nerobol	7	0,722 ± 0,08*	1582 ± 100*	126,2 ± 5,1*	1,88 ± 0,08*	−260,1
CCl_4 (control)		0,630 ± 0,04*	1774 ± 54	118,9 ± 6,3*	2,09 ± 0,09	−257,9
CCl_4 + ecdysterone	14	0,489 ± 0,04**	2064 ± 64**	89,4 ± 3,3**	2,26 ± 0,08	−253,0
CCl_4 + turkesterone	14	0,476 ± 0,02**	2144 ± 68**	86,2 ± 4,1**	2,42 ± 0,06**	−251,6
CCl_4 + cyasterone	14	0,525 ± 0,03	1949 ± 64	90,4 ± 2,7**	2,19 ± 0,05	−253,6
CCl_4 + Nerobol	14	0,527 ± 0,03	2013 ± 60**	102,3 ± 4,4*,**	2,15 ± 0,07	−255,5

cation. In control animals, total bile output over 4 h decreased by 30% following CCl_4 administration (Fig. 5.1).

In contrast, bile secretion declined by only 6.2–10.6% in phytoecdysteroid-treated animals, whereas the reduction reached 19.3% in the Nerobol group. By day 7, bile secretion in all treated groups had recovered to levels observed in intact animals. This normalization was maintained under phytoecdysteroid treatment until the end of the experiment. In contrast, bile output in Nerobol-treated rats exceeded intact values by 27.7% at this time point. Analysis of bile composition revealed a unidirectional and favorable effect of all tested compounds in toxic liver injury, primarily through improvement of the hepatic component of bilirubin metabolism (Fig. 5.2). In control animals, bilirubin content in bile, particularly during the early period following CCl_4 administration, was markedly reduced. In contrast, in experimental animals treated with phytoecdysteroids, bilirubin levels from the first day of therapy differed only slightly from those of intact controls. A similar normalization was observed in Nerobol-treated animals, but only after 1 week of treatment. Total bile acid content in rats receiving the investigated compounds was consistently higher than in control animals at all observation time points. By day 14, bile acid levels in these experimental groups exceeded those of intact animals by 24.5–36.5%. Treatment was also

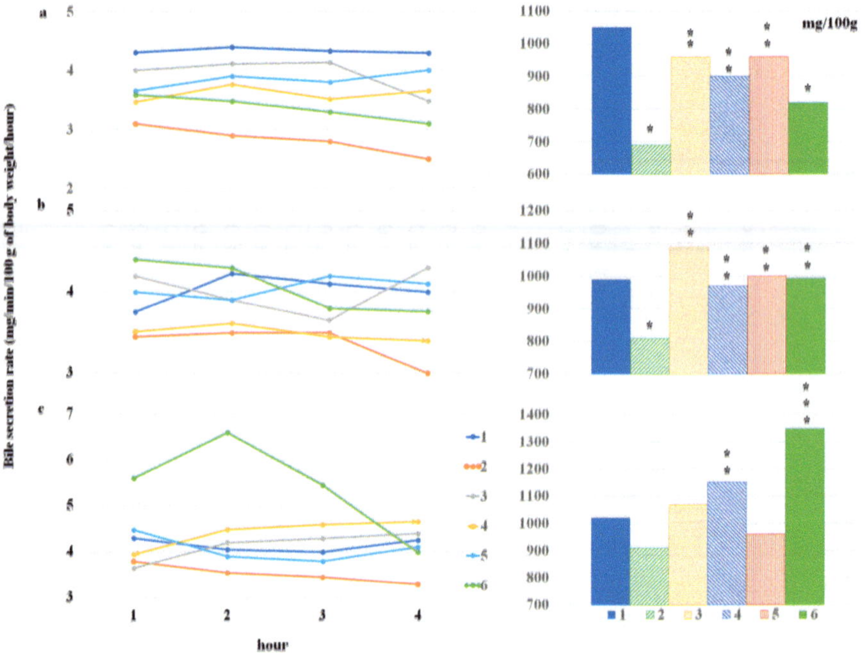

Fig. 5.1 The effect of the studied drugs on the rate of bile secretion (in mg/min per 100 g of body weight for each hour of the experiment, left) and the total amount of bile excreted in 4 h (in mg per 100 g, right) in rats with acute toxic hepatitis on the 3rd (**a**), 7th (**b**), and 14th (**c**) days of their introduction (n = 8). 1—intact; 2—CCl_4; 3—CCl_4 + ecdysterone; 4—CCl_4 + turkesterone; 5—CCl_4 + cyasterone; 6—CCl_4 + Nerobol

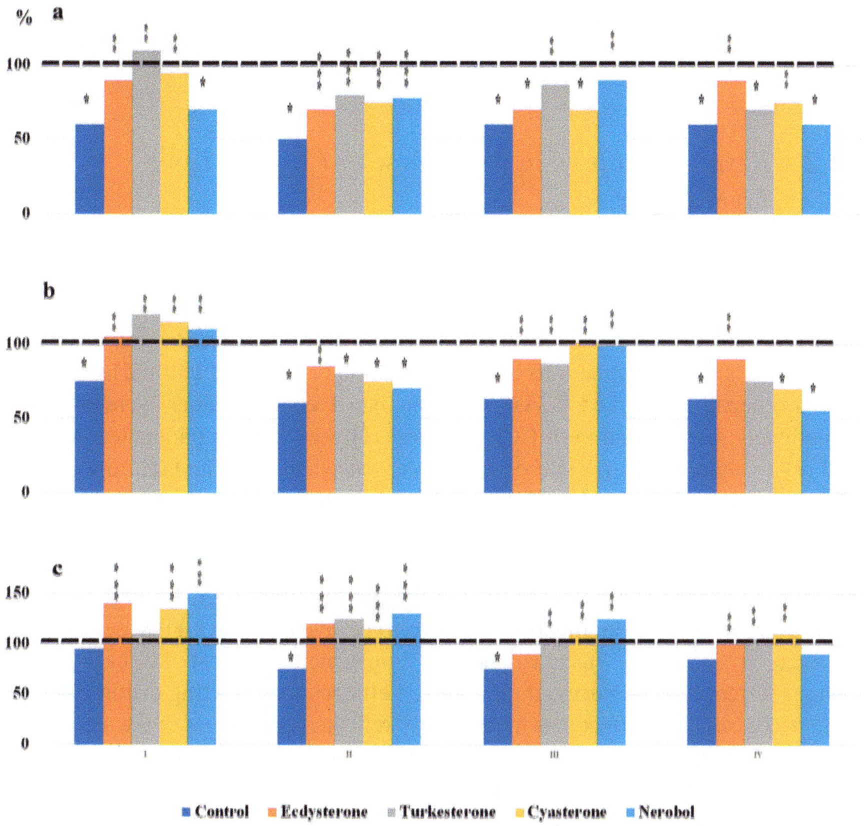

Fig. 5.2 The effect of the studied drugs on the content of bilirubin (I), bile acids (II), cholesterol (III), and the value of the cholate-to-cholesterol coefficient (IV) in bile in rats with acute toxic hepatitis (in % of the corresponding indicators of intact animals) on the 3rd (**a**), 7th (**b**), and 14th (**c**) days of their introduction (n = 8). *The significance of differences ($p < 0.05$) with the intact control; **with the corresponding group of untreated animals

associated with accelerated recovery of the hepatic cholesterol-excretory function (Fig. 5.2).

At the same time, clear differences were observed in the qualitative composition of bile. In rats treated with phytoecdysteroids, a favorable balance between bile acids and cholesterol was maintained. As a result, the cholate-to-cholesterol coefficient remained higher than in the corresponding control groups throughout the experiment and, by day 14, did not differ from that of intact animals. In contrast, a distinctly different pattern was observed in rats with toxic hepatitis treated with Nerobol. In these animals, although bile concentrations of both bile acids and cholesterol were higher than in controls, their concentration ratio remained significantly lower than in intact animals throughout the study. Taken together, these experimental data indicate that phytoecdysteroids improve the course of acute carbon tetrachloride-induced toxic liver injury more effectively than Nerobol. This therapeutic advantage appears to be mediated by favorable metabolic shifts in hepatocytes, leading to restoration of the functional state of the damaged liver.

5.2.2 *Results Obtained in Animals with Chronic Toxic Liver Injury*

In a model of chronic toxic hepatitis induced by prolonged administration of low doses of carbon tetrachloride (CCl_4), the beneficial effects of phytoecdysteroids were more pronounced than in the acute liver dystrophy model. This difference reflects fundamental distinctions in the progression of acute and chronic liver injury. In acute hepatitis, marked deviations from intact control values were observed immediately after cessation of CCl_4 exposure. However, most parameters normalized relatively rapidly over subsequent days, consistent with the high regenerative capacity of hepatic tissue. In contrast, during chronic hepatitis, most functional and biochemical parameters remained persistently altered or continued to deteriorate throughout ongoing CCl_4 administration. Notably, in animals exposed to CCl_4 for 7 weeks—a duration sufficient to induce chronic hepatitis (Bardin 1978)—no substantial recovery of liver function was observed even 15 days after cessation of intoxication, indicating sustained hepatic impairment. Against this background, phytoecdysteroids demonstrated pronounced efficacy in correcting disturbed metabolic processes. This effect was most clearly reflected in changes in the serum protein profile. In control rats, total serum protein levels remained reduced by 8–10% throughout the experiment, unlike in acute hepatitis. Moreover, 15 days after CCl_4 withdrawal, serum protein levels in control animals continued to decline, reaching reductions of up to 14%. (Table 5.4).

A markedly different pattern was observed in animals treated with the investigated compounds. Under phytoecdysteroid treatment, total serum protein levels remained consistently higher than in control animals throughout the experimental period and were almost completely restored to intact values within 2 weeks after cessation of carbon tetrachloride (CCl_4) administration. This increase in total protein was primarily attributable to sustained elevation of serum albumin levels, indicating preservation and stimulation of the hepatic protein-synthetic function. In control rats, serum albumin concentrations progressively declined during CCl_4 intoxication, reaching 2.18 g% by the end of exposure (a 26.1% reduction relative to intact animals), and decreased further to 1.96 g% (−33.6%) 15 days after CCl_4 withdrawal. In contrast, rats receiving phytoecdysteroids during the development of chronic toxic hepatitis maintained significantly higher serum albumin levels at all observation time points. Following cessation of CCl_4 exposure, this effect became even more pronounced, with albumin concentrations exceeding intact values. At this stage, serum albumin levels under the influence of ecdysterone, turkesterone, and cyasterone were 59.7%, 55.1%, and 51.5% higher, respectively, than those of control animals. Analysis of serum globulin fractions further supported the hepato-

Table 5.4 Effects of phytoecdysteroids and Nerobol on serum total protein and protein fractions in rats with chronic toxic hepatitis (M ± m, n = 6–8)

Experimental groups	Administration period (days)	Total protein (g%)	Albumin (g%)	Globulins (g%)			
				α_1	α_2	β	γ
Intact animals	–	7,30 ± 0,06	2,95 ± 0,09	1,76 ± 0,08	0,66 ± 0,06	1,10 ± 0,07	0,77 ± 0,07
CCl_4 (**C**)	The 28th	6,73 ± 0,08*	2,66 ± 0,05*	1,18 ± 0,05*	0,54 ± 0,06	1,47 ± 0,08*	0,85 ± 0,09
CCl_4 + **E**	28 + 7 days	7,08 ± 0,22	3,09 ± 0,15**	1,50 ± 0,09**	0,55 ± 0,04	1,20 ± 0,06**	0,74 ± 0,10
CCl_4 + **T**	28 + 7 days	7,18 ± 0,15**	3,09 ± 0,09**	1,60 ± 0,11**	0,53 ± 0,04	1,18 ± 0,07**	0,76 ± 0,10
CCl_4 + **C**	28 + 7 days	6,83 ± 0,31	2,88 ± 0,19	1,38 ± 0,07*,**	0,51 ± 0,03*	1,20 ± 0,13	0,87 ± 0,08
CCl_4 + **N**	28 + 7 days	6,93 ± 0,09*	3,02 ± 0,13**	1,47 ± 0,07*,**	0,43 ± 0,04*	1,34 ± 0,03*	0,66 ± 0,06
CCl_4 (**C**)	The 35th	6,61 ± 0,11*	2,28 ± 0,08*	1,05 ± 0,08*	0,67 ± 0,05	1,55 ± 0,07*	1,01 ± 0,03*
CCl_4 + **E**	35 + 14 days	7,06 ± 0,09**	3,26 ± 0,10*,**	1,54 ± 0,07**	0,47 ± 0,02*,**	1,10 ± 0,06**	0,73 ± 0,06**
CCl_4 + **T**	35 + 14 days	7,16 ± 0,16**	3,16 ± 0,15**	1,53 ± 0,12**	0,49 ± 0,04*,**	1,16 ± 0,07**	0,84 ± 0,05**
CCl_4 + **C**	35 + 14 days	6,78 ± 0,28	2,82 ± 0,17**	1,18 ± 0,09*	0,56 ± 0,06	1,26 ± 0,07**	0,96 ± 0,10
CCl_4 + **N**	35 + 14 days	7,06 ± 0,09**	2,94 ± 0,09**	1,25 ± 0,12*	0,59 ± 0,05	1,38 ± 0,05*	0,89 ± 0,06
CCl_4 (**C**)	The 50th	6,58 ± 0,06*	2,18 ± 0,06*	1,03 ± 0,06*	0,78 ± 0,06	1,50 ± 0,08*	1,07 ± 0,06*
CCl_4 + **E**	50 + 30 days	6,83 ± 0,15*	2,77 ± 0,10**	1,23 ± 0,08*	0,59 ± 0,05**	1,42 ± 0,06*	0,75 ± 0,07**
CCl_4 + **T**	50 + 30 days	6,90 ± 0,08*,**	2,72 ± 0,09**	1,25 ± 0,07*,**	0,63 ± 0,07	1,33 ± 0,04*	0,95 ± 0,06
CCl_4 + **C**	50 + 30 days	6,83 ± 0,05*,**	2,62 ± 0,10*,**	1,23 ± 0,04*,**	0,69 ± 0,07	1,30 ± 0,04*,**	0,97 ± 0,05*
CCl_4 + **N**	50 + 30 days	6,88 ± 0,09*,**	2,90 ± 0,08**	1,11 ± 0,04*	0,72 ± 0,07	1,24 ± 0,05**	0,82 ± 0,04**
CCl_4 (**C**)	The 50th	6,30 ± 0,17*	1,96 ± 0,14*	0,90 ± 0,08*	0,71 ± 0,05	1,46 ± 0,06*	1,23 ± 0,16*
CCl_4 + **E**	50 + 45 days	7,20 ± 0,09**	3,13 ± 0,10**	1,36 ± 0,06*,**	0,74 ± 0,05	1,03 ± 0,07**	0,93 ± 0,10
CCl_4 + **T**	50 + 45 days	7,28 ± 0,05**	3,04 ± 0,08**	1,76 ± 0,06**	0,54 ± 0,04**	1,14 ± 0,06**	0,80 ± 0,07**
CCl_4 + **C**	50 + 45 days	7,10 ± 0,11**	2,97 ± 0,06**	1,80 ± 0,07**	0,52 ± 0,05**	0,95 ± 0,08**	0,82 ± 0,082**
CCl_4 + **N**	50 + 45 days	6,90 ± 0,15*,**	2,89 ± 0,09**	1,44 ± 0,15**	0,64 ± 0,06	1,02 ± 0,07**	0,92 ± 0,07

Note: *The significance of differences ($p < 0.05$) with the intact control; **with the corresponding group of untreated animals. (C)—control, E—ecdysterone, T—turkesterone, C—cyasterone, N—Nerobol protective activity of phytoecdysteroids. All tested compounds induced a sustained increase in α_1-globulin levels throughout the experiment, with maximal elevations of 51–100% at the final time point, whereas this fraction was markedly reduced in control animals. Conversely, phytoecdysteroid treatment normalized α_2-, β-, and γ-globulin fractions, which were substantially elevated in rats with chronic hepatitis (Table 5.4).

Overall, the changes observed in the serum protein profile of rats with chronic hepatitis treated with phytoecdysteroids were largely comparable to those induced by Nerobol under these pathological conditions (Table 5.4). Equally important indicators of improvement in chronic hepatic pathology, similar to those observed in acute liver dystrophy, were the reductions in serum alanine aminotransferase (ALT) and aspartate aminotransferase (AST) activities. In control rats subjected to prolonged carbon tetrachloride (CCl_4) administration, ALT and AST activities remained persistently elevated throughout the experiment (Fig. 5.3). Although some fluctuations were noted at individual time points, enzyme activities consistently exceeded intact control values, albeit without reaching the extreme elevations typical of acute liver injury. Discontinuation of CCl_4 exposure resulted in only minimal normalization of transaminase activity in control animals, particularly with respect to ALT. In contrast, animals treated with phytoecdysteroids exhibited a clear reduction in enzyme activity as early as 7 days after initiation of treatment. At this time point, ALT activity was reduced by 35.8%, 39.0%, and 42.2%, and AST activity by 26.9%, 27.9%, and 28.9%, respectively, compared with control animals. Similar trends were

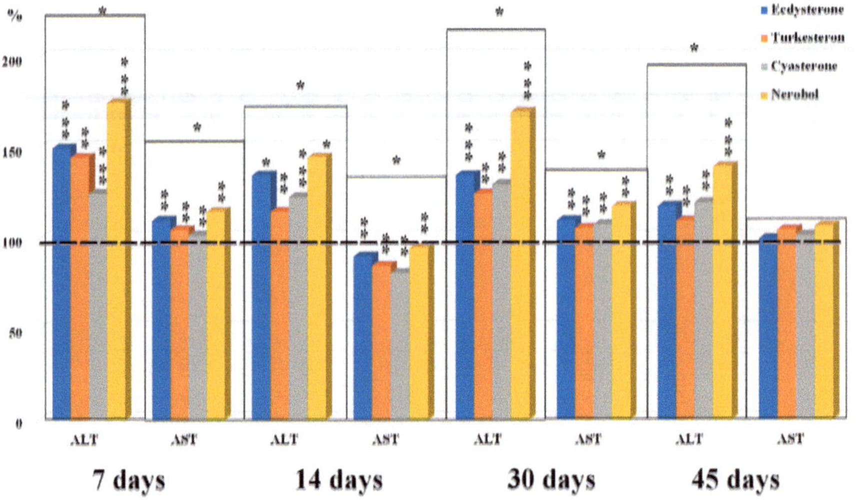

Fig. 5.3 ALT and AST activity in the blood serum of rats with chronic toxic hepatitis (light columns) and treated with ecdysterone, turkesterone, cisterone, and Nerobol; n = 6–8. The abscissa axis shows the timing of drug administration (in days), and the ordinate axis shows transaminase activity (mmol of PVC per 1 ml of serum per 1 h of incubation), expressed as % of the corresponding values of intact rats. *The significance of differences ($p < 0.05$) with the intact control; **with the corresponding group of untreated animals

observed at subsequent observation periods. Notably, continued administration of phytoecdysteroids after cessation of CCl_4 injections led to a further pronounced decrease in transaminase activity. As a result, ALT activity in treated animals exceeded intact control values by only 11.0%, 6.2%, and 19.7%, differences that were not statistically significant, while AST activity did not differ appreciably from intact levels. In this respect, the effect of Nerobol was substantially weaker (Fig. 5.3).

Given that ALT and AST activities are widely regarded as markers of hepatocellular cytolysis (Bluger 1970), the more pronounced effect of phytoecdysteroids observed in both chronic and acute hepatitis appears to be closely related to their membrane-stabilizing properties (Syrov et al. 2023). The development of chronic toxic hepatitis under the present experimental conditions was accompanied by severe disturbances in carbohydrate metabolism, which became particularly pronounced toward the end of CCl_4 exposure and exceeded those observed in acute hepatitis (Table 5.5). Recovery processes were also markedly impaired. Most notably, hepatic glycogen content remained persistently reduced. Four weeks after the onset of CCl_4 intoxication, hepatic glycogen levels were decreased by 44.4% relative to intact animals maintained under identical vivarium conditions. Glycogen content continued to decline progressively, reaching its lowest level after 12 weeks, when it amounted to only 31.8% of intact control values. Fifteen days after discontinuation of CCl_4 administration, hepatic glycogen content showed only a slight increase and remained markedly reduced, being 64.1% lower than intact values. Administration of phytoecdysteroids, as well as Nerobol, even for a period of 1 week, significantly attenuated the decline in hepatic glycogen content. In animals receiving the investigated compounds for 7 days, glycogen levels were reduced by only 14–24%. After 15 and 30 days of treatment, hepatic glycogen content was markedly higher than in control animals. Following cessation of CCl_4 exposure, continued administration of phytoecdysteroids or Nerobol for an additional 15 days further promoted restoration of hepatic glycogen stores. Among the tested compounds, turkesterone demonstrated the highest efficacy, under which hepatic glycogen levels were virtually indistinguishable from those observed in intact control animals.

The pronounced beneficial effect of phytoecdysteroids on carbohydrate metabolism in the injured liver, similar to that observed in acute hepatitis, was also reflected in changes in the lactate/pyruvate (L/P) ratio and the redox index of the L/P system. These parameters indicate improvement of aerobic oxidative processes in hepatocytes and a metabolic shift toward more efficient energy production. To further characterize the effects of phytoecdysteroids on hepatic bioenergetics, we examined the influence of ecdysterone and turkesterone, in comparison with Nerobol, on mitochondrial respiration and oxidative phosphorylation in isolated liver mitochondria. In addition, the activity and stability of multienzyme systems associated with mitochondrial membranes were assessed. Administration of carbon tetrachloride (CCl_4) to rats caused pronounced uncoupling of mitochondrial respiration and oxidative phosphorylation in liver cells. When α-ketoglutarate or succinate was used as an oxidation substrate, respiration in the phosphorylating state (V_3) remained unchanged or was suppressed, whereas respiration in the nonphosphorylating state

Table 5.5 Effects of phytoecdysteroids and Nerobol on hepatic malondialdehyde levels and selected indicators of carbohydrate metabolism in rats with chronic toxic hepatitis ($M \pm m$, $n = 6–8$)

Experimental groups	Administration period (days)	Glycogen (mg%)	Lactic acid (mg%)	PVC (mg%)	ORP MK/PVC (mV)
Intact animals	–	2012 ± 177	$87{,}5 \pm 4{,}7$	$2{,}28 \pm 0{,}12$	−252,6
CCl_4, end of seeding (control)	The 28th	$1119 \pm 95^{*}$	$127{,}6 \pm 10{,}6^{*}$	$1{,}63 \pm 0{,}11^{*}$	−262,1
CCl_4 + ecdysterone	28 + 7 days	$1586 \pm 133^{**}$	$111{,}8 \pm 5{,}4^{*}$	$2{,}00 \pm 0{,}06^{**}$	−257,6
CCl_4 + turkesterone	28 + 7 days	$1733 \pm 59^{**}$	$105{,}2 \pm 4{,}1^{*}$	$2{,}03 \pm 0{,}09^{**}$	−256,6
CCl_4 + cyasterone	28 + 7 days	$1722 \pm 94^{**}$	$110{,}8 \pm 5{,}6^{*}$	$1{,}97 \pm 0{,}07^{*,**}$	−257,7
CCl_4 + Nerobol	28 + 7 days	$1528 \pm 57^{*,**}$	$118{,}7 \pm 4{,}5^{*}$	$2{,}01 \pm 0{,}12^{**}$	−258,4
CCl_4 (control)	The 35th	$701 \pm 59^{*}$	$160{,}6 \pm 5{,}4^{*}$	$1{,}41 \pm 0{,}09^{*}$	−267,1
CCl_4 + ecdysterone	35 + 14 days	$1803 \pm 112^{**}$	$104{,}4 \pm 8{,}8^{**}$	$1{,}89 \pm 0{,}09^{*,**}$	−257,5
CCl_4 + turkesterone	35 + 14 days	$1859 \pm 91^{**}$	$100{,}6 \pm 7{,}0^{**}$	$2{,}04 \pm 0{,}10^{**}$	−255,9
CCl_4 + cyasterone	35 + 14 days	$1763 \pm 52^{**}$	$112{,}0 \pm 8{,}4^{*,**}$	$1{,}86 \pm 0{,}07^{*,**}$	−258,6
CCl_4 + Nerobol	35 + 14 days	$1835 \pm 105^{**}$	$124{,}4 \pm 5{,}7^{*,**}$	$1{,}96 \pm 0{,}06^{*,**}$	−259,3
CCl_4 (control)	The 50th	640 ± 45	$163{,}0 \pm 8{,}1^{*}$	$1{,}37 \pm 0{,}15^{*}$	−267,7
CCl_4 + ecdysterone	50 + 30 days	$1674 \pm 58^{**}$	$126{,}6 \pm 5{,}6^{*,**}$	$1{,}80 \pm 0{,}12^{*,**}$	−260,7
CCl_4 + turkesterone	50 + 30 days	$1752 \pm 112^{**}$	$122{,}1 \pm 7{,}8^{*,**}$	$1{,}88 \pm 0{,}08^{*,**}$	−259,6
CCl_4 + cyasterone	50 + 30 days	$1431 \pm 141^{*,**}$	$139{,}8 \pm 5{,}6^{*,**}$	$1{,}75 \pm 0{,}07^{*,**}$	−262,4
CCl_4 + Nerobol	50 + 30 days	$1716 \pm 107^{**}$	$134{,}0 \pm 9{,}5^{*,**}$	$1{,}79 \pm 0{,}08^{*,**}$	−261,5
CCl_4 (control)	The 50th	723 ± 28	$142{,}1 \pm 7{,}9^{*}$	$1{,}42 \pm 0{,}08^{*}$	−265,4
CCl_4 + ecdysterone	50 + 45 days	$1893 \pm 72^{**}$	$106{,}7 \pm 6{,}6^{*,**}$	$2{,}03 \pm 0{,}07^{**}$	−256,8
CCl_4 + turkesterone	50 + 45 days	$1938 \pm 43^{**}$	$94{,}7 \pm 2{,}9^{**}$	$2{,}06 \pm 0{,}07^{**}$	−255,0
CCl_4 + cyasterone	50 + 45 days	$1675 \pm 75^{**}$	$116{,}9 \pm 5{,}4^{*,**}$	$1{,}88 \pm 0{,}11^{*,**}$	−259,06
CCl_4 + Nerobol	50 + 45 days	$1769 \pm 92^{**}$	$114{,}6 \pm 4{,}8^{*,**}$	$1{,}94 \pm 0{,}08^{*,**}$	−258,4

Note: *The significance of differences ($p < 0.05$) with the intact control; **with the corresponding group of untreated animals

(V_4) increased markedly. As a result, the respiratory control ratio (RCR) decreased by 44–47%, and the ADP/O ratio declined by 38–39%, indicating a substantial reduction in the efficiency of oxidative phosphorylation. Administration of phytoecdysteroids during ongoing CCl_4 intoxication prevented, to varying degrees, the development of mitochondrial dysfunction induced by this hepatotropic toxin. After 7 days of treatment, ecdysterone exerted a moderate protective effect. However, after 14 days, it almost completely restored the efficiency of oxidative phosphorylation in liver mitochondria (Table 5.6). Under the influence of ecdysterone, the rate of phosphorylating oxidation of α-ketoglutarate increased by 41%, while oxidation of succinate increased by 83% compared with control animals. These changes resulted in near-complete normalization of both the respiratory control ratio and the ADP/O ratio.

An even more pronounced protective effect was observed following turkesterone administration. Its normalizing influence on the evaluated bioenergetic parameters

Table 5.6 Effect of phytoecdysteroids and Nerobol on respiration and oxidative phosphorylation of liver mitochondria in rats with chronic toxic hepatitis (M ± m, n = 8)

Experimental groups	Day 7: Respiration rate (nanoatom $O_2 \cdot min^{-1} \cdot mg^{-1}$ protein) V_3	V_4	Respiratory control ratio (RCR)	ADP/O	Day 14: Respiration rate (nanoatom $O_2 \cdot min^{-1} \cdot mg^{-1}$ protein) V_3	V_4	RCR	ADP/O
α- ketoglutarate								
Intact animals	60 ± 4	20 ± 2	3,0 ± 0,10*	2,7 ± 0,09	55 ± 10	18 ± 3	3,0 ± 0,10	2,6 ± 0,08
Cl_4 (**C**)	60 ± 6	30 ± 3*	2,0 ± 0,12*	2,0 ± 0,12*	48 ± 4	29 ± 4*	1,6 ± 0,12*	1,6 ± 0,10*
CCl_4 + **E**	72 ± 5	34 ± 5*	2,1 ± 0,11*	2,1 ± 0,08*	68 ± 2**	24 ± 1	2,8 ± 0,09**	2,5 ± 0,09**
CCl_4 + **T**	62 ± 4	22 ± 2**	2,8 ± 0,13**	3,0 ± 0,10*,**	113 ± 17*,**	29 ± 3*	3,9 ± 0,13*,**	2,6 ± 0,13**
CCl_4 + **N**	65 ± 6	31 ± 4*	2,1 ± 0,09*	2,1 ± 0,08*	75 ± 2**	24 ± 1	3,1 ± 0,11**	2,4 ± 0,08**
Succinate								
Intact animals	110 ± 5	38 ± 3	2,9 ± 0,08	1,8 ± 0,07	108 ± 8	36 ± 3	3,0 ± 0,09	1,8 ± 0,09
CCl_4 (**C**)	120 ± 6	58 ± 7*	2,1 ± 0,09*	1,3 ± 0,09*	87 ± 4*	51 ± 2*	1,7 ± 0,09*	1,1 ± 0,08*
CCl_4 + **E**	164 ± 9*,**	76 ± 7*	2,1 ± 0,10*	1,4 ± 0,10*	159 ± 8*,**	51 ± 7	3,1 ± 0,10**	1,8 ± 0,10**
CCl_4 + **T**	113 ± 4	38 ± 2**	3,0 ± 0,11**	1,9 ± 0,07**	220 ± 7*,**	65 ± 13	3,4 ± 0,14*,**	1,6 ± 0,12**
CCl_4 + **N**	128 ± 6	57 ± 4*	2,2 ± 0,06*	1,4 ± 0,08*	130 ± 5*,**	43 ± 4	3,0 ± 0,12**	1,8 ± 0,09**

Note: *The significance of differences ($p < 0.05$) with the intact control; **with the corresponding group of untreated animals. (C)—control, E—ecdysterone, T—turkesterone, N—Nerobol became apparent as early as day 7 of treatment. At this time point, the respiratory control ratio (RCR) increased relative to control values by 40% when α-ketoglutarate was used as the oxidation substrate and by 43% when succinate was used. Simultaneously, the ADP/O ratio increased by 50% and 46%, respectively. By day 14, both RCR and ADP/O values remained elevated, primarily due to an increased rate of phosphorylating substrate oxidation. Notably, under turkesterone treatment, unlike ecdysterone, the normalization of coupling between respiration and oxidative phosphorylation via the NAD-dependent pathway was more pronounced than via the succinate-dependent pathway. Accordingly, the ADP/O ratio increased by 62.5% in the NAD-dependent system and by 45% in the succinate-dependent system. Overall, comparison of the investigated compounds revealed no substantial differences between ecdysterone, turkesterone, and Nerobol in their general effects on mitochondrial respiration and oxidative phosphorylation. Ecdysterone demonstrated efficacy comparable to that of Nerobol, whereas turkesterone exerted a more pronounced positive effect on mitochondrial bioenergetic parameters. Chronic hepatitis induced by CCl_4 administration was also associated with marked alterations in the activity of membrane-bound enzymes in liver mitochondria. These changes were characterized by increases in the activities of succinate dehydrogenase, succinate oxidase, rotenone-sensitive NADH oxidase, and NADH dehydrogenase by 18–76%. In contrast, the activities of rotenone-insensitive NADH oxidase and the cytochrome c oxidase system did not differ significantly from normal values. In rats receiving ecdysterone, turkesterone, or Nerobol concomitantly with CCl_4 (data shown for 15 days of treatment), the activities of the investigated mitochondrial enzymes differed markedly from those observed in the control group, indicating a substantial corrective effect of these compounds on mitochondrial membrane-associated enzymatic systems (Table 5.7).

Under the influence of ecdysterone, either a clear tendency toward normalization or complete normalization was observed in the activities of NADH dehydrogenase, as well as in rotenone-sensitive and rotenone-insensitive NADH oxidase systems. Turkesterone, while normalizing the activity of rotenone-insensitive NADH oxidase, simultaneously promoted a marked increase in the activities of succinate oxidase and rotenone-sensitive NADH oxidase, by approximately 1.5- and 1.4-fold, respectively. Nerobol reduced the activities of NADH dehydrogenase and NADH oxidase components of the mitochondrial respiratory chain to levels close to those of intact animals. However, in contrast to phytoecdysteroids, Nerobol induced a more pronounced increase in the activities of succinate dehydrogenase and succinate oxidase compared with the control group. The elevation of activity in several mitochondrial polyenzyme systems observed during toxic liver injury, together with the unidirectional effects produced by phytoecdysteroids and Nerobol, likely reflects the combined action of multiple processes occurring in hepatocytes. On the one hand, membrane damage may facilitate substrate access to the active sites of polyenzyme complexes.

On the other hand, membrane-stabilizing effects, particularly characteristic of phytoecdysteroids, as well as their anabolic activity, contribute to the observed enzymatic changes. For this reason, assessment of mitochondrial status based solely

Table 5.7 Effects of phytoecdysteroids and Nerobol on the activity of polyenzyme systems of liver mitochondrial membranes in rats with chronic toxic hepatitis (M ± m, n = 6–8)

Enzyme	Intact animals	Animals with hepatitis: Hepatitis control (CCl_4)	+ Ecdysterone	+ Turkesterone	+ Nerobol
Succinate dehydrogenase	29,0 ± 1,3	34,4 ± 1,5	33,3 ± 1,3	36,4 ± 1,6	42,2 ± 2,0
		p < 0,02	$p_2 < 0,5$	$p_2 < 0,25$	$p_2 < 0,01$
Succinate oxidase	200 ± 25	260 ± 21	268 ± 27	400 ± 70	270 ± 20
		$p_1 < 0,05$	$p_2 < 0,5$	$p_2 < 0,05$	$p_2 < 0,5$
NADH dehydrogenase	7,3 ± 0,7	12,9 ± 1,1	11,6 ± 1,4	12,1 ± 1,6	10,0 ± 1,0
		$p_1 < 0,001$	$p_2 < 0,25$	<0,5	$p_2 < 0,05$
Rotenone-sensitive NADH oxidase	79,4 ± 5,4	83,2 ± 5,5	67,2 ± 4,8	113,3 ± 11,2	69,2 ± 4,4
		$p_1 < 0,5$	$p_2 < 0,05$	$p_2 < 0,05$	$p_2 < 0,05$
Rotenone-sensitive NADH oxidase	19,2 ± 1,3	24,4 ± 1,3	20,1 ± 1,5	20,0 ± 1,6	17,6 ± 0,9
		$p_1 < 0,02$	$p_2 < 0,05$	$p_2 < 0,05$	$p_2 < 0,001$
Cytochrome c oxidase	300 ± 27	291 ± 50	283 ± 24	284 ± 35	260 ± 26
		$p_1 < 0,5$	$p_2 < 0,5$	$p_2 < 0,5$	$p_2 < 0,5$

Note: p_1_given in relation to intact animals; p_2—in relation to control

on measurements of polyenzyme system activity cannot, under these conditions, be considered a sufficient criterion for evaluating either the severity of hepatic pathology or the therapeutic efficacy of phytoecdysteroids and Nerobol. More informative insight into mitochondrial membrane integrity was therefore obtained by examining mitochondrial degradation under the influence of external stressors, including thermal stress and exposure to controlled concentrations of exogenous lytic enzymes. Experiments using this approach demonstrated that the kinetics of changes in the activity of membrane-bound enzymes in liver mitochondria isolated from treated animals under thermal stress differed markedly from those observed in untreated controls and closely resembled the behavior characteristic of intact animals. Oxidases and dehydrogenases of the inner mitochondrial membrane in rats with established hepatitis lost activity at a relatively rapid rate. In contrast, mitochondria isolated from animals receiving the investigated compounds exhibited significantly greater resistance to thermal inactivation. Of particular interest is the observation that incubation of liver mitochondria from phytoecdysteroid-treated animals at 36 °C resulted in a pronounced activation phase of oxidase systems. In animals treated with Nerobol, a comparable activation phenomenon was observed only for NADH dehydrogenase, indicating a more limited protective effect on mitochondrial membrane-associated enzyme systems (Fig. 5.4).

Calculation of the half-life of NADH oxidase inactivation in mitochondria isolated from the livers of control animals revealed a markedly reduced value of 40 min, compared with approximately 115 min under normal physiological conditions. In animals with liver pathology treated with ecdysterone, turkesterone, or Nerobol, the half-life of inactivation increased to 130, 115, and 80 min, respectively. Similar trends were observed for other oxidase and dehydrogenase systems of the mitochondrial respiratory chain. Increased thermal stability of membrane-bound enzymes is known to be closely associated with strong interactions between phospholipids and membrane proteins. Such interactions restrict enzyme accessibility to phospholipases and proteases and thereby enhance membrane stability (Almatov et al. 1982; Rakhimov and Almatov 1977). Support for this mechanism was obtained from our analysis of the structural integrity of liver mitochondrial membranes in animals treated with the investigated compounds. As shown in Fig. 5.4, incubation of liver mitochondria from treated animals with phospholipase A_2 resulted in a substantially smaller decline in the activities of NADH dehydrogenase and succinate dehydrogenase than that observed in mitochondria from untreated rats over the same incubation period. In control animals, NADH dehydrogenase activity decreased twofold within 20 min of phospholipase A_2 exposure. In contrast, in animals receiving ecdysterone, turkesterone, or Nerobol under conditions of hepatitis, the time required to achieve a comparable reduction was prolonged by 50, 60, and 86 min, respectively (under normal physiological conditions, this parameter is approximately 75 min). After 40 min of incubation with phospholipase A_2, NADH dehydrogenase activity in liver mitochondria from untreated animals declined by 60%. In contrast, ecdysterone markedly attenuated this decrease, while turkesterone and Nerobol exerted an even stronger protective effect, maintaining NADH dehydrogenase activity close to its initial level (Fig. 5.5).

Succinate dehydrogenase exhibited a pattern similar to that described above. During mitochondrial incubation, an initial activation phase was followed by inhibition of enzyme activity, with the most rapid decline observed in liver mitochondria isolated from untreated animals. In control rats, succinate dehydrogenase activity in the presence of phospholipase A_2 decreased twofold after 230 min of incubation. In contrast, in animals treated with ecdysterone, turkesterone, or Nerobol, the half-life of succinate dehydrogenase inactivation increased to 320, 430, and 360 min, respectively, approaching the normal physiological value of approximately 440 min. Comparable results were obtained when the proteolytic enzyme trypsin was used instead of phospholipase A_2 (Fig. 5.5). These findings further indicate that the normalizing effects of phytoecdysteroids on hepatic metabolism under conditions of CCl_4-induced liver injury extend to lipid metabolism. In our experiments, ecdysterone, turkesterone, and cyasterone, similar to Nerobol, reduced fatty infiltration of the liver. As early as 7 days after initiation of treatment, total hepatic lipid content was 20–24% lower than in control animals. By the end of the observation period, this difference increased to 31–36% in animals treated with ecdysterone, turkesterone, and Nerobol, whereas in the cyasterone-treated group, it remained approximately unchanged. In all experimental groups, the reduction in total hepatic lipid content was primarily attributable to decreased triglyceride accumulation. In contrast, phospholipid levels in the livers of rats with chronic toxic hepatitis treated with phytoecdysteroids or Nerobol remained relatively high throughout the experimental period, indicating preservation of membrane lipid components (Fig. 5.6).

A critically important indicator of the pharmacotherapeutic efficacy of phytoecdysteroids in chronic toxic hepatitis, as in acute liver dystrophy, was the reduction in hepatic malondialdehyde (MDA) levels, a terminal product of lipid peroxidation. This effect is particularly relevant in chronic hepatitis because carbon tetrachloride

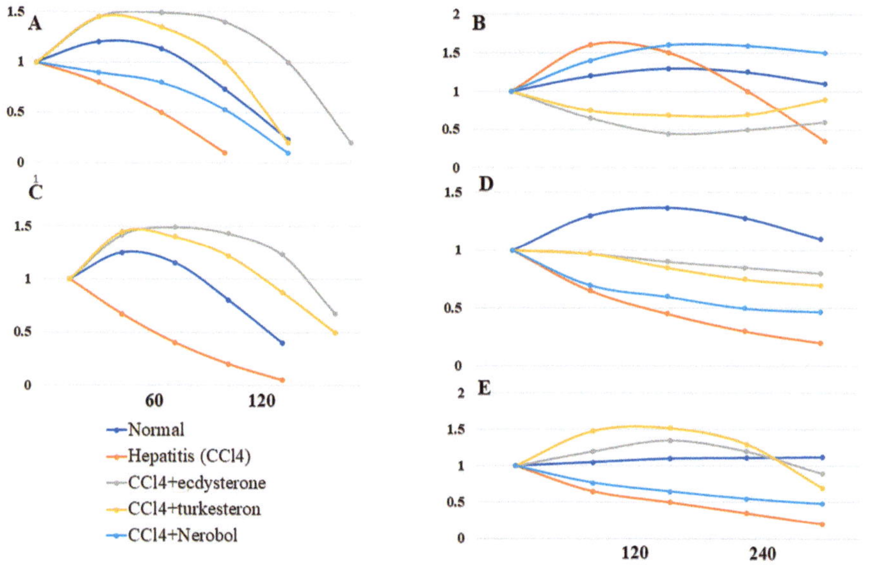

Fig. 5.4 The effect of phytoecdysteroids and nerobol on the thermal stability (36 °C) of polyenzyme systems of liver mitochondrial membranes in rats with chronic toxic hepatitis (n = 6–8). A—rotenone-soluble NADH oxidase; B—NADH dehydrogenase; C—succinate oxidase; D—succinate dehydrogenase; D—cytochrome C oxidase. 1—normal (intact control); 2—hepatitis (CCl_4); 3—CCl_4+ecdysterone; 4—CCl_4+turkesterone; 5—CCl_4+Nerobol. A_0—activity before incubation; A—during incubation; A/A_0—relative activity

(CCl_4), a potent prooxidant (Vladimirov and Archakov, 1972), acts during prolonged exposure as a persistent inducer of peroxidation of unsaturated fatty acids in membrane phospholipids. Sustained lipid peroxidation under these conditions suppresses oxidative processes in mitochondria and the endoplasmic reticulum, destabilizes lysosomal membranes, and promotes the release of necrotic hydrolases. In this context, the antioxidant effects of phytoecdysteroids and Nerobol observed in chronic hepatitis (Table 5.8) should be regarded as one of the key mechanisms underlying their hepatoprotective action.

The favorable metabolic shifts induced by phytoecdysteroids and Nerobol in hepatocytes of animals with chronic toxic hepatitis, similar to those observed in acute hepatitis, also translated into marked improvements in liver-specific functions. These included bile secretion, bilirubin metabolism, bile acid synthesis, and cholesterol excretion. Corresponding data are presented in Tables 5.9 and 5.10. As shown by these results, rats with chronic hepatitis exhibited a significant reduction in bile secretion compared with intact animals, particularly during the third and fourth hours of observation. At all time points, total bile output over a 4-h period was reduced by 7–12%, while at the first observation point (4 weeks after initiation of CCl_4 administration), the reduction reached 25%. Marked disturbances were also detected in the hepatic component of bilirubin metabolism. These alterations were accompanied by deterioration of the bile acid-to-cholesterol balance, resulting in a

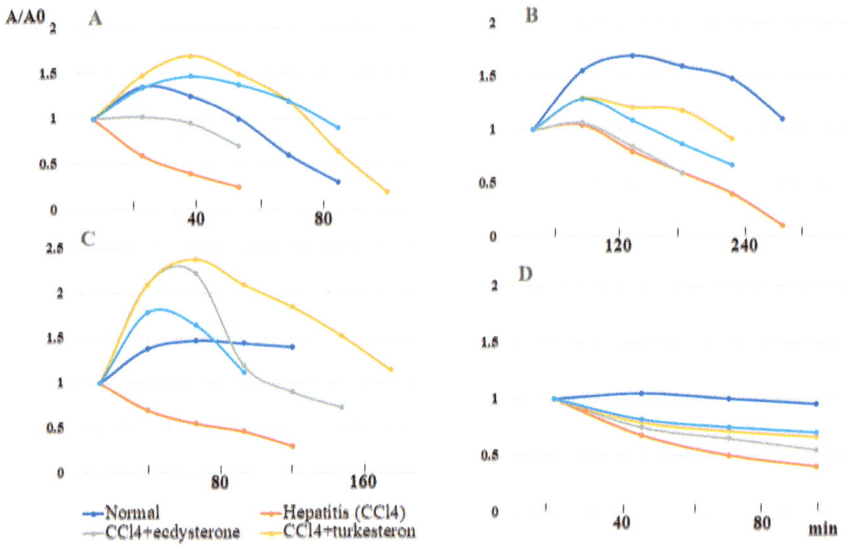

Fig. 5.5 Effect of phospholipase A2 (A, B) and trypsin (C, D) on the activity of NADH-dehydrogenase (A, B) and succinate dehydrogenase (C, D) of liver mitochondrial membranes in rats with chronic toxic hepatitis (n = 6–8). A0 is the activity of dehydrogenases up to, A—after the addition of lytic enzymes. The remaining values are the same as in Fig. 5.4

persistently reduced cholate-to-cholesterol coefficient throughout the experiment. In contrast, in rats receiving phytoecdysteroids during prolonged CCl_4 exposure, bile secretion was not suppressed and, in most cases, remained close to, or even exceeded, values observed in intact animals. The only exceptions were groups treated with ecdysterone or turkesterone for a short duration of 1 week, in which bile secretion had not yet fully normalized.

The same conclusion applies to the severity of hyperbilirubinemia, bile acid synthesis, and cholesterol excretion. Overall, the effects of phytoecdysteroids in this experimental series were largely comparable to those of Nerobol. At certain observation time points, particularly with respect to bile secretion rate and bile composition, the anabolic steroid produced a somewhat more pronounced response. However, as in the acute hepatitis model, animals treated with Nerobol consistently exhibited a lower cholate-to-cholesterol ratio in bile throughout the experiment compared with rats receiving phytoecdysteroids. In addition, in Nerobol-treated animals, this ratio differed only slightly from values observed in control groups, indicating limited normalization of bile acid–cholesterol balance. In contrast, treatment with phytoecdysteroids was associated with a more favorable correction of this parameter. Thus, phytoecdysteroids—ecdysterone, turkesterone, and cyasterone—which demonstrated superior efficacy compared with Nerobol in acute toxic liver injury, produced effects comparable to those of this anabolic steroid in the chronic hepatitis model. In both experimental settings, phytoecdysteroid activity was reflected in normalization of the full spectrum of investigated parameters

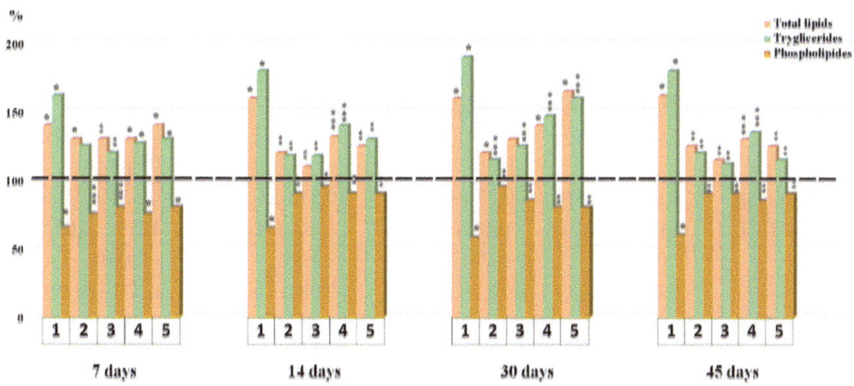

Fig. 5.6 The effect of phytoecdysteroids and Nerobol on the content of total lipids, triglycerides, and phospholipids in the liver of rats with chronic toxic hepatitis n = 6–8. The abscissa axis shows the timing of drug administration (in days), and the ordinate axis shows the content of total lipids, triglycerides, and phospholipids (in % of intact control values). 1—control (CCl_4), 2—CCl_4+ecdysterone, 3—CCl_4+turkesterone, 4—CCl_4+cyasterone, and 5— CCl_4+Nerobol. *The significance of differences ($p < 0.05$) with the intact control; **with the corresponding group of untreated animals

characterizing metabolic and functional disturbances of the liver. Importantly, in both acute and chronic models, phytoecdysteroids, unlike Nerobol, promoted a more rapid restoration of the cholate-to-cholesterol ratio to values characteristic of intact animals. Taken together, these findings indicate that phytoecdysteroids represent a more promising class of potential therapeutic agents than steroidal anabolic analogs for the treatment and prevention of toxic liver injury and its long-term consequences.

5.3 Pharmacotherapeutic Evaluation of the Effects of Phytoecdysteroids in Toxic Kidney Injury

Given their stimulatory effects on protein biosynthesis, phytoecdysteroids may also be of interest as agents with beneficial effects in certain renal pathologies. This assumption is supported by earlier observations that many anabolic compounds exert a normalizing influence on renal morphofunctional status following nephrotoxic injury and reduce manifestations of uremic intoxication in chronic renal failure (Rosenfeld 1977). Accordingly, we investigated the nephroprotective properties of phytoecdysteroids in comparison with nerobol in a rat model of chronic nephritis induced by a uranyl–glycerol nephrotoxic mixture. Administration of the nephrotoxic mixture resulted in pronounced hyperazotemia. In control rats, by the end of the first week, serum urea levels increased by 170% and residual nitrogen by 124% relative to baseline. Although these parameters declined slightly at later time points, they remained significantly elevated throughout most of the experiment. Serum

Table 5.8 Effects of phytoecdysteroids and Nerobol on malondialdehyde content in the liver of rats with chronic toxic hepatitis (M ± m, n = 6–8)

Experiment groups	MDA (nmol/mg protein) 28 days CCl_4 + 7 days treatment	35 days CCl_4 + 14 days treatment	50 days CCl_4 + 30 days treatment	50 days CCl_4 + 45 days treatment
Intact animals	0,515 ± 0,044	0,530 ± 0,024	0,518 ± 0,042	0,524 ± 0,024
Control (CCl_4)	1195 ± 0,079	1389 ± 0,079	1568 ± 0,078	1215 ± 0,058
p_1	<0,001	<0,001	<0,001	<0,001
CCl_4+ ecdysterone	0,869 ± 0,079	0,775 ± 0,077	0,984 ± 0,087	0,731 ± 0,060
p_1	<0,01	<0,02	<0,001	<0,01
p_2	<0,02	<0,001	<0,001	<0,001
CCl_4+ turkesteron	0,891 ± 0,057	0,800 ± 0,078	1093 ± 0,03	0,692 ± 0,076
p_1	<0,001	<0,01	<0,001	<0,05
p_2	=0,01	<0,001	<0,001	<0,001
CCl_4 + cyasterone	0,905 ± 0,064	1031 ± 0,093	1418 ± 0,085	0,869 ± 0,063
p_1	<0,001	<0,001	<0,001	<0,001
p_2	<0,02	<0,02	<0,01	<0,01
CCl_4 + Nerobol	0,927 ± 0,069	0,881 ± 0,092	1085 ± 0,100	0,832 ± 0,071
p_1	<0,001	<0,01	<0,001	<0,01
p_2	<0,05	<0,002	<0,01	<0,002

Note: p_1—calculated in relation to the corresponding value of intact animals; p_2—in relation to the control

creatinine levels also remained markedly increased (Table 5.11). Rats exposed to the nephrotoxic mixture additionally exhibited sustained polyuria over the four-week observation period. Daily urinary excretion of urea decreased by 57.0%, 54.3%, 37.6%, 17.3%, and 9.4% from the first week to the end of the experiment, with statistically significant reductions during the first 20 days. Creatinine excretion decreased by 40.5%, 23.8%, 16.7%, 9.5%, and 7.2%, with significant differences observed during the first 14 days. This is because serum urea and creatinine levels remained elevated during this period, and reduced urinary excretion indicates a marked decline in glomerular filtration rate. Severe renal injury was further confirmed by persistent proteinuria, detected from the first days after nephrotoxic exposure until the end of the experiment. Urinary sediment analysis revealed abundant erythrocytes, leukocytes, casts, and renal epithelial cells. In contrast, rats treated with phytoecdysteroids or Nerobol showed less pronounced increases in serum urea and residual nitrogen by the end of the first week (Table 5.11).

At later time points, these parameters remained consistently lower than in controls. From day 28 in phytoecdysteroid-treated rats and day 21 in Nerobol-treated rats, serum urea and residual nitrogen levels no longer differed significantly from baseline values. During the same periods, serum creatinine concentrations also returned to normal.

Table 5.9 Effects of phytoecdysteroids and Nerobol on bile formation in rats with chronic toxic hepatitis (M ± m, n = 8)

Experimental groups	Day/Treatment	Bile secretion rate ($mg \cdot min^{-1} \cdot 100\ g^{-1}$ body weight)				Total bile output (4 h), mg/100 g
		1st	2nd	3rd	4th	
Intact animals	–	4,2 ± 0,20	4,3 ± 0,16	4,2 ± 0,18	4,1 ± 0,18	1010 ± 37
CCl_4, end of seeding (control)	The 28th	3,4 ± 0,28*	3,4 ± 0,28*	3,1 ± 0,26*	2,7 ± 0,23*	757 ± 59*
CCl_4 + ecdysterone	28th + 7	3,4 ± 0,24*	3,9 ± 0,22	3,6 ± 0,32	3,4 ± 0,35	861 ± 50*
CCl_4 + turkesterone	28th + 7	3,2 ± 0,10*	3,7 ± 0,12*	3,7 ± 0,10*	3,6 ± 0,13*,**	850 ± 18*
CCl_4 + cyasterone	28th + 7	5,0 ± 0,10*,**	5,9 ± 0,10*,**	4,8 ± 0,14*,**	4,9 ± 0,39**	1243 ± 22*,**
CCl_4 + Nerobol	28th + 7	4,8 ± 0,26**	4,9 ± 0,10*,**	4,9 ± 0,40**	4,6 ± 0,35**	1155 ± 32*,**
CCl_4 (control)	The 35th	3,8 ± 0,17	3,9 ± 0,17	3,7 ± 0,17	3,5 ± 0,17*	938 ± 34
CCl_4 + ecdysterone	35th +14	3,9 ± 0,30	4,7 ± 0,29**	4,5 ± 0,26**	4,1 ± 0,26	1033 ± 45
CCl_4 + turkesterone	35th +14	4,4 ± 0,10**	4,4 ± 0,33	4,6 ± 0,17**	4,6 ± 0,15**	1075 ± 41**
CCl_4 + cyasterone	35th +14	3,8 ± 0,10	4,1 ± 0,30	4,6 ± 0,35**	3,6 ± 0,18	973 ± 45
CCl_4 + Nerobol	35th +14	5,8 ± 0,32*,**	4,9 ± 0,42	4,8 ± 0,39**	4,3 ± 0,44	1191 ± 85**
CCl_4 (control)	The 50th	3,6 ± 0,22	3,7 ± 0,12*	3,8 ± 0,23	3,4 ± 0,20*	887 ± 36*
CCl_4 + ecdysterone	50th + 30	4,6 ± 0,33**	4,3 ± 0,33	4,5 ± 0,43	3,7 ± 0,37	1024 ± 84
CCl_4 + turkesterone	50th + 30	3,5 ± 0,27	3,7 ± 0,28	4,5 ± 0,43	4,1 ± 0,16**	945 ± 55
CCl_4 + cyasterone	50th + 30	4,8 ± 0,48**	4,8 ± 0,46**	4,4 ± 0,41	4,2 ± 0,36	1089 ± 80**
CCl_4 + Nerobol	50th + 30	6,7 ± 0,24*,**	6,3 ± 0,36*,**	5,7 ± 0,34*,**	4,9 ± 0,48**	1424 ± 59*,**
CCl_4 (control)	The 50th	3,8 ± 0,36	4,2 ± 0,33	3,6 ± 0,32	3,1 ± 0,22*	882 ± 44*
CCl_4 + ecdysterone	50th + 45	4,6 ± 0,50	5,2 ± 0,62	4,8 ± 0,32**	4,7 ± 0,34**	1218 ± 82*,**
CCl_4 + turkesterone	50th + 45	4,6 ± 0,52	4,5 ± 0,46	4,7 ± 0,30**	4,3 ± 0,41**	1051 ± 56**
CCl_4 + cyasterone	50th + 45	4,6 ± 0,59	4,2 ± 0,52	4,2 ± 0,21	4,1 ± 0,26**	1021 ± 95
CCl_4 + Nerobol	50th + 45	4,2 ± 0,10	4,1 ± 0,10	4,1 ± 0,11	4,1 ± 0,11**	993 ± 18**

Note: *The significance of differences ($p < 0.05$) with the intact control; **with the corresponding group of untreated animals

Table 5.10 Effects of phytoecdysteroids and Nerobol on the chemical composition of bile in rats with chronic toxic hepatitis (M ± m, n = 8)

Experimental groups	Day/Treatment	Total amount in mg/100 g of body weight in 4 h			Cholate-to-cholesterol ratio
		Bilirubin	Bile acids	Cholesterol	
Intact animals	–	0,140 ± 0,01	9614 ± 0,42	0,195 ± 0,02	64,2 ± 5,4
CCl_4, end of seeding (control)	The 28th	0,067 ± 0,008*	5045 ± 0,38*	0,132 ± 0,01*	43,8 ± 6,6*
CCl_4 + ecdysterone	28th + 7	0,137 ± 0,02**	7042 ± 0,42*,**	0,202 ± 0,02**	33,7 ± 3,4*
CCl_4 + turkesterone	28th + 7	0,115 ± 0,02**	8520 ± 0,48**	0,245 ± 0,03**	47,4 ± 4,8*
CCl_4 + cyasterone	28th + 7	0,143 ± 0,01**	8210 ± 0,56**	0,268 ± 0,02*,**	30,2 ± 3,2*
CCl_4 + Nerobol	28th + 7	0,109 ± 0,01*	7759 ± 0,38*,**	0,227 ± 0,02**	35,6 ± 3,6*
CCl_4 (control)	The 35th	0,107 ± 0,01*	5658 ± 0,54*	0,146 ± 0,01	33,9 ± 3,4*
CCl_4 + ecdysterone	35th +14	0,130 ± 0,03	8839 ± 0,56**	0,164 ± 0,01	53,5 ± 7,2**
CCl_4 + turkesterone	35th +14	0,121 ± 0,02	8890 ± 0,62**	0,184 ± 0,01**	48,6 ± 4,6**
CCl_4 + cyasterone	35th +14	0,115 ± 0,01	7684 ± 0,70*,**	0,169 ± 0,02	42,8 ± 4,4*
CCl_4 + Nerobol	35th +14	0,137 ± 0,02	9428 ± 0,78**	0,245 ± 0,02**	38,6 ± 4,0*
CCl_4 (control)	The 50th	0,099 ± 0,008*	5157 ± 0,46*	0,116 ± 0,01*	37,2 ± 3,6*
CCl_4 + ecdysterone	50th + 30	0,115 ± 0,01	9304 ± 0,68**	0,184 ± 0,02**	52,3 ± 5,2**
CCl_4 + turkesterone	50th + 30	0,124 ± 0,01	9449 ± 0,54**	0,186 ± 0,02**	53,2 ± 5,8**
CCl_4 + cyasterone	50th + 30	0,187 ± 0,03**	11,707 ± 0,62*,**	0,201 ± 0,03**	58,2 ± 6,4**
CCl_4 + Nerobol	50th + 30	0,193 ± 0,02*,**	14,796 ± 0,82*	0,310 ± 0,03*,**	49,4 ± 5,2
CCl_4 (control)	The 50th	0,107 ± 0,01*	5743 ± 0,54*	0,114 ± 0,01*	49,4 ± 4,8
CCl_4 + ecdysterone	50th + 45	0,224 ± 0,03*,**	10,360 ± 0,62**	0,189 ± 0,02**	55,7 ± 5,2
CCl_4 + turkesterone	50th + 45	0,122 ± 0,02	8260 ± 0,48**	0,150 ± 0,01**	54,9 ± 3,8
CCl_4 + cyasterone	50th + 45	0,121 ± 0,02	6592 ± 0,36*	0,134 ± 0,01*	63,4 ± 4,0**
CCl_4 + Nerobol	50th + 45	0,184 ± 0,02**	8179 ± 0,38*,**	0,159 ± 0,01**	50,0 ± 3,6

Note: *The significance of differences ($p < 0.05$) with the intact control; **with the corresponding group of untreated animals

Table 5.11 Effects of phytoecdysteroids and Nerobol on the course of experimental nephritis in rats (mean ± SEM, n = 8)

Sample	Parameter (mg%)	Experimental group	Initial values	Days of observation				
				7th	14th	21st	28th	35th
1	2	3	4	5	6	7	8	9
BLOOD	Urea	Control	28,4 ± 1,5	76,7 ± 7,9*	63,2 ± 5,4*	54,4 ± 5,2*	45,0 ± 3,9*	37,3 ± 3,2*
		Ecdysterone	27,4 ± 1,4	66,9 ± 4,1*	48,4 ± 3,7*	37,2 ± 2,5*,**	29,1 ± 1,6**	28,2 ± 1,6**
		Turkesterone	29,3 ± 0,9	61,9 ± 4,2*	44,5 ± 2,5*,**	35,7 ± 2,8*,**	28,0 ± 1,4**	28,9 ± 1,7**
		Nerobol	28,7 ± 2,3	57,6 ± 4,4*	41,8 ± 2,6*,**	32,6 ± 2,1**	27,8 ± 1,5**	27,0 ± 1,6**
	Residual nitrogen	Control	37,4 ± 2,4	84,1 ± 7,6*	70,9 ± 7,2*	62,6 ± 6,0*	53,4 ± 4,7*	46,0 ± 4,0
		Ecdysterone	36,5 ± 2,4	75,4 ± 4,9*	56,8 ± 3.6*	45,8 ± 3,6*,**	38,1 ± 2,6**	37,2 ± 3,6
		Turkesterone	38,3 ± 3,5	69,7 ± 7,8*	52,9 ± 3,8*,**	44,5 ± 4,2**	37,0 ± 3,8**	37,9 ± 3,4
		Nerobol	37,7 ± 4,8	65,6 ± 8,4*	50,4 ± 4,8**	41,5 ± 4,2**	36,8 ± 3,4**	36,1 ± 3,0
	Creatinine	Control	1,3 ± 0,19	2,5 ± 0,37*	2,4 ± 0,23*	1,8 ± 0,07*	1,9 ± 0,09*	1,7 ± 0,06
		Ecdysterone	1,2 ± 0,05	1,7 ± 0,08*	1,6 ± 0,05*,**	1,5 ± 0,06*,**	1,2 ± 0,09**	1,3 ± 0,04**
		Turkesterone	1,3 ± 0,09	1,8 ± 0,11*	1,7 ± 0,06*,**	1,5 ± 0,07*,**	1,2 ± 0,08**	1,2 ± 0,06**
		Nerobol	1,2 ± 0,12	1,9 ± 0,12*	1,4 ± 0,12**	1,3 ± 0,07**	1,1 ± 0,08**	1,2 ± 0,08**
URINE	Diuresis daily (ml)	Control	7,3 ± 0,49	15,8 ± 1,1*	14,8 ± 1,0*	13,0 ± 1,0*	9,6 ± 0,49*	8,5 ± 0,57
		Ecdysterone	7,0 ± 0,38	13,8 ± 0,70*	11,2 ± 0,59*,**	9,0 ± 0,33*,**	8,0 ± 0,32**	7,3 ± 0,37
		Turkesterone	7,6 ± 0,42	13,2 ± 0,52*	10,8 ± 0,64*,**	8,9 ± 0,44**	8,1 ± 0,39**	7,0 ± 0,30**
		Nerobol	6,8 ± 0,45	12,8 ± 1,1*	9,6 ± 0,8*,**	7,4 ± 0,7**	6,8 ± 0,6**	6,4 ± 0,4**
	Creatinine (mg)	Control	4,2 ± 0,34	2,5 ± 0,19*	3,2 ± 0,17*	3,3 ± 0,13	3,8 ± 0,27	3,4 ± 0,22
		Ecdysterone	4,1 ± 0,29	3,1 ± 0,13*,**	3,5 ± 0,17	4,2 ± 0,35**	4,1 ± 0,30	4,3 ± 0,34
		Turkesterone	4,0 ± 0,30	3,4 ± 0,32**	3,7 ± 0,29	4,1 ± 0,23**	4,5 ± 0,29	4,2 ± 0,19
		Nerobol	4,0 ± 0,28	3,5 ± 0,24**	3,9 ± 0,26**	4,2 ± 0,4**	3,7 ± 0,25	3,9 ± 0,15
	Protein (mg)	Control	0,65 ± 0,07	10,8 ± 1,2*	8,2 ± 1,0*	6,0 ± 1,07*	3,4 ± 0,53*	1,9 ± 0,33*
		Ecdysterone	0,63 ± 0,05	4,7 ± 0,67*,**	2,1 ± 0,50*,**	0,78 ± 0,06**	0,61 ± 0,04**	0,57 ± 0,03**
		Turkesterone	0,54 ± 0,04	3,1 ± 0,38*,**	1,6 ± 0,26*,**	0,63 ± 0,05**	0,52 ± 0,04**	0,46 ± 0,05**
		Nerobol	0,60 ± 0,06	3,4 ± 0,44*,**	1,3 ± 0,2*,**	0,67 ± 0,05**	0,55 ± 0,04**	0,47 ± 0,08**

Note: *The reliability of differences ($p < 0.05$) with the initial values; **with the corresponding indicator in the group of control animals

Administration of phytoecdysteroids, similar to Nerobol, promoted a more rapid resolution of polyuria. Proteinuria was less pronounced in treated animals and was largely confined to the first 2 weeks of observation. It should be noted that intact rats in our experiments also exhibited a minor degree of spontaneous proteinuria. After 1 week of treatment, urinary urea excretion decreased by 40% under ecdysterone, 39% under turkesterone, and 38.3% under Nerobol relative to baseline ($p < 0.05$). After 3 weeks, reductions in urea excretion (12–9%) were no longer statistically significant. Creatinine excretion was fully restored in all treated groups by day 21 and remained at baseline levels until the end of the experiment (Table 5.11). These findings indicate that administration of phytoecdysteroids prevents early impairment of glomerular filtration in animals with chronic renal failure induced by a uranyl–glycerol mixture. From the second week onward, urinary sediment from rats treated with phytoecdysteroids or Nerobol contained virtually no erythrocytes, casts, or renal epithelial cells. The therapeutic effects of ecdysterone and turkesterone were generally similar in direction to those observed with Nerobol, although for some parameters, the anabolic steroid produced a somewhat more pronounced effect. Overall, the data demonstrate that phytoecdysteroids exert a pronounced hypoazotemic effect in severe renal pathology associated with marked elevations in serum urea and residual nitrogen. Under their influence, a broad spectrum of indices reflecting renal functional status improved significantly. Nevertheless, with respect to overall therapeutic efficacy in this experimental model, phytoecdysteroids were somewhat less effective than Nerobol.

5.4 Clinical Relevance, Mechanisms of Action, and Translational Prospects of Phytoecdysteroids

The clinical potential of phytoecdysteroids extends beyond their established applications in cardiology, hepatology, and nephrology. Their future use may include adjunctive therapy for metabolic syndrome, supportive treatment of viral hepatitis and drug-induced liver injury, and rehabilitation after myocardial infarction or surgical interventions, owing to their favorable effects on energy metabolism, membrane stability, and antioxidant defense. Phytoecdysteroids may also be valuable as adaptogenic agents in the aging population, where they enhance stress resistance, limit oxidative damage, and improve recovery capacity. Despite these promising perspectives, further randomized controlled clinical trials are required to define optimal dosing regimens, long-term safety, and comparative efficacy relative to established pharmacotherapies. Nevertheless, currently available experimental and clinical data allow several important conclusions to be drawn. Preparations such as Ecdysten and related formulations have demonstrated efficacy in cardiovascular and hepatic disorders, with consistent improvements in biochemical parameters including transaminase activity, lipid profiles, and bilirubin levels under both experimental and clinical conditions (Syrov 2015).

Enhanced physical and functional performance has been documented during postinfarction rehabilitation and in patients with chronic myocardial dystrophy. Importantly, phytoecdysteroids exhibit a markedly superior safety profile compared with synthetic anabolic steroids, supporting their potential use in metabolic syndrome, viral hepatitis, and geriatric medicine. Extensive experimental evidence demonstrates that phytoecdysteroids exert pronounced cardioprotective and hepatoprotective effects through the integration of multiple biological mechanisms. In the myocardium, they preserve glycogen and ATP stores, limit necrosis, maintain sarcomere and mitochondrial ultrastructure, stabilize ionic balance and ECG parameters, suppress lipid peroxidation, and enhance antioxidant defenses, ultimately improving contractile function and recovery. In the liver, phytoecdysteroids reduce serum transaminase and bilirubin levels, prevent hepatocellular necrosis, preserve lobular architecture, normalize lipid metabolism by reducing cholesterol and triglyceride accumulation, stimulate bile secretion, and stabilize mitochondrial respiration with preservation of cristae structure and bioenergetic capacity. Clinical experience with phytoecdysteroid-based preparations, including Ecdysten and Eksumid, corroborates these findings. Their use has been associated with improved biochemical markers, functional outcomes, and overall well-being in patients with cardiovascular and hepatic diseases, without the androgenic or hepatotoxic adverse effects characteristic of synthetic anabolic steroids. Taken together, these data identify phytoecdysteroids as multifunctional natural agents that combine therapeutic efficacy with an exceptional safety profile. This unique combination strongly supports their translational potential as protective and restorative agents for vital organs under conditions of metabolic, toxic, ischemic, and age-related stress.

References

Albano GD, Amico F, Cocimano G, Liberto A, Maglietta F, Esposito M, Rosi GL, Di Nunno N, Salerno M, Montana A (2021) Adverse Effects of Anabolic-Androgenic Steroids: A Literature Review. Healthcare 9:97. https://doi.org/10.3390/healthcare9010097

Almatov KT, Agzamov Kh, Rakhimov MM, Turakulov YaK (1982) Oxidative phosphorylation and activity of polyenzyme systems of rat liver mitochondria membranes during starvation Vopr Med Chem 4:50–56 (in Russ)

Bakhteev RR (1987) Effect of retabolil on hyperuricemia in chronic heart failure. Kazan Med J 68:102–103. https://doi.org/10.17816/kazmj96023

Bardin EV (1978) Experimental substantiation and clinical use of anabolic steroids in myocardial infarction Zdravookhr Belarus 1:19–21 (in Russ)

Báthori M, Tóth N, Hunyadi A, Márki Á, Zador E (2008) Phytoecdysteroids and anabolic-androgenic steroids-structure and effects on humans. Curr Med Chem 15:75–91. https://doi.org/10.2174/092986708783330674

Belenky EE, Sokolov IK, Lantsberg LA et al (1973) Prevention of experimental adrenalin damage to the myocardium with potassium orotate. Pharmacol and toxicol 3:301–304 (in Russ)

Bin Sayeed MS, Karim SMR, Sharmin T, Morshed MM (2016) Critical analysis on characterization, systemic effect and therapeutic potential of beta-sitosterol: a plant-derived orphan phytosterol. Medicines 3:29. https://doi.org/10.3390/medicines3040029

Bluger AF (1970) Viral hepatitis and its outcomes Riga: Zvaigzne Publ p 546

Buniam J, Chukijrungroat N, Rattanavichit Y, Surapongchai J, Weerachayaphorn J, Bupha-Intr T, Saengsirisuwan V (2020) 20-Hydroxyecdysone ameliorates metabolic and cardiovascular dysfunction in high-fat highfructose-fed ovariectomized rats (BMC) Compl Med Ther 20:140–152. https://doi.org/10.1186/s12906-020-02936-1

Cahlíková L, Macáková K, Chlebek J, Hošt'álková A, Kulhánková A, Opletal L (2011) Ecdysterone and its activity on some degenerative diseases. Nat Prod Commun 6:1934578X1100600527. https://doi.org/10.1177/1934578X1100600527

Cevostyanov AE, Sokolov VA, Darmogray VN (2006) Prospects for the use of phytoecdysteroids in ophthalmology. Russian Medical and Biological Bulletin 1:71–79 (in Russ)

Gorelick-Feldman J, MacLean D, Ilic N, Poulev A, Lila MA, Cheng D, Raskin I (2008) Phytoecdysteroids increase protein synthesis in skeletal muscle cells. J of Agric Food Chem 56:3532–3537. https://doi.org/10.1021/jf073059z

Grebenchikov OA, Zabelina TS, Filippovskaya JS, Gerasimenko ON, Plotnikov EYu, Likhvantsev VV (2016) Molecular mechanisms of oxidative stress. Bull Inten Ther named after AI Saltanova 3:13–21

Hilchuk MA (2013) Metabolic changes in toxic liver damage and the possibility of their correction (experimental study). Doct dissert (in Russ)

Lafont R, Dinan L (2003) Practical uses for ecdysteroids in mammals including humans: and update. J Ins Sci 3:7. https://doi.org/10.1093/jis/3.1.7

Mamarasulov B, Davranov K, Jabborova D (2020) Phytochemical, pharmacological and biological properties of *Ajuga turkestanica* (Rgl.) Brig (*Lamiaceae*). Ann Phytomed 9:44–57. https://doi.org/10.21276/ap.2020.9.1.6

Meyers TA and Townsend D (2019) Cardiac Pathophysiology and the Future of Cardiac Therapies in Duchenne Muscular Dystrophy. Inter J Mol Sci 20:4098. https://doi.org/10.3390/ijms20174098

Nemeth E, Bernath J (2008) Biological activities of yarrow species (Achillea spp.) Curr Pharm Des 14:3151–3167. https://doi.org/10.2174/138161208786404281

Palladino A, Papa AA, Morra S, Russo V, Ergoli M, Rago A, Orsini C, Nigro G, Politano L (2019) Are there real benefits to implanting cardiac devices in patients with end-stage dilated dystrophinopathic cardiomyopathy? Review of literature and personal results. Acta Myol 38:1–7.

Rakhimov MM, Almatov KT (1977) Some features of degradation of liver mitochondrial enzyme systems in rats exposed to thermal influences. Biochemistry 42:1852–1863 (in Russ)

Rosenfeld JB (1977) Androgens in nephrology. Pharmacol Ther Part C: Clin Pharm Therap 2:57–63. https://doi.org/10.1016/S0362-5486(77)80012-5

Solovyova AG, Yerimbetov KT, Accusing OV, Fedorova AV, Mikhailov VV (2021) Physiological mechanisms of action and prospects for the use of phytoecdysteroids in biomedical technologies. Probl Biol Prod Anim 1:26–40. https://doi.org/10.25687/1996-6733.prodanimbio1.2021.1.26-40

Syrov VN (2015) Experimental and clinical results of evaluating the efficacy of ecdysten as a metabolic-type drug. Byul Pharm 3–4:111–117 (in Russ)

Syrov VN, Gusakova SD, Khushbaktova ZA, Egamova FR, Khidoyatova SK, Sagdullaev SS (2023) Hepatoprotective Efficacy of a New Phytocomposition of Essential Phospholipids with Glycyrrhizic Acid, Ecdysterone, and Lycopene in Experimental Chronic Hepatitis Compared to Phosphogliv. Pharm Chem J 56:1433–1438. https://doi.org/10.1007/s11094-023-02811-6

Syrov VN, Islamova JI, Egamova FR, Yuldasheva NH, Khushbaktova ZA (2014) Stress-protective properties of phytoecdysteroids Experim and Clin Pharmacol 77:35–38. https://doi.org/10.30906/0869-2092-2014-77-7-35-38 (in Russ)

Syrov VN, Nabiyev AN, Sultanov MB (1986) Effect of phytoecdysteroids on the bile-secreting function of the liver under normal conditions and in experimental hepatitis. J Pharm and Toxicol 3:100–103 (in Russ)

Vladimirov YuA, Archakov AI (1972) Lipid peroxidation in biological membranes. Moscow: Nauka Publ p 252 (in Russ)

Yusupova UY, Ramazonov NS, Syrov VN, Sagdullaev SS (2022b) Biological Activities of Phytoecdisteroids In: Phytoecdysteroids Springer, Singapore. https://doi.org/10.1007/978-981-16-6711-4_6

Yusupova UY, Ramazonov NSh, Syrov VN, Sagdullaev SS (2022a) Technologies for Producing Drugs and Biologically Active Additives Based on Ecdysteroids from Plants *Rhaponticum carthamoides*, *Silene praemixta*, and *Ajuga turkestanica*. In: Phytoecdysteroids Springer, Singapore. https://doi.org/10.1007/978-981-16-6711-4_7

Yusupova U, Sasmakov S, Usmanov D, Ramazonov N, Azimova S and Sagdullaev S (2020) Phytoecdysteroids from Silene claviformis and their antibacterial and antifungal activities. Pol. J. Natur. Sc. 35(3):313–321.

Yusupova UY, Bobakulov KM, Girault JP, Usmanov DA, Ramazonov NS and Abdullaev ND (2022) New Phytoecdysteroid from Silene popovii. Chem Nat Compd 58(6):1082–1084. https://doi.org/10.1007/s10600-022-03874-9.

Yusupova UY, Bobakulov KM, Khurramov AR, Syrov VN, Egamova FR, Karuth A, Usmanov DA, Quadir M and Rasulev B (2025) Phytochemical constituents isolated from Silene popovii Schischk. Med Chem Res 34(2):423–431. https://doi.org/10.1007/s00044-024-03345-6.

Yusupova UY, Ramazonov NS, Bobakulov KM, Egamova FR, Syrov VN and Usmanov DA (2021) Biological active components of silene tomentella and their pharmacological properties. Khim. Rastit. Syr′ya 1:197–202. https://doi.org/10.14258/jcprm.2021018323.

Yusupova UY, Usmanov DA and Ramazonov NS (2019) Phytoecdysteroids from the Plant Dianthus helenae. Chem Nat Compd 55(2):393–394. https://doi.org/10.1007/s10600-019-02701-y.

Chapter 6
Phytoecdysteroids in Wound Healing, Bone Regeneration, and Tissue Repair

Vladimir Nikolayevich Syrov and Egamova Feruza

Abstract Phytoecdysteroids (PEs) are naturally occurring, plant-derived polyhydroxylated steroids with pronounced regenerative potential. In addition to their well-documented anabolic and adaptogenic properties, a growing body of experimental and clinical evidence indicates that phytoecdysteroids actively promote tissue repair and regeneration across multiple organ systems. These compounds stimulate protein and collagen synthesis, support angiogenesis, preserve mitochondrial integrity, and protect cellular membranes from oxidative damage. In experimental wound-healing models, phytoecdysteroids accelerate re-epithelialization, attenuate inflammatory responses, and improve the structural restoration of both epidermal and dermal layers. During bone regeneration, phytoecdysteroids enhance osteoblast activity, improve calcium and phosphorus metabolism, and promote fracture healing by increasing mineralization and biomechanical strength. Their membrane-stabilizing and antioxidant properties also extend to skeletal muscle, tendons, and internal organs, thereby facilitating functional recovery following traumatic or ischemic injury. Clinical observations further support these experimental findings. Phytoecdysteroid-based preparations have been associated with improved outcomes in postoperative rehabilitation, fracture healing, and tissue recovery, while maintaining an excellent safety profile. Importantly, unlike synthetic anabolic steroids, phytoecdysteroids do not exhibit androgenic or organ-toxic effects during prolonged use. Taken together, these data position phytoecdysteroids as a promising class of natural regenerative agents with significant translational potential in regenerative and restorative medicine, offering a safe and multifaceted strategy for enhancing tissue repair and functional recovery.

Keywords Phytoecdysteroids · Wound healing · Bone regeneration · Tissue repair · Collagen synthesis · Fracture healing

---

V. N. Syrov (✉) · E. Feruza
Department of Pharmacology and Toxicology, Institute of the Chemistry of Plant Substances, Academy of Science of the Republic of Uzbekistan, Tashkent, Uzbekistan

V. N. Syrov et al. (eds.), *Phytoecdysteroids II*,
https://doi.org/10.1007/978-981-95-8423-9_6

Wound healing, bone regeneration, and tissue repair are complex processes that rely on the coordinated action of inflammation, angiogenesis, collagen synthesis, osteogenesis, mitochondrial energy production, and tissue remodeling. Pharmacological agents that support these processes without causing adverse effects are therefore of high clinical importance, particularly in traumatology, surgery, sports medicine, and chronic degenerative diseases. PEs are naturally occurring polyhydroxylated steroids widely distributed in plants and have emerged as promising regenerative agents. Originally identified as insect molting hormones, phytoecdysteroids were later shown to exert diverse biological effects in mammals, including anabolic, adaptogenic, cardioprotective, hepatoprotective, and regenerative activities (Zádor 2025; Cheng 2010; (Yusupova et al. 2019; Yusupova et al. 2020; Yusupova et al. 2022; Yusupova et al. 2025; Yusupova et al. 2021). Their ability to stimulate protein and nucleic acid synthesis, enhance cellular energy metabolism, stabilize biological membranes, and protect tissues against oxidative damage provides a strong mechanistic basis for their application in tissue repair and regenerative medicine. Experimental studies show that phytoecdysteroids accelerate wound closure, promote re-epithelialization, and reduce necrotic changes in models of skin injury.

During bone regeneration, compounds such as ecdysterone and turkesterone stimulate osteoblast proliferation, normalize calcium and phosphorus metabolism, and improve the biomechanical strength of healing bone. Their membrane-stabilizing and antioxidant effects also extend to skeletal muscle and connective tissues, facilitating faster and more complete functional recovery after injury. Importantly, unlike synthetic anabolic steroids, which are often associated with androgenic and hepatotoxic side effects, phytoecdysteroids exhibit a superior safety profile in long-term experimental and clinical studies (Badawy et al. 2021; Bajguz et al. 2015). This combination of efficacy and safety makes them promising candidates for clinical use in wound healing, bone repair, and tissue regeneration. The following sections examine the molecular mechanisms, experimental evidence, and clinical perspectives underlying the use of phytoecdysteroids in regenerative medicine, with particular focus on wound healing, fracture repair, and restoration of tissue integrity. The regenerative effects of phytoecdysteroids arise from the coordinated activation of multiple molecular pathways that support cell proliferation, tissue repair, and functional recovery. These include stimulation of protein and collagen synthesis, enhancement of mitochondrial energy metabolism, antioxidant, and membrane-stabilizing actions, modulation of metabolic and hormonal signaling, and regulation of angiogenesis and osteogenesis (Wang et al. 2024; Li et al. 2025).

PEs exert their regenerative effects through a combination of complementary mechanisms. One of the earliest and most extensively studied actions of these compounds is stimulation of protein and collagen biosynthesis. Administration of ecdysterone and turkesterone increases the incorporation of labeled amino acids into cellular proteins in hepatocytes, cardiomyocytes, and fibroblasts. In wound-healing models, this response promotes fibroblast proliferation and enhances collagen deposition, thereby providing a stable extracellular matrix framework that supports tissue remodeling and bone regeneration. At the same time, phytoecdysteroids preserve mitochondrial function and ensure an adequate energy supply during tissue repair. Phytoecdysteroids support oxidative phosphorylation and increase the adenosine diphosphate-to-oxygen (ADP/O) ratio, a key indicator of mitochondrial coupling efficiency, in stressed liver and muscle mitochondria. They stabilize the activity of essential mitochondrial enzymes, including cytochrome c oxidase, succinate dehydrogenase, and glutamate dehydrogenase, thereby sustaining adenosine triphosphate (ATP) production required for active protein synthesis and cell proliferation. In parallel, phytoecdysteroids exhibit pronounced antioxidant and membrane-stabilizing properties. By reducing malondialdehyde levels and suppressing lipid peroxidation, they protect cellular membranes, preserve mitochondrial integrity, and maintain the viability of newly formed cells. These effects collectively protect fibroblasts, osteoblasts, hepatocytes, and cardiomyocytes from apoptosis and necrosis under hypoxic or oxidative stress. Together, these coordinated actions create optimal metabolic and structural conditions for efficient tissue repair and accelerated regeneration.

6.1 Hormonal and Metabolic Modulation

Unlike synthetic anabolic steroids, phytoecdysteroids exert a balanced regulatory effect on endocrine and metabolic pathways. They normalize carbohydrate and lipid metabolism, improve nitrogen balance, and support calcium–phosphorus homeostasis, all of which are essential for effective bone regeneration and tissue repair (Devi et al. 2023; Parr et al. 2015). Through their involvement in bone metabolism, phytoecdysteroids enhance osteoblast activity and promote mineralization, resulting in accelerated fracture healing and improved biomechanical strength. Growing evidence also indicates that phytoecdysteroids stimulate angiogenesis, thereby improving oxygen and nutrient delivery to regenerating tissues. Increased capillary density and enhanced microcirculatory blood flow have been documented in ischemia and wound-healing models treated with ecdysterone (Omanakuttan et al. 2016; Chen et al. 2008; Vágvölgyi et al. 2024). Adequate vascularization is a key requirement for successful tissue repair, particularly during skin wound healing and bone regeneration, where sustained blood supply is essential. The regenerative effects of phytoecdysteroids can be explained by an integrated, multi-level mechanism. First, stimulation of protein and collagen synthesis enhances fibroblast and osteoblast activity, thereby accelerating structural tissue repair. Second, preservation of mitochondrial function and optimization of cellular energy metabolism ensure sufficient ATP production to support reparative and proliferative processes. Third, antioxidant and membrane-stabilizing actions protect cells from oxidative and mechanical damage, creating a favorable microenvironment for regeneration. In addition, regulatory effects on hormonal and metabolic balance help maintain calcium–phosphorus homeostasis and systemic metabolic stability. Finally, induction of angiogenesis ensures adequate perfusion of damaged and recovering tissues. Together, these complementary mechanisms provide a solid molecular basis for the regenerative effects of phytoecdysteroids observed in experimental and clinical studies.

6.2 Effects of Phytoecdysteroids on Erythropoiesis Under Anemic Conditions

In describing the anabolic effects of phytoecdysteroids in mammals, we have previously highlighted their pronounced stimulation of erythropoiesis in intact animals. Experimental evidence shows that 20-hydroxyecdysone enhances hematopoietic activity, promotes regenerative processes, and increases both erythrocyte number and hemoglobin concentration in the blood. Stimulation of blood cell regeneration is a characteristic feature of anabolic agents. However, unlike synthetic anabolic steroids, phytoecdysteroids do not exert undesirable effects on androgen-dependent organs and tissues, indicating a fundamentally different and more favorable biological profile. At the same time, phytoecdysteroids increase total blood protein content, predominantly through an elevation of albumin levels, and promote an increase in erythrocyte number and hemoglobin concentration. In rats with experimental hemolytic anemia induced by a 5% phenylhydrazine solution, administration of phytoecdysteroids (ecdysone and 2-deoxyecdysterone) at a dose of 5 mg/kg increased erythrocyte counts by 6.9–8.1% on day 7 and by 11.9–13.8% on day 14 of the experiment (Syrov 1994).

Stimulation of erythropoiesis, enhancement of regenerative capacity, and elevation of erythrocyte and hemoglobin levels were also observed following administration of 20-hydroxyecdysterone. In addition, phytoecdysteroids were shown to increase the activity of protective components of the blood system, including lymphocytes and neutrophils, and to enhance phagocytic activity. Acetate derivatives and other modified forms of 20-hydroxyecdysterone have also been shown to stimulate DNA biosynthesis in human and animal lymphocytes activated by polyclonal mitogens. In addition, phytoecdysteroids exhibit preventive and therapeutic effects in experimental models of anemia and leukopenia, as well as in disorders of reproductive function and climacteric syndrome associated with age-related dysregulation of the reproductive cycle (Vasiliev 2012; Sidorova 2014). This chapter presents evidence demonstrating the high efficacy of phytoecdysteroids in experimental phenylhydrazine-induced anemia. In rats with phenylhydrazine-induced anemia, recovery of erythrocyte count and hemoglobin concentration occurred relatively slowly in the control group. Erythrocyte numbers declined from 5.7 ± 0.19 to 2.7 ± 0.32 million/mm^3, while hemoglobin levels decreased from 12.5 ± 0.47 to 6.2 ± 0.35 g%.

Only after 2 weeks did these parameters begin to approach baseline values; however, even at this time point, erythrocyte counts remained approximately 15% lower than pre-intoxication levels. In contrast, treatment with ecdysterone markedly accelerated hematological recovery. An increase in erythrocyte numbers was evident as early as day 6 of treatment and proceeded significantly faster than in control animals. From day 9 onward, erythrocyte counts no longer differed from baseline values, and complete normalization was achieved by the end of the experiment. A similar but even more pronounced effect was observed for hemoglobin concentration. By day 9, hemoglobin levels had nearly returned to baseline, while on days 12 and 15 they exceeded initial values, reaching 13.9 ± 0.47 and 14.7 ± 0.40 g%, respectively. These corresponded to increases of 11.2% ($p > 0.05$) and 17.6% ($p < 0.05$) relative to the baseline hemoglobin concentration of 12.5 ± 0.47 g%. Regeneration of erythrocytes under treatment with sileneoside followed a pattern similar to that observed with ecdysterone. However, both erythrocyte counts and hemoglobin levels were slightly higher in anemic rats treated with sileneoside A, indicating a somewhat more pronounced stimulatory effect on hematopoietic recovery (Fig. 6.1).

As a result, by the end of the experiment, the erythrocyte count in treated animals exceeded the baseline level by 7.1%, increasing from 5.6 ± 0.39 million/mm^3 at the beginning to 6.0 ± 0.27 million/mm^3 at the end of the experiment. This value was 23.2% higher than that observed in the control group ($p < 0.05$). At the same time point, hemoglobin concentration in rats treated with sileneoside A reached 15.6 ± 0.69 g%, exceeding the initial level by 21.9% ($p < 0.05$). In anemic rats treated with turkesterone, stimulation of erythropoiesis was even more pronounced, closely approaching the response observed in intact animals and clearly exceeding

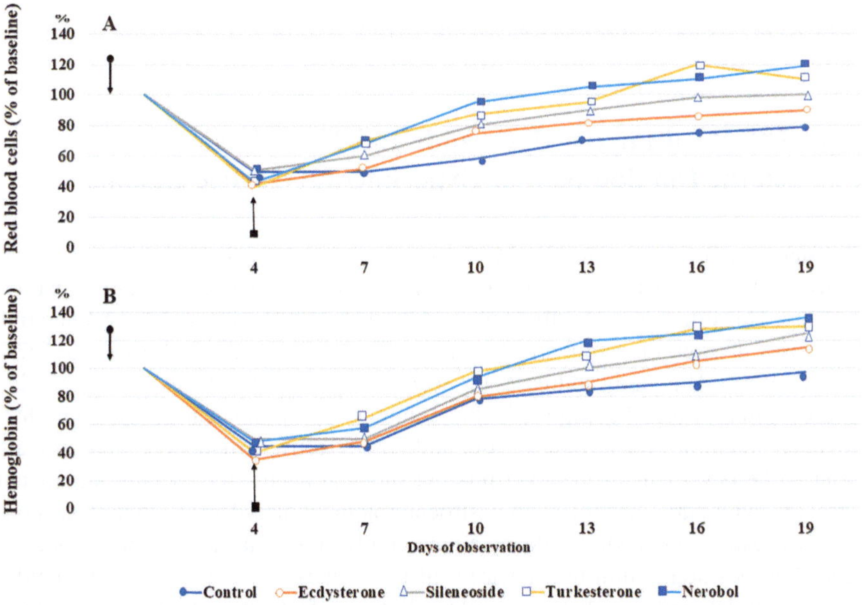

Fig. 6.1 Effect of phytoecdysteroids and Nerobol on red blood cell regeneration in rats with phenylhydrazine-induced anemia ($n = 6$). Arrows indicate the time of phenylhydrazine administration and the onset of treatment with the studied drugs, respectively

the effects of ecdysterone and sileneoside A. An increase in erythrocyte count was evident as early as day 3 of treatment and persisted throughout the entire experimental period. By the end of the experiment, erythrocyte levels reached 6.7 ± 0.26 million/mm^3, exceeding baseline values by 24.7% ($p < 0.05$) and surpassing those of the control group by 37% ($p < 0.05$). Changes in hemoglobin concentration were even more marked. By day 3 of turkesterone treatment, hemoglobin levels had already increased markedly, reaching 10.4 ± 0.55 g%. In contrast, hemoglobin concentrations in rats treated with ecdysterone or sileneoside A differed only slightly from control values. This value was clearly higher than in the control group and did not differ significantly from the baseline level of 11.8 ± 0.52 g%. From day 6 onward, mean hemoglobin content in this group (12.3 ± 0.65 g%) slightly exceeded the intact level and remained the highest among all treated groups until the end of the experiment. On days 9, 12, and 15, hemoglobin concentrations in turkesterone-treated rats exceeded baseline values by 22.0%, 35.6%, and 33.0%, respectively ($p < 0.05$ for all time points) and were significantly higher than those observed in the control group. Ecdysterone and sileneoside A were less effective than the synthetic anabolic steroid Nerobol in promoting recovery of erythrocyte number and hemoglobin concentration. In contrast, turkesterone produced a response comparable to that of Nerobol (Fig. 6.1). Thus, phytoecdysteroids administered to animals with phenylhydrazine-induced hematotoxic anemia markedly accelerated the regeneration of red blood cells compared with untreated controls. Within the investigated series, the erythropoiesis-stimulating activity of phytoecdysteroids in anemic rats increased in the order ecdysterone < sileneoside A < turkesterone, with turkesterone exhibiting efficacy comparable to that of Nerobol.

6.3 Effects of Phytoecdysteroids and Their Analogues on the Development of Experimental Gastric Lesions

Substances that improve metabolic processes, particularly those that activate protein oxidation in organs and tissues, often also exhibit pronounced anti-inflammatory activity. This effect is generally explained by their ability to stimulate regenerative processes at sites of pathological damage (Eming et al. 2017; Uzbekova 1974; Kasich 1963). This study addresses the search for new plant sources of ecdysteroids and ecdysteroid-based compounds that combine high biological activity with low toxicity, suitability for long-term use, and cost-effective production. The ultimate goal is the development of novel pharmacological agents for socially significant diseases, with particular focus on gastroenterological disorders. Among these conditions, gastric and duodenal ulcers remain highly prevalent and continue to occupy a leading position within the spectrum of gastrointestinal diseases. Global epidemiological data indicate that peptic ulcer disease affects up to 16% of the population, predominantly individuals of working age. Despite substantial advances in medical technologies and therapeutic approaches, the risk of severe complications and hospital mortality remains significant. This is associated both with the frequently latent course of ulcer disease and with the widespread, often uncontrolled, use of non-steroidal anti-inflammatory drugs. In this context, the search for safe and effective gastroprotective agents remains highly relevant.

Long-term studies involved chemical and pharmacological screening of plant species from the genera *Silene* and *Lychnis* (family Caryophyllaceae), which are known producers of phytoecdysteroids. Available literature summarizes the occurrence of secondary metabolites in these plants, their traditional medicinal uses, pharmacological properties, and chemical composition. Chromatographic analysis of seed extracts identified new sources of biologically valuable compounds, leading to the structural elucidation of seven previously unknown ecdysteroids. Subsequently, the pharmacological activity of extracts, isolated fractions, and individual compounds obtained from several *Silene* and *Lychnis* species was evaluated. Particular attention was given to the anti-ulcer activity of extracts from ecdysteroid-containing plants, including *Lychnis chalcedonica* L., *Silene viridiflora* L., and *Silene frivaldszkyana* Hampe, using experimental models of neurogenic and aspirin-induced ulcerogenesis. The results demonstrated a clear and reproducible gastroprotective effect of *S. viridiflora* and *L. chalcedonica* extracts, comparable to that of famotidine. Repeated intragastric administration significantly reduced the formation of ulcerative lesions in all applied models. These findings indicate that such phytocomplexes represent promising sources for the development of novel gastroenterological drugs (Zibareva et al. 2021; Krylova et al. 2014). Since many of the investigated phytoecdysteroids demonstrated detoxifying effects comparable to those of known anabolic agents, their pharmacological activity was further assessed using additional experimental models of gastric injury. Phytoecdysteroids significantly prevented the formation of destructive lesions in the secretory region of the stomach in reserpine-treated mice (Table 6.1). Ecdysterone reduced the number of

Table 6.1 Effect of phytoecdysteroids and Nerobol on reserpine-induced gastric ulceration in mice (M ± m, n = 10)

	Gastric lesions (number per animal)			
Experimental conditions	Punctate	Focal	Liner lesions	Total
Control (reserpine)	10,7 ± 0,95	2,3 ± 0,30	1,7 ± 0,23	14,7 ± 1,34
Ecdysterone	6,9 ± 2,1*	1,2 ± 0,32*	1,0 ± 0,20*	9,1 ± 1,16*
Viticosterone E	7,0 ± 1,9*	1,1 ± 0,40*	1,1 ± 0,34*	9,2 ± 1,19*
2-Deoxy ecdysterone	8,4 ± 1,04	1,6 ± 0,47	1,3 ± 0,30	11,3 ± 1,74
2-Deoxy-α-ecdysone	10,5 ± 0,83	2,0 ± 0,36	1,5 ± 0,37	14,0 ± 0,81
Cyasterone	6,6 ± 0,70*	1,4 ± 0,30*	0,9 ± 0,22*	8,9 ± 0,90*
Turkesterone	6,0 ± 0,49*	0,9 ± 0,34*	0,8 ± 0,24*	7,7 ± 0,88*
Integristerone A	9,4 ± 0,96	1,6 ± 0,40	1,2 ± 0,38	12,2 ± 1,23
24 (28)-Dehydromakisterone	7,7 ± 0,61*	1,1 ± 0,27*	0,9 ± 0,21*	9,7 ± 0,83*
Nerobol	7,8 ± 0,89*	1,3 ± 0,33*	1,1 ± 0,34	10,2 ± 1,34*

Note: *reliable values of the indicated values ($p < 0.05$) were noted in relation to the control gastric lesions per animal by 38.1%. A comparable effect was observed for 24(28)-deyhdromakisterone, which reduced lesion number by 34.1%. The anti-ulcer activity of cyasterone and viticosterone E reached 39.5% and 37.4%, respectively. The most pronounced gastroprotective effect was observed for turkesterone. Under its influence, the number of gastric lesions decreased by 47.6% compared with control animals and by 15.4% compared with the ecdysterone-treated group.

In addition, turkesterone produced the most substantial reduction (53%) in the proportion of severe linear lesions. In contrast, integristerone A exhibited a considerably weaker effect, reducing gastric lesion formation by only 17.1%. Compounds such as 2-deoxyecdysterone showed moderate anti-ulcer activity (23.2%), whereas 2-deoxy-α-ecdysone displayed the weakest protective effect against reserpine-induced gastric injury. In these experiments, the anabolic steroid Nerobol reduced ulcer formation by 30.6%. Convincing evidence for the anti-ulcer activity of phytoecdysteroids was also obtained in a model of severe, slowly healing chronic gastric ulcer induced by subserosal injection of a 10% acetic acid solution into the glandular region of the anterior stomach wall. Gastric lesions were detected in 100% of control rats on day 10 of the experiment. In seven animals, macroscopically visible round ulcers with diameters of 0.4–0.5 cm and depths of 0.3–0.4 cm were observed, accompanied by pronounced perifocal inflammation. In the remaining three animals, ulcer perforation was detected (Table 6.2).

After an additional 10 days, the majority of rats in this group continued to exhibit round gastric ulcers with diameters of 0.3–0.4 cm and depths of 0.2–0.3 cm. Only two animals displayed very shallow ulcers with a diameter of approximately 0.1 cm. By the end of the experiment, five of the sacrificed rats still presented with small residual ulcers measuring about 0.2 cm in diameter and 0.1 cm in depth. In the remaining five animals, the ulcer sites had been replaced by coarse scar tissue, and the folds of the gastric mucosa were markedly flattened. In contrast, a distinctly different pattern of ulcer healing was observed in the experimental groups in which the investigated phytoecdysteroids were administered immediately after surgery. On day 10 of the experiment, among rats treated with ecdysterone, turkesterone, and cyasterone (10 animals examined in each group), complete epithelialization of gastric mucosal defects was observed in two, three, and one animals, respectively. In these treatment groups, complete healing of gastric ulcers was achieved after 35.7, 33.3, and 32.5 days, respectively.

The wound-healing effect of the investigated phytoecdysteroids was generally not inferior to that of the anabolic steroid Nerobol, and in several cases, ecdysterone and cyasterone even exhibited a slightly more pronounced therapeutic effect (Table 6.3).

The activation of regenerative processes in the skin, manifested as accelerated wound closure in rats treated with phytoecdysteroids and Nerobol, appears to be associated with optimization of metabolic processes in the skin as well as in other organs and tissues under the influence of these compounds. According to literature data, substances with anabolic activity enhance DNA and RNA biosynthesis in the skin and increase its protein content (Uglyanitsa 1980). In our experiments, morphological examination of the skin following complete healing of wound defects

Table 6.2 Effect of phytoecdysteroids and Nerobol on experimental chronic gastric ulcers in rats (macroscopic evaluation)

Experimental groups	Observation period (days)								
	10th			20th			30th		
	n	n_1	% of animals with affected	n	n_1	% of animals with affected	n	n_1	% of animals with affected
Control	10	10	100	10	10	10	10	10	100
EChU+ecdysterone	10	8	80	10	6	60	10	3	30
EChU+cyasterone	10	9	90	10	6	60	10	4	40
EChU+turkesterone	10	7	70	10	5	50	10	2	20
EChU+Nerobol	10	8	80	10	5	50	10	2	20

Note: EChU—an experimental chronic ulcer; n—the total number of decapitated animals; n, n_1 - number of animals with lesions

Table 6.3 Effect of phytoecdysteroids and Nerobol on wound area reduction and healing time (M ± m, n = 8–10)

Experimental groups	Average wound area (cm^2)					Healing time	
	Day 7	Day14	Day 21	Day 28	Day 35	In days	The effect in %
Control	2,6 ± 0,18	1,8 ± 0,18	1,3 ± 0,14	0,7 ± 0,10	0,4 ± 0,08	41,1 ± 0,7	–
Integristerone A	2,4 ± 0,17	1,9 ± 0,14	1,2 ± 0,09	0,5 ± 0,10	0,2 ± 0,05*	35,7 ± 1,2*	13,1
Ecdysterone	2,5 ± 0,16	1,6 ± 0,13	0,9 ± 0,11*	0,4 ± 0,06*	0,1 ± 0,03*	33,3 ± 1,0*	19,0
Cyasterone	2,4 ± 0,14	11,5 ± 0,11	0,9 ± 0,09*	0,4 ± 0,08*	0,1 ± 0,05*	32,5 ± 1,1*	20,9
Nerobol	2,5 ± 0,17	1,9 ± 0,15	1,0 ± 0,10	0,3 ± 0,05*	0,2 ± 0,05*	34,3 ± 1,0*	16,6

revealed no significant structural abnormalities. In the gastric mucosa, healed round ulcers with scar formation were observed in two rats in each group treated with ecdysterone and turkesterone, and in one rat in the cyasterone-treated group. In the remaining animals of these groups, gastric ulcers were still present; however, they were not accompanied by signs of perifocal inflammation.

The diameter of these lesions ranged from 0.2 to 0.3 cm, and their depth from 0.1 to 0.2 cm. In rats that received the tested compounds for 20 days, an even higher proportion of animals exhibited no visible macroscopic changes in the gastric mucosa. In those animals in which ulcers were still detected, the lesions were small, measuring approximately 0.1–0.2 cm in diameter with a comparable depth. In experimental groups sacrificed on day 30 after initiation of treatment with ecdysterone, turkesterone, and cyasterone, healed ulcers were identified in three, two, and four animals, respectively, while the remaining animals showed no marked disruption of gastric mucosal integrity. Notably, the overall effects of phytoecdysteroids were nearly comparable to those of Nerobol (Table 6.2). This observation was further confirmed in an additional series of experiments in which phytoecdysteroids and the anabolic steroid were compared using models of atophan- and butadione-induced

Table 6.4 Effect of phytoecdysteroids and Nerobol on atophan- and butadione-induced gastric mucosal injury in rats (M ± m, n = 10)

Experimental groups	Gastric lesions (number per animal) Small lesions (d < 1.5 mm)	Medium lesions (d < 1.5–2.5 mm)	Large lesions (d < 2.5–5 mm)	Liner lesions	Total lesions
Control (atophan-induced)	21,3 ± 1,26	6,1 ± 0,50	7,6 ± 0,54	2,8 ± 0,63	37,8 ± 0,05
Ecdysterone	12,4 ± 0,96*	2,2 ± 0,42*	1,7 ± 0,52*	0,7 ± 0,26*	17,0 ± 1,44*
Turkesterone	11,4 ± 0,46*	2,1 ± 0,45*	1,6 ± 0,47*	0,6 ± 0,26*	15,7 ± 1,33*
Nerobol	11,7 ± 1,14*	2,0 ± 0,54*	1,6 ± 0,58*	1,3 ± 0,52	16,6 ± 2,43*
Control (butadione-induced)	26,4 ± 2,2	5,4 ± 0,59	5,6 ± 0,41	2,4 ± 0,29	39,8 ± 2,4
Ecdysterone	15,5 ± 1,63*	2,2 ± 0,46*	2,0 ± 0,47*	1,1 ± 0,35*	20,8 ± 2,7*
Turkesterone	12,3 ± 1,6*	1,6 ± 0,43*	1,1 ± 0,48*	1,2 ± 0,44*	16,2 ± 2,5*
Nerobol	18,3 ± 1,7*	2,1 ± 0,41*	2,1 ± 0,60*	1,5 ± 0,54	24,0 ± 3,0*

Note: *reliable values of the indicated values ($p < 0.05$) were noted in relation to the control

gastric injury, both of which are characterized by toxic–dystrophic lesions of the stomach. Phytoecdysteroids, similarly to Nerobol, significantly reduced the total number of gastric lesions in these experimental settings (Table 6.4).

In the atophan-induced injury model, the protective effects of ecdysterone, turkesterone, and Nerobol amounted to 55.1%, 58.5%, and 56.1%, respectively. In the butadion-induced model, the corresponding protective effects were 47.7%, 59.3%, and 39.7%. Taken together, these data demonstrate that phytoecdysteroids act as effective anti-ulcer agents, with activity comparable to that of the anabolic drug Nerobol, based on macroscopic evaluation of gastric lesions. These conclusions are further supported by similarities observed in certain aspects of their proposed mechanisms of anti-ulcer action. In a subset of animals treated with butadione, the content of nucleic acids in gastric tissues was analyzed. Extensive gastric injury was associated with a decrease in RNA content by 17.8% and DNA content by 10.2% (Fig. 6.2). In contrast, in animals pretreated with phytoecdysteroids or Nerobol, RNA and DNA levels did not decrease and, in some cases, even exceeded those observed in intact control animals.

The results of this study demonstrate that phytoecdysteroids also hold strong potential as wound-healing agents. From the first weeks of treatment, animals in the experimental groups showed a clear reduction in edema and active formation of granulation tissue. In parallel with connective tissue growth and increased vascularization of the wound bed, marked epithelialization of the defect was observed. Based on clinical observation of the wound-healing process and planimetric analysis (Table 6.4), it can be concluded that phytoecdysteroids, which display marked protein-anabolic activity in other experimental models, also promoted accelerated wound closure under these conditions. Treatment with integristerone A resulted in a statistically significant reduction in mean wound area only by day 35 compared with

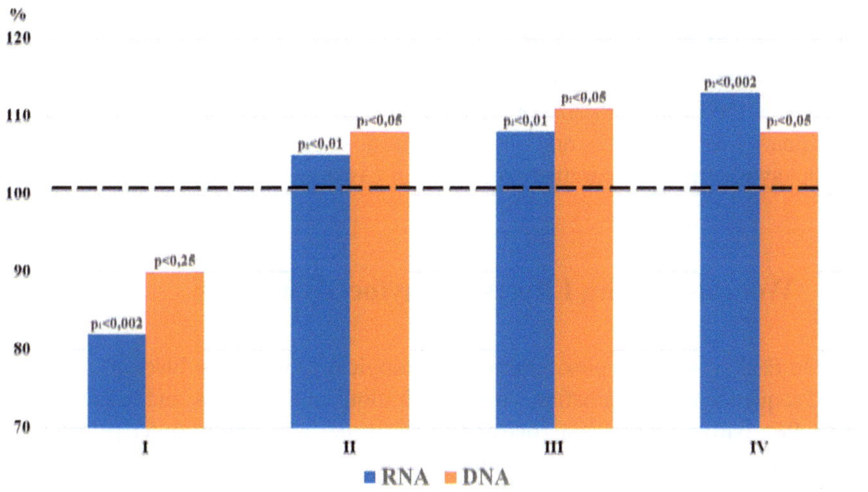

Fig. 6.2 Effect of ecdysterone (II), turkesterone (III), and Nerobol (IV) on nucleic acid content in rat gastric tissue (n = 6) after administration of butadione (I) (expressed as % of intact control). p_1 indicates statistical significance versus the intact control, and p_2 indicates statistical significance versus butadione-treated animals

control animals. In contrast, rats treated with ecdysterone and cyasterone exhibited a significant effect already by day 21. At this time point, the mean skin wound area in animals receiving ecdysterone and cyasterone was 30.8% smaller than that observed in the control group. By day 28 of observation, this difference became even more pronounced, reaching 42.9%. It should be noted that on day 28, the wound defect in rats treated with ecdysterone accounted for only 16.0% of the initial wound area, and in the cyasterone-treated group, for 16.6%, whereas in control animals, the wound surface still represented 26.9% of the area created at the time of injury.

A similar trend was observed on day 35: in control animals, the wound area corresponded to 15.4% of the initial size; in the integristerone A-treated group, it was reduced to 8.3%; while in rats treated with ecdysterone and cyasterone, it decreased further to 4.0% and 4.2%, respectively. Complete wound healing in the control group was observed only by day 41 of the experiment. Treatment with integristerone A, ecdysterone, and cyasterone was accompanied by favorable metabolic changes in the skin, including normalization of carbohydrate and lipid metabolism. Specifically, in the skin of rats treated with ecdysterone and Nerobol, glycogen content was higher than in control animals by 27.3% ($p < 0.05$) and 20.6% ($p < 0.05$), respectively. At the same time, skin cholesterol levels decreased by 18.1% and 17.2% ($p < 0.05$), while phospholipid content—an important determinant of skin elasticity—increased by 20.3% and 22.3% ($p < 0.05$) compared with control values. Overall, phytoecdysteroids significantly accelerated the healing of clean incisional skin wounds. When the time required for complete wound closure was compared with that in control animals, the therapeutic effects of ecdysterone and cyasterone were even more pronounced than those of the reference anabolic steroid Nerobol (Table 6.4). Thus, the experimental data presented in this section indicate that phytoecdysteroids, similarly to Nerobol, exert a pronounced anti-ulcer effect irrespective of the ulcerogenic factor applied. This gastroprotective activity appears to be primarily driven by restoration of nucleic acid synthesis and normalization of plastic (synthetic and reparative) processes in gastric tissues.

6.4 Wound-Healing Effects of Phytoecdysteroids

Wound healing is a complex, multistage biological process that involves a coordinated sequence of inflammation, cell proliferation, angiogenesis, and tissue remodeling. PEs, particularly ecdysterone and turkesterone, have demonstrated pronounced wound-healing activity by accelerating fibroblast proliferation, collagen deposition, vascularization, and epithelial regeneration (Khushbaktova et al. 2012). These effects are mediated through a combination of molecular mechanisms, including stimulation of protein biosynthesis, antioxidant protection, and optimization of metabolic processes (Mukhin et al. 1988; Ermishina 1986; Kholodova 1979; Reitman and Frankel 1957; Darmogray et al. 1996).

In official medical practice, ecdysteroid-containing natural formulations are already used for the treatment of wounds and ulcers, as well as for the management of burns (Darmogray et al. 1996; Meybeck et al. 1997; Lafont et al. 2021). Experimental studies have shown that *Rhaponticum carthamoides*, a plant widely regarded in traditional Eastern medicine as a source of physical strength and vitality, synthesizes more than 50 individual phytoecdysteroids. These include highly active compounds such as rapisterone D, dachryhainansterone, 24(28)-dehydromakisterone A, 22-benzoate-ecdysterone, 5-deoxy-kaladasterone, ecdysterone (20-hydroxyecdysone), polypodine B, ajugasterone C, makisterone A and C, integristerone, leuzeasterone, carthamosterone, coronatasterone, and others. Preparations derived from *R. carthamoides* are used to enhance physical endurance, stimulate sexual activity, exert tonic and restorative effects, accelerate wound healing, display antitumor activity, and promote anabolic processes (Timofeev 2005). In experimental studies investigating the wound-healing activity of phytoecdysteroids, Nerobol was employed as a reference compound, as in earlier investigations. Anabolic steroid preparations are known to be widely used for these purposes, particularly in cases where rapid epithelialization of skin wounds is required (Mukhin et al. 1988; Uglyanitsa 1980).

In addition, a retrospective analysis of plant-based remedies used for the treatment and prevention of dental and periodontal diseases was conducted (Mosaddad et al. 2023). These studies demonstrated the promising therapeutic potential of phytoecdysteroids, molting and metamorphosis hormones of insects and other arthropods, which, although not functioning as hormones in warm-blooded organisms, exert a wide spectrum of biological and pharmacological effects. A comprehensive review of patents and experimental studies highlighted their high efficacy, fundamental biological significance, and relevance for the development of medicinal and pharmaceutical products. Importantly, phytoecdysteroids were shown to be practically non-toxic and did not induce adverse effects even during long-term administration, making them attractive candidates for the development of import-substituting therapeutic agents. Consequently, the need for further exploration of plant and marine sources of phytoecdysteroids has been emphasized. Studies conducted by university research groups and dental clinics demonstrated beneficial effects of phytoecdysteroids and phytoecdysteroid-containing plant mixtures in periodontal and dental diseases, with additional bioactive compounds from other plant families modulating and enhancing their therapeutic efficacy (Darmogray et al. 2021).

Overall, the accumulated research revealed a wide range of pronounced biological effects induced by both individual phytoecdysteroids and their combined preparations. In particular, a strong wound-healing activity was consistently observed across various models of tissue damage, including thermal and chemical burns, cryogenic injuries, trophic ulcers, and pyodermal lesions. In these models, treatment resulted in the formation of flat scars without signs of keloid formation or hypertrophic scarring, underscoring the high regenerative potential of phytoecdysteroids (Darmogray et al. 2004).

In the combined treatment of chemical esophageal burns, administration of an oil-based phytoecdysteroid formulation together with guidewire-assisted esophageal bougienage resulted in clinically meaningful benefits. Within 5 days of therapy initiation, patients showed marked improvement in general condition, and dysphagia resolved completely by day five. The required number of bougienage procedures was reduced, leading to a shortening of hospital stay by an average of four bed-days compared with standard treatment. In patients with diabetic ulcers, topical application of a phytoecdysteroid-based ointment increased epithelial regeneration by an average of 19%. Cytological analysis of surface imprints revealed a 13.7% rise in cellular elements, including fibroblasts, leukocytes, lymphocytes, and macrophages. Subjectively, 83% of patients reported a reduction in pain intensity. In patients with purulent diseases of the lungs and pleura, evaluation of the integral index of regulatory system activity (PARS) demonstrated that oral administration of an alcoholic phytoecdysteroid tincture reduced PARS values from 6.0, reflecting severe functional strain, to 4.8, corresponding to moderate strain. This improvement was accompanied by faster subjective recovery and more rapid normalization of radiographic findings. In contrast, patients receiving conventional therapy exhibited a smaller PARS reduction, from 6.0 to 5.4, indicating persistent regulatory system stress (Mikheev et al. 2004).

Clinical observations also support the use of phytoecdysteroids in patients with gastric and duodenal ulcer disease complicated by gastrointestinal bleeding and ulcer perforation. In one group of patients receiving an oral formulation consisting of hawthorn tincture (50.0), calendula tincture (50.0), and ecdysterone (0.01), administered at a dose of 65–70 drops three times daily, a statistically significant increase in erythrocyte count and hemoglobin concentration was observed by day five of treatment. Evaluation of PARS in patients with perforated ulcers revealed that, at hospital admission, both study groups exhibited moderate functional strain of regulatory systems. By day ten of treatment with an ecdysterone-containing solution, PARS values decreased from 4.5 to 3.4, whereas in the control group, they declined from 4.7 to 3.8. Morphometric analysis further demonstrated that inclusion of phytoecdysteroids in therapeutic regimens resulted in a significant reduction in segmented neutrophils and a concomitant increase in fibroblast numbers within the ulcer zone. This cellular shift reflects stimulation of regenerative processes through transition from the exudative phase to the proliferative phase of inflammation (Landén et al. 2016).

Additional evidence of regenerative activity was obtained in studies evaluating Serpisten ointment, derived from phytoecdysteroids isolated from *Serratula coronata* L. In a model of linear aseptic skin wounds, Serpisten exerted a pronounced regenerative effect comparable to that of reference medications and was manifested by a statistically significant increase in scar tensile strength on days 5 and 7 compared with control animals ($p \leq 0.05$). In experimental models of thermal burns, pharmaceutical formulations containing Serpisten reduced eschar rejection time by approximately 5 days, prevented wound infection and animal mortality, and increased the frequency of complete wound healing by two- to threefold relative to controls. Importantly, topical formulations containing Serpisten were not inferior in efficacy to widely used international products, including Bepanthen ointment and Actovegin cream (Lipin 2015). Elucidating the mechanisms underlying ecdysteroid activity is a key focus of modern biomedical research. When integrated with contemporary biotesting strategies and three-dimensional computational modeling, these studies help define ligand–receptor interactions within structure–activity relationships, enabling prediction of optimal molecular structures and conformations for the rational design of highly active compounds. In vivo application of non-purified extracts from ecdysteroid-containing plants allows simultaneous genomic and non-genomic effects, together with the action of other primary and secondary plant metabolites. Such synergistic interactions support the development of novel pharmaceutical formulations based on whole medicinal plant materials, characterized by high adaptogenic and pharmacological efficacy at low doses. On this basis, a new class of pharmaceuticals and dietary supplements has been developed, including Bioinfuzin, BCL-Phyto, Lipolite, and Rapontik™, produced from selected aerial biomass of *Rhaponticum carthamoides* cultivated under agropopulation conditions using technologies designed to enhance the biosynthesis and preservation of highly active phytoecdysteroids during raw material processing (Timofeev 2005).

The conducted studies demonstrated that PEs actively interact with membrane phospholipids and are readily incorporated into liposomes and highly dispersed emulsions. These compounds exhibit pronounced membranotropic properties, manifested by an increase in erythrocyte membrane microviscosity and enhanced resistance to osmotic, thermal, and acid-induced hemolysis. Phytoecdysteroids induce minor conformational rearrangements on the protein surface of erythrocyte membranes and exert a protective effect against both ultraviolet-induced hemolysis and hemolysis mediated by lipid peroxidation. In this context, they display significant antioxidant activity through effective inhibition of lipid peroxidation (LPO). In addition to these membrane-related effects, phytoecdysteroids increase hemoglobin concentration and the color index, enhance neutrophil phagocytic activity, elevate total protein and γ-globulin levels, and reduce elevated cholesterol concentrations. They promote the accumulation of hepatic glycogen, exhibit pronounced hepatoprotective properties, and stimulate protein biosynthesis.

Furthermore, phytoecdysteroids are capable of activating multiple components of the immune system, including mechanisms of innate resistance as well as antimicrobial and antitoxic immunity, without exerting direct suppressive effects on microbial viability in vitro. Additional biological activities include nootropic effects and a pronounced wound-healing and regenerative capacity in tissue damage caused by chemical, thermal, cryogenic, and mechanical factors (Raza et al. 2024). Ecdysten, a phytoecdysteroid-based preparation isolated from *Rhaponticum carthamoides*, has demonstrated notable clinical potential in the treatment of gastric and duodenal ulcer disease. In patients receiving Ecdysten as part of combined therapeutic regimens together with antacids and antispasmodics over a 24-day treatment period, accelerated ulcer scarring was observed along with improvement of laboratory parameters, including increased hemoglobin concentration and erythrocyte counts (Raza et al. 2024). Ecdysten is widely recognized for its strong stimulatory effect on regenerative processes in the organism (Syrov 2000, 2009, 2022). Preliminary phytochemical investigations of *Stachys hissarica* confirmed the presence of phytoecdysteroids and enabled the isolation and identification of its principal ecdysteroids, including 20-hydroxyecdysone (0.33% of dry plant weight), polypodine B (0.22%), 2-deoxy-20-hydroxyecdysone (0.20%), integristerone A (0.20%), and 2-deoxyecdysone (0.01%).

The isolated individual ecdysteroids were structurally characterized and evaluated for biological activity. A total ecdysteroid preparation (TEP) obtained from *S. hissarica* was subsequently tested in a series of biological experiments designed to assess its wound-healing efficacy. The experimental results demonstrated that administration of TEP significantly accelerated the healing of linear incisional wounds in rats compared with untreated control animals. By day 7 following TEP administration, complete detachment of the scab formed within 1–2 days after injury was observed, revealing a well-organized and mechanically stable scar. In contrast, in control animals, complete scab separation occurred only by day 12, and the resulting scar tissue was markedly less robust. Mechanical testing of scar tissue (Table 6.5) further confirmed the superior regenerative effect of TEP. The tensile force required to rupture the scar in control animals was 382.4 g, whereas in rats treated with TEP from *S. hissarica*, scar tensile strength increased by 48.7%. Notably, the wound-healing activity of the total ecdysteroid preparation exceeded that of methyluracil, a widely used reference compound known for its regenerative properties.

Under these experimental conditions, phytoecdysteroids markedly activated all key processes involved in skin wound healing. On day 10 of observation in rats with alloxan-induced diabetes, the force required to rupture the postoperative scar was 286.4 g in untreated animals, whereas treatment with the total ecdysteroid preparation (TEP) increased this value to 520.6 g. Thus, the wound-healing effect of TEP derived from *Stachys hissarica* was substantially superior to that of the reference drug methyluracil and was even more pronounced than the effect observed in intact animals (Table 6.5). These results indicate a high regenerative potential of the ecdysteroid-rich preparation obtained from *S. hissarica*. The present study therefore provides experimental confirmation of the pronounced wound-healing activity of an ecdysteroid-containing preparation from *S. hissarica*. In parallel, detailed chemical and spectroscopic analyses were performed to identify the principal bioactive constituents responsible for the observed synergistic regenerative effect. The major components were identified as 20-hydroxyecdysone, polypodine B, 2-deoxy-20-hydroxyecdysone, integristerone A, and 2-deoxy-α-ecdysone. Repeated oral administration of the whole-plant extract at a dose of 10 mg/kg significantly accelerated healing of linear skin wounds in rats. Notably, the wound-healing efficacy of the *S. hissarica* extract exceeded that of the widely used reference drug methyluracil, particularly in animals with alloxan-induced diabetes—a model characterized by impaired regenerative capacity (Ramazanov et al. 2016).

In summary, the available experimental and clinical evidence clearly demonstrates that phytoecdysteroids exhibit strong, reproducible wound-healing activity. These compounds accelerate all major stages of skin repair, including granulation tissue formation, angiogenesis, epithelialization, and scar maturation. Their effects are mediated through stimulation of protein and collagen biosynthesis, stabilization of cellular membranes, antioxidant protection, and optimization of carbohydrate and lipid metabolism in damaged tissues. Phytoecdysteroid-containing preparations, particularly those derived from *Rhaponticum, Silene,* and *Stachys* significantly increased scar tensile strength and shortened healing time under both normal

Table 6.5 Tensile strength of wound scars formed after linear incisional skin wounds in rats on day 10 following treatment with a total ecdysteroid preparation from *Stachys hissarica* and methyluracil (M ± m, n = 6)

Experimental conditions	Tensile strength of post operative scar (g)	Difference in tensile strength of a scar in com paring to control (%)	p1 – to the test I p2 – to the test II
Test I (normal animals)	382.4 ± 12.6	–	–
TEP from *Stachys hissarica*	568.8 ± 18.4	48.7	p1 < 0.001
Methyluracil	522.2 ± 16.8	36.5	p1 < 0.001
Test II (diabetes)	286.4 ± 23.4	–	p1 < 0.01
Diabetes + TEP from *Stachys hissarica*	520.6 ± 32.6	81.8	p1 < 0.01 p2 < 0.001
Diabetes + methyluracil	413.2 ± 29.4	44.3	p1 < 0.50 p2 < 0.01

Note: p_1 denotes statistical significance compared with the corresponding control group (Test I or Test II). p_2 denotes statistical significance compared with the reference treatment (methyluracil) within the same experimental condition. Statistical differences were considered significant at $p < 0.05$

and metabolically compromised conditions, including experimental diabetes. In several experimental models, their efficacy was comparable to or exceeded that of established reference drugs such as Nerobol and methyluracil, while maintaining a more favorable safety profile. Taken together, these findings indicate that phytoecdysteroids represent a promising class of natural regenerative agents with high translational potential for the treatment of acute and chronic wounds, burns, ulcers, and conditions associated with impaired tissue repair, thereby supporting their further development for clinical and pharmaceutical applications.

6.5 Effect of Ecdysterone on Bone Fracture Healing

Throughout the history of medicine, substantial attention has been devoted to the search for pharmacological agents capable of stimulating osteogenesis and accelerating bone repair. In recent decades, particular interest has focused on compounds with anabolic activity, which—through selective stimulation of protein biosynthesis—not only improve the general physiological condition of the organism following trauma but also significantly accelerate the consolidation of bone fragments (Rolevich et al. 1975). Within this context, phytoecdysteroids represent a promising class of bioactive substances, as they have demonstrated pronounced corrective effects on metabolic processes in various pathological models, often comparable to those of synthetic anabolic steroids, yet without the adverse effects typically associated with steroid therapy. In the present study, one of the most widespread phytoecdysteroids, ecdysterone, was evaluated in a rat model of experimental bone fracture. As in previous investigations, the anabolic steroid Nerobol was used as a reference compound.

The experimental results demonstrated that bone fracture was accompanied by marked systemic metabolic disturbances when compared with intact animals. During the first 5–7 days following injury, body-weight gain decreased by 5–10% ($p < 0.05$). Concurrently, serum total protein and albumin concentrations declined by 9% and 11%, respectively ($p < 0.05$). While the α_1-globulin fraction remained largely unchanged, α_2- and γ-globulin levels increased by 25–50% throughout the observation period ($p < 0.05$). In addition, serum calcium concentrations in control animals increased by 10–17% ($p < 0.05$). Administration of ecdysterone or Nerobol after fracture resulted in a clear improvement in the general physiological and metabolic status of the animals. Body-weight gain in treated rats did not differ from that of intact animals from the early stages of treatment and subsequently exceeded the values observed in untreated fractured controls. Serum total protein concentrations remained 5–9% higher than in control animals with fractures, while albumin levels increased by 13–14% ($p < 0.05$). The α_1-globulin fraction showed a progressive increase and exceeded control values by 9–18% by the end of the experiment ($p < 0.05$). In contrast, α_2-globulin levels declined most noticeably on days 7, 21, 36, and 42, with reductions of 8–13% ($p < 0.05$).

Levels of β- and γ-globulins exhibited variable changes but tended to decrease during the later stages of bone healing, with reductions of 4–12% and 14–24%, respectively ($p < 0.05$). Serum calcium and phosphorus concentrations in treated animals were reduced at specific time points—calcium by 8–9% and phosphorus by 18–20% ($p < 0.05$)—which likely reflects enhanced mineral utilization at the fracture site during periods of active osteogenesis and bone remodeling. In parallel with systemic metabolic improvements, radiographic analysis (Fig. 6.3) revealed favorable local changes, including progressive restoration of bone integrity and activation of osteogenic processes at the fracture site. Comparative assessment showed that although ecdysterone and Nerobol produced similar normalizing effects on whole-body protein metabolism, their local actions at the site of bone injury differed. Radiographs obtained 42 days after fracture indicated that callus formation was less pronounced in animals treated with ecdysterone than in those receiving Nerobol. Thus, while phytoecdysteroids effectively support systemic anabolic and

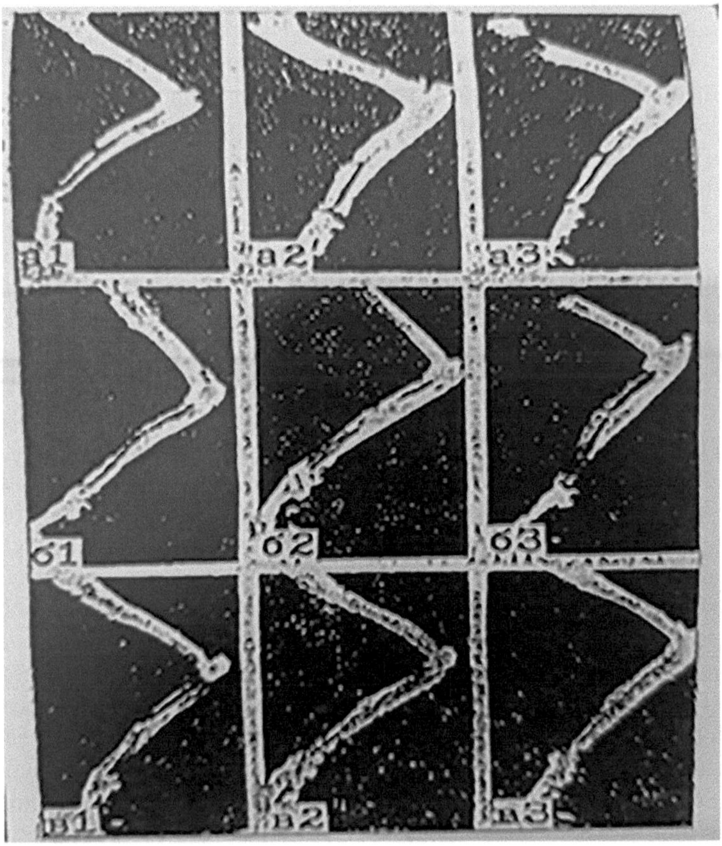

Fig. 6.3 Radiographs of the front left paw of rats taken on days 7 (a), 28 (б), and 42 (в) after fracture

metabolic recovery following bone injury, their direct stimulatory effect on local callus formation appears to be somewhat weaker than that of synthetic anabolic steroids. Overall, these findings indicate that phytoecdysteroids, including ecdysterone, promote recovery after bone trauma primarily through normalization of protein metabolism and improvement of the general physiological state of the organism. Although their local osteogenic activity may be less pronounced, phytoecdysteroids offer a clear advantage due to their superior safety profile and absence of steroid-associated adverse effects. Consequently, they may serve as valuable complementary agents in the pharmacological management of bone fractures. Taken together, the data presented in this chapter demonstrate that phytoecdysteroids function as potent metabolic regulators. By enhancing anabolic and energy-dependent processes across multiple tissues, they increase resistance to adverse factors and facilitate recovery in a broad range of pathological conditions, including traumatic bone injury.

With respect to adaptive capacity, physical performance enhancement, and fatigue prevention, phytoecdysteroids clearly outperform both Eleutherococcus extract and saparal. In addition, they offer several important advantages over the synthetic anabolic steroid Nerobol in conditions such as stress syndrome, myocardial dystrophy, hepatitis, and other metabolically driven pathologies. Many of these disorders arise primarily from impaired metabolic regulation rather than from direct damage to steroid-sensitive organs such as skeletal muscle or the kidneys. In this context, the therapeutic potential of phytoecdysteroids is particularly attractive, as their beneficial effects are achieved without the hormonal side effects characteristic of synthetic anabolic steroids. Phytoecdysteroids lack androgenic activity and are not associated with steroid-related toxicity, substantially widening their therapeutic window and enabling safe use across a broad range of clinical settings. Taken together, these properties support the use of phytoecdysteroids as agents of choice for enhancing physical performance and preventing fatigue, especially under conditions of pre-pathological functional overload. Their value is particularly evident in diseases in which disruption of metabolic homeostasis represents the central pathogenic mechanism. By targeting systemic metabolism rather than exerting narrow organ-specific effects, phytoecdysteroids address the fundamental drivers of disease development and progression.

Ecdysterone and structurally related phytoecdysteroids promote recovery after bone fracture primarily through restoration of metabolic balance, stimulation of protein biosynthesis, and improvement of overall physiological status (Syrov et al. 1986). Although their direct stimulation of local callus formation is somewhat less pronounced than that of synthetic anabolic steroids, the absence of hormonal imbalance and organ-specific toxicity represents a decisive therapeutic advantage. By simultaneously supporting adaptive capacity, physical performance, and resistance to fatigue, phytoecdysteroids emerge as safe and effective metabolic regulators with broad regenerative potential. Collectively, these characteristics justify the consideration of phytoecdysteroids as valuable complementary or alternative agents in fracture management and in a wide range of pathological conditions in which impaired metabolism underlies tissue damage, delayed regeneration, and incomplete functional recovery.

6.6 Clinical and Translational Applications

Clinical observations provide strong evidence for the efficacy of phytoecdysteroids (PEs) in postoperative recovery, burn management, and treatment of chronic wounds (Sivamani et al. 2012). Patients receiving phytoecdysteroid-based preparations, including Ecdysten, showed faster postoperative healing, formation of stronger and more organized scars, and a lower incidence of inflammatory complications. In sports medicine and rehabilitation, phytoecdysteroids accelerate recovery of muscles and tendons, enabling a safe return to physical activity without adverse effects on endocrine regulation or hepatic function. By combining anabolic stimulation with antioxidant protection, angiogenic activation, and membrane stabilization, phytoecdysteroids act as multifunctional natural regenerative agents.

Accordingly, they exert a broad restorative effect by accelerating skin wound closure, promoting formation of a dense and well-organized collagen matrix, enhancing protein synthesis, and reducing creatine kinase leakage during skeletal muscle regeneration. This clinically validated, pleiotropic profile clearly differentiates phytoecdysteroids from synthetic anabolic steroids and supports their application in wound healing, orthopedics, reconstructive surgery, and rehabilitation medicine (Rice 2025; Goslen 2025). They also improve the stability of glycosaminoglycans and extracellular matrix components during cartilage and connective tissue repair. These clinical benefits have been consistently confirmed in postoperative rehabilitation. The data summarized in this chapter demonstrate that phytoecdysteroids possess strong regenerative potential by integrating anabolic, adaptogenic, and cytoprotective properties into coordinated biological actions that accelerate wound healing, bone regeneration, and tissue repair.

In wound-healing models, phytoecdysteroids promote rapid epithelialization, increase collagen deposition, stimulate angiogenesis, and reduce oxidative stress and inflammatory infiltration, resulting in faster wound closure and restoration of normal skin architecture. During bone regeneration, they enhance osteoblast activity, stimulate protein and matrix synthesis, normalize calcium–phosphorus metabolism, and accelerate fracture consolidation, thereby improving the strength and biomechanical stability of newly formed bone. At the cellular and tissue levels, phytoecdysteroids preserve mitochondrial integrity, stabilize cell membranes, and activate endogenous antioxidant defenses, limiting necrosis and supporting recovery of muscles, tendons, and internal organs. Clinical observations further confirm that ecdysterone- and turkesterone-based preparations improve postoperative recovery, fracture healing, and soft-tissue repair without the androgenic or hepatotoxic effects typical of synthetic anabolic steroids (Kuchkarova et al. 2020).

Taken together, the evidence presented in this chapter identifies phytoecdysteroids as a distinct class of natural metabolic regulators with strong regenerative potential. By coordinately enhancing protein biosynthesis, mitochondrial energy metabolism, angiogenesis, and antioxidant defenses, these compounds accelerate wound healing, promote fracture consolidation, and support soft-tissue recovery while maintaining an excellent safety profile. Unlike synthetic anabolic steroids, phytoecdysteroids act predominantly through non-hormonal mechanisms, providing substantial biological efficacy without disrupting endocrine homeostasis. Their consistent activity in experimental models and clinical observations highlights their translational relevance for regenerative medicine, orthopedics, postoperative rehabilitation, and sports medicine and supports the need for further controlled clinical trials and targeted pharmaceutical development.

References

Badawy MT, Sobeh M, Xiao J, Farag MA (2021) Androstenedione (a Natural Steroid and a Drug Supplement): A Comprehensive Review of Its Consumption, Metabolism, Health Effects, and Toxicity with Sex Differences. Molecules 26:6210. https://doi.org/10.3390/molecules26206210

Bajguz A, Bakala I, Talarek M (2015) Ecdysteroids in plants and their pharmacological effects in vertebrates and humans. Stud Nat Prod Chem 45:121–145. https://doi.org/10.1016/B978-0-444-63473-3.00005-8

Chen Z, Zhu G, Zhang JH, Liu Z, Tang W, Feng H (2008) Ecdysterone-sensitive smooth muscle cell proliferation stimulated by conditioned medium of endothelial cells cultured with bloody cerebrospinal fluid. In Cerebral Vasospasm: New Strategies in Research and Treatment 104:183–187. https://doi.org/10.1007/978-3-211-75718-5_36

Cheng DM (2010) Phytoecdysteroid accumulation in plants and bioactivities in animal models. Doct disser p 146

Darmogray SV, Guskov AV, Zimankov DA, Yerofeeva NS, Darmogray VN (2021) Possible use of phytoecdysteroids in dental practice. Pharmacy 70:9–14. https://doi.org/10.29296/25419218-2021-06-02

Darmogray VN, Ukhov YuI, Petrov VK (2004) On the mechanism of action of phytoecdysteroids. Modern science-intensive technologies 5:100–101 (in Russ)

Darmogray VN, Ukhov YuI, Sysykin AA, Petrov VK, Ryvkin OM, Potekhinsky SM (1996) Efficacy of phytoecdysteroid ointment for chemical burns. Ryazan: Inform leaflet Ryazan 4:54–96 (in Russ)

Devi S, Kumar P, Kaur G (2023) Biological Significance of Steroids. Steroids and their Medicinal Potential. p 98–124. https://doi.org/10.2174/97898150493361230101

Eming SA, Wynn TA, Martin P (2017) Inflammation and metabolism in tissue repair and regeneration. Science 356:1026–1030. https://doi.org/10.1126/science.aam792

Ermishina OA (1986) The effect of ecdysterone and nerobol on the course of experimental myocardial infarction in rats. Ischemic heart disease p 36–40 (in Russ)

Goslen KH (2025) Effect of phytoecdysteroids on protein synthesis and Akt/mTOR signaling after downhill running in skeletal muscle of mice. Doct disser p 17. https://doi.org/10.71889/5fylantbak.29862371

Kasich AM (1963) Clinical evaluation of nandrolone phenylpropionate in patients with gastrointestinal disease. Amer J Gastroent 41:628–633 (in Russ)

Kholodova YuD (1979) Phytoecdysones—biologically active polyoxyhydroxylated sterides. Ukr Biochem J 51:560–575 (in Russ)

Khushbaktova ZA, Saruk AV, Gusakov VM, Syrov VN (2012) Experimental evaluation of the effects of phytoecdysteroids on lipid peroxidation processes in rat liver during in vitro and in vivo studies. Theor Appl Ecol 1:31–34 (in Russ)

Krylova SG, Zueva EP, Zibareva LN, Amosova EN, Razina TG (2014) Antiulcer activity of extracts of ecdysteroid-containing plants of the genera *Lychnis* and *Silene* of the family *Caryophyllaceae.* Bull Exp Biol Med 158:190–194. https://doi.org/10.1007/s10517-014-2728-1

Kuchkarova LS, Rokhimova SO, Syrov VN (2020) Effect of turkesterone on the Pancreas histology and function in diabetic rats. Inter J Curr Res Rev 12:1–5. https://doi.org/10.31782/IJCRR.2020.12216

Lafont R, Balducci C, Dinan L (2021) Ecdysteroids. Encyclopedia 1:1267–1302. https://doi.org/10.3390/encyclopedia1040096

Landén NX, Dongqing Li, Mona Ståhle (2016) Transition from inflammation to proliferation: a critical step during wound healing. Cell Mol Life Sci 73:3861–3885. https://doi.org/10.1007/s00018-016-2268-0

Li J, Liu Y, Li Y. Hua Y, Jiang X, Zhou M (2025) Harnessing Energy Metabolism for Enhanced Bone Defect Repair: Mechanisms and Therapeutic Strategies. Authorea. https://doi.org/10.22541/au.175698028.84284546/v1

Lipin DE (2015) Development of technology and standardization of soft dosage forms containing phytoecdysteroids *Serratula coronata* L. PhD Disser p 189

Meybeck A, Bonte F, Redziniak G (1997) Use of an ecdysteroid for the preparation of cosmetic or dermatological compositions intended, in particular, for strengthening the water barrier function of the skin or for the preparation of a skin cell culture medium, as well as to the compositions. US Patent 5:609–873.

Mikheev NN, Zharikova MV, Makarova TP, Taiber GS (2004) Kombinirovannaya stress–ehokardiografiya s dipiridamolom i chrespishchevodnoi elektrostimulyatsiei predserdii v diagnostike stenoziruyushchego porazheniya koronarnykh arterii (Combined stress echocardiography with dipyridamole and transesophageal atrial pacing in the diagnosis of stenotic coronary artery disease). In Proceedings of the Fifth Scientific and Practical Conference "Diagnosis and Treatment of Cardiovascular System Disorders". Russ Acad Agricult Sci 4:134136

Mosaddad SA, Hussain A, Tebyaniyan H (2023) Green Alternatives as Antimicrobial Agents in Mitigating Periodontal Diseases: A Narrative Review. Microorganisms 11:1269. https://doi.org/10.3390/microorganisms11051269

Mukhin EA, Dubchenko VV, Gonchar VI, Gasani VB (1988) Hormones and oxygen under high pressure. Chisinau: Stiinca p 134 (in Russ)

Omanakuttan A, Bose C, Pandurangan N, Kumar GB, Banerji A, Nair BG (2016) Nitric Oxide and ERK mediates regulation of cellular processes by Ecdysterone. Experimental cell research 346:167–175. https://doi.org/10.1016/j.yexcr.2016.07.019

Parr MK, Botrè F, Naß A, Hengevoss J, Diel P (2015) Ecdysteroids: A novel class of anabolic agents? Biology of sport 32:169–173. https://doi.org/10.5604/20831862.1144420

Ramazanov NSh, Bobayev ID, Yusupova UYu, Aliyeva NK, Egamova FR, Yuldasheva NKh, Syrov VN (2016) Phytoecdysteroids-containing extract from *Stachys hissarica* plant and its wound-healing activity. Nat Prod Res 31:593–597. https://doi.org/10.1080/14786419.2016.1205058

Raza A, Chohan TA, Zaidi SHH, Hai A, Alzahrani AR, Abida Saleem H (2024) A Systematic Review on Biochemical Perspectives on Natural Products in Wound Healing: Exploring Phytochemicals in Tissue Repair and Scar Prevention. Chemistry & Biodiversity 21:e202400615. https://doi.org/10.1002/cbdv.202400615

Reitman S and Frankel S (1957) A colorimetric method for the determination of serum glutamic oxalacetic and glutamic pyruvic transaminases. Amer J Clin Pathol 28:56–63. https://doi.org/10.1093/ajcp/28.1.56

Rice TT (2025) Effects of phytoecdysteroids on skeletal muscle contractile function and protein synthesis after eccentric muscle damage. Doct disser p 45. https://doi.org/10.71889/5fylantbak.29862167

Rolevich IV, Petruchuk IV, Vologovskaya OA (1975) Comparative characteristics of the effect of nerobol and methylandrostenediol on the formation of bone marrow. Act probl theory and wedge honey Min p 379-381 (in Russ)

Sidorova YuS (2014) Physiological and biochemical in vivo assessment of adaptogenic properties of phytoecdysteroid-containing extract of crowned serpukha. PhD disser p 27 (in Russ)

Sivamani RK, Ma BR, Wehrli LN, Maverakis E (2012) Phytochemicals and naturally derived substances for wound healing. Advances in wound care 1:213–217. https://doi.org/10.1089/wound.2011.0330

Syrov VN (1994) Phytoecdysteroids: biological effects in an organism of the higher animals, and prospects for the use in medicine. Exp Clin Pharmacol 57:61–66 (in Russ)

Syrov VN (2000) The comparative study of anabolic activity of phytoecdysteroids and steranabols in experiment. J Chem Pharm 4:31–34. https://doi.org/10.1007/BF02524596

Syrov VN (2009) Medications and dietary supplements based on phytoecdysteroids (new approaches to pharmacorrection of disturbed metabolic processes in the organism). Ufa: Abstracts p 24

Syrov VN (2022) Ecdisten and Ecsumid as Effective Means of Preventing Negative Shifts in Certain Metabolic Processes of the Myocardium, Caused by Immobilization Stress. Amer J Med and Med Sci 12:179–182. https://doi.org/10.5923/j.ajmms.20221202.26

Syrov VN, Matveyev SV, Kurmukov AG, Islambekov US (1986) Effect of ecdysterone and nerobol on the course of bone fracture healing in an experimental model. Med J Uzb 3:67–69 (in Russ)

Timofeev NP (2005) Phytoecdysteroids: pharmacological use and activity (Review). J Med Sci 4:26–66 (in Russ)

Uglyanitsa KN (1980) Vliyanie retabolima na metabolicheskie protsessy v kozhnykh ranakh (Influence of retabolil on metabolic processes in skin wounds). Zdravookhranenie v Belorussii 1:53–55 (in Russ)

Uzbekova DG (1974) Anabolic steroid drugs (literature review). Pharmacol and toxicol 4:484–490 (in Russ)

Vágvölgyi M, Laczkó D, Santa-Maria AR, Vigh JP, Walter FR, Berkecz R, Hunyadi A (2024) 17-Oxime ethers of oxidized ecdysteroid derivatives modulate oxidative stress in human brain endothelial cells and dose-dependently might protect or damage the blood-brain barrier. Plos one 19:e0290526. https://doi.org/10.1371/journal.pone.0290526

Vasiliev AS (2012) Pharmacological effects of extracts of ecdysteroid-containing plants in the conditions of models of the syndrome of increased blood viscosity. Doct disser p 49 (in Russ)

Wang Sh, Liu J, Zhou L, Xu H, Zhang D, Zhang X, Wang Q, Zhou Q (2024) Research progresses on mitochondrial-targeted biomaterials for bone defect repair. Regener Biomat 11:rbae082. https://doi.org/10.1093/rb/rbae082

Yusupova U, Sasmakov S, Usmanov D, Ramazonov N, Azimova S and Sagdullaev S (2020) Phytoecdysteroids from Silene claviformis and their antibacterial and antifungal activities. Pol. J. Natur. Sc. 35(3):313–321.

Yusupova UY, Bobakulov KM, Girault JP, Usmanov DA, Ramazonov NS and Abdullaev ND (2022) New Phytoecdysteroid from Silene popovii. Chem Nat Compd 58(6):1082–1084. https://doi.org/10.1007/s10600-022-03874-9.

Yusupova UY, Bobakulov KM, Khurramov AR, Syrov VN, Egamova FR, Karuth A, Usmanov DA, Quadir M and Rasulev B (2025) Phytochemical constituents isolated from Silene popovii Schischk. Med Chem Res 34(2):423–431. https://doi.org/10.1007/s00044-024-03345-6.

Yusupova UY, Ramazonov NS, Bobakulov KM, Egamova FR, Syrov VN and Usmanov DA (2021) Biological active components of silene tomentella and their pharmacological properties. Khim. Rastit. Syr′ya 1:197–202. https://doi.org/10.14258/jcprm.2021018323.

Yusupova UY, Usmanov DA and Ramazonov NS (2019) Phytoecdysteroids from the Plant Dianthus helenae. Chem Nat Compd 55(2):393–394. https://doi.org/10.1007/s10600-019-02701-y.

Zádor E (2025) Molecular Targets of 20-Hydroxyecdysone in Mammals, Mechanism of Action: Is It a Calorie Restriction Mimetic and Anti-Aging Compound? Cells 14:431. https://doi.org/10.3390/cells14060431

Zibareva LN, Amosova EN, Nesterova YuV, Poveteva TN, Afanaseva OG, Erst AA, Suslov NI (2021) Plants of the genera *Silene* L. and *Lychnis* L. (*Caryophyllaceae*): Composition of chemical components and biological activity. Book p 495 (in Russ)

Chapter 7
Pharmacokinetic and Safety Profiles of Phytoecdysteroids in Mammals

Laurence Dinan and René Lafont

Abstract Owing to the wide range of pharmaceutical effects which have been ascribed to ecdysteroids in mammals and the general consensus on their low toxicity in these species, there is currently considerable interest in developing these compounds as drugs to treat a wide range of conditions (as described elsewhere in this book) and as nutritional supplements. Many of the early data were indicative of the considerable potential of ecdysteroids as pharmaceutical agents but were not supported by thorough clinical trials. Some of these are now being performed. A major part of the process (preclinical and clinical) is to establish the pharmacokinetic and safety profiles, along with the best form of delivery of the candidate drug. In this chapter, we give a brief overview of what is entailed in assessing whether the candidate compound and its formulation have the required properties to become an effective drug. We then review the data which are currently available for ecdysteroids in mammalian systems and finally consider what might be done to overcome the low bioavailability of 20-hydroxyecdysone.

Keywords Absorption · ADMET · Bioavailability · Cyclodextrin · Distribution · Ecdysteroid · Excretion · Pharmacokinetics · Safety · Steroid metabolism

7.1 Introduction

Although ecdysteroids were first identified as the steroid hormones of insects and other arthropods where they control moulting, development, and reproduction (summarised in Koolman 1989; Smagghe 2009), it was only after the discovery of much larger amounts of ecdysteroid analogues (phytoecdysteroids) in certain species of plants (Dinan 2001; Dinan et al., 2021a; other chapters in this book) that it became possible to test the effects of ecdysone, 20-hydroxyecdysone, and a few other analogues on mammals, including humans. It is fair to say that this has become

L. Dinan · R. Lafont (✉)
Sorbonne University, Paris, France

V. N. Syrov et al. (eds.), *Phytoecdysteroids II*,
https://doi.org/10.1007/978-981-95-8423-9_7

the main current focus of research in ecdysteroid studies. Initial results indicated that ecdysteroids might possess a plethora of predominantly beneficial effects on mammals, but it is only recently that thorough clinical trials have started to be carried out to substantiate the promising initial findings. The development of ecdysteroids as effective and safe pharmaceutical agents requires a thorough understanding of the pharmacology and toxicology of ecdysteroids in mammals, initially in models such as rodent species, and then in humans.

The registration of a new molecule as a medicine requires, among other things, the completion of a regulatory dossier which includes an extensive set of pharmacokinetic data, including data on its toxicity, the "ADMET" (Absorption, Distribution, Metabolism, Excretion and Toxicity). Such properties are established by means of various experiments in vitro and in vivo. On the other hand, it is also possible to predict, more or less accurately, these properties with an array of informatic tools (reviewed in Fuart Gatnik and Worth 2010), e.g. TOPKAT by Discovery Studio 2016 (Accelrys Inc., San Diego, CA, USA) or SwissADME (Molecular Modelling Group of the Swiss Institute of Bioinformatics, Lausanne, Switzerland; Daina et al. 2017). These tools have been recently used with sets of ecdysteroids in the context of Structure-Activity Relationships (SAR) studies (Mamadalieva et al. 2022; Shafiq et al. 2024). In addition to pharmacokinetic parameters, predictions may concern carcinogenic, mutagenic, or irritant properties. Nevertheless, direct experimental analyses remain necessary to obtain accurate parameters.

In this chapter, we shall consider the general background to establishing the pharmacokinetic and safety profiles of a candidate drug, review the data currently available for ecdysteroid pharmacokinetics and toxicity in mammals, and critically discuss how to improve the bioavailability. We shall also summarise and discuss the few 'negative' toxicological findings, which stand in stark contrast to the vast majority of 'positive' pharmaceutical effects of ecdysteroids seen in mammals.

7.2 General Aspects

7.2.1 Physicochemical Properties

A molecule's ADME properties are the direct consequence of its physicochemical properties, among which water solubility, molecular size, and polarity are key parameters, and can be used to describe its behaviour in a specific biological system (if experimentally determined) or to predict its behaviour generally in biological systems (if calculated). The delivery of the drug can result in improved efficacy by incorporation of appropriate solubilising excipients (organic solvents, non-ionic surfactants, water-insoluble lipids, organic liquids/semisolids or phospholipids) into the formulation (Strickley 2004), which act as an interface between the drug and the complex biological tissues to provide optimal interaction.

7.2.2 *Absorption/Distribution/Metabolism/Elimination/Toxicity (ADMET) Studies*

The different component stages of ADME studies are depicted in Fig. 7.1. The aim is to determine the oral bioavailability, then the distribution of the absorbed molecule between tissues, metabolic conversions (in tissues and/or the microbiome), and its elimination (in urine and faeces). It is also important to determine whether some portion of the active pharmaceutical ingredient (API) is stored (e.g. in liver or adipocytes). This is best determined by the administration of a radioactively labelled (preferably ^{14}C) molecule to check whether the administered molecule is completely eliminated (as the original molecule or any metabolites), which allows the establishment of a so-called mass balance study (i.e. proportion of the applied API accounted for by distributed, sequestered, and excreted API and metabolites).

7.2.3 *Mode of Delivery*

There are many possible ways for the administration of a molecule: oral (solutions, pills, capsules), transdermal (patches), injections (subcutaneous, intravenous, intraperitoneal), buccal sprays, and intranasal drops. Amongst these, oral delivery is the most widely used.

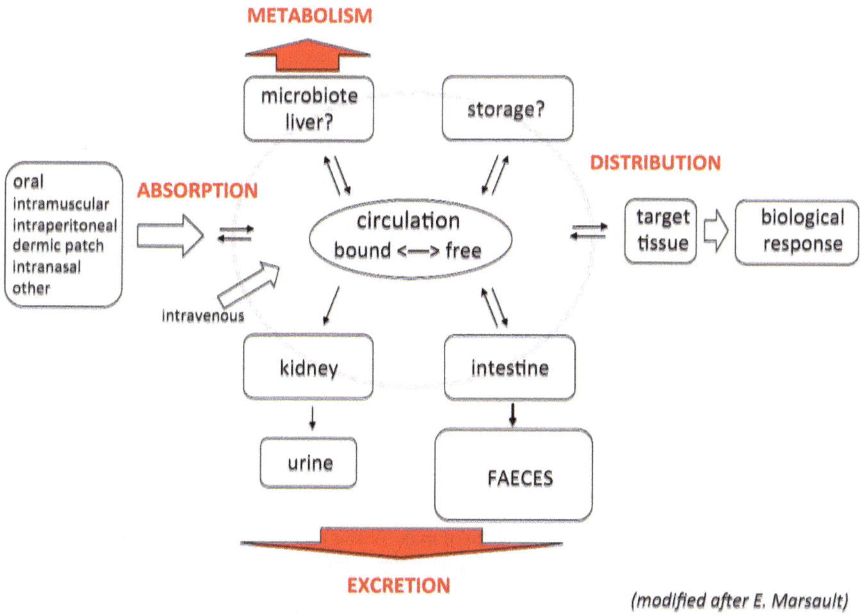

Fig. 7.1 General representation of ADME (Absorption, Distribution, Metabolism, and Excretion) (modified after Marsault (2011), Pharmacology Institute of Sherbrooke, http://pages.usherbrooke.ca/ydory/pdf/BCM-400_ADME.pdf, downloaded on 20/03/2020)

7.2.4 Bioavailability

The bioavailability of a drug (Hetal et al. 2010; Danish and Lubhan, 2016; Stielow et al. 2023; Wu et al. 2025) is the proportion of an administered dose of the drug which reaches the site of action in an unchanged state; in practice, the proportion reaching the systemic circulation is used in place of the site of action, because the precise site(s) and locations may not be known. The **absolute bioavailability** (f) is determined by comparing the profiles of unchanged drug in the plasma over time to give the Area Under the Curve (AUC)/dose after a specified administration method relative to the AUC/dose after intravenous injection of the drug (see Fig. 7.2 for an example). Thus, absolute bioavailability (f) can have a value between 0 (low) and 1 (high), but it is often expressed as a percentage (F = 100xf). The **relative bioavailability** is used to compare two administration methods (e.g. formulation 1 versus formulation 2), i.e. (AUC1/dose1)/(AUC2/dose2). **Bioequivalence** indicates that the API when given to the same patient in the same dosage regimen by different routes/formulations results in equivalent drug concentrations in the plasma and tissues. **Therapeutic equivalence** means that the drug when given to the same patient in the same dosage regimen by different formulations/routes has the same therapeutic and adverse effects (Le and Vivian 2025). The **therapeutic window** is the difference between the dose required to bring about a therapeutic effect and a (higher) dose which starts to bring about adverse effects.

There are many factors which contribute to the bioavailability of a drug: the physicochemical properties of the drug (polarity, solubility, molecular size, pKa,

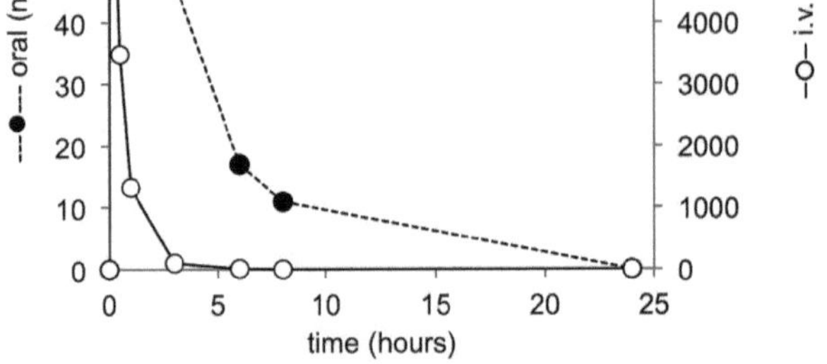

Fig. 7.2 Pharmacokinetics of 20E in plasma of male Wistar rats were performed after intravenous (5 mg/kg in 0.9% NaCl) or per os (50 mg/kg in water) application of the compound. 20E levels were quantified by HPLC/MS (after Dinan et al. 2021a). From these data, the bioavailability (F) is calculated to be only 1%

etc.), the drug formulation and administration route, the drug's pharmacokinetic parameters (distribution, metabolism, sequestration, and rate of elimination), interactions with other drugs/nutrients (induction/inhibition of relevant enzyme activities), the patient's age and health/disease status, circadian rhythms, and individual genetic variation (Hetal et al. 2010; Danish and Lubhan 2016; Stielow et al. 2023).

Drugs are classified into 4 classes depending on their solubilities and permeabilities (Biopharmaceutics Classification System):

BCS Class I: high solubility, high permeability.
BCS Class II: low solubility, high permeability.
BCS Class III: high solubility, low permeability.
BCS Class IV: low solubility, low permeability.

Owing to its high aqueous solubility but low membrane permeability (low bioavailability), 20E falls into BCS Class III (Dilda et al. 2025).

7.3 Pharmacokinetics of Ecdysteroids in Mammals

The available data up to 5 years ago has been reviewed (Dinan et al. 2021b) and this will be summarised along with more recent findings below:

7.3.1 Physicochemical Properties of 20E and Other Phytoecdysteroids

7.3.1.1 Solubility

20-Hydroxyecdysone is highly soluble in alcohols (methanol and ethanol) and DMSO, but owing to its high number of hydroxyl groups (when compared with vertebrate steroid hormones), it is also soluble to a significant extent in water. Its maximal solubility is in the range of 10 mg/mL, i.e. ca. 20 mM (Table 7.1).

7.3.1.2 Polarity

The polarity of a molecule is often expressed as its LogP. This parameter is measured by its partition coefficient in a 2-phase system (octanol:water): LogP = Log $(C_{octanol})/C_{water})$.

Table 7.1 Reported solubilities of 20E in water

Solubility (mg/mL)	Temperature	Source/reference
0.084	–	Temirgaziyev et al. (2019a)
0.42	–	ALOGPS*
0.9	–	OSADHI**
4.8	–	Abcam Ltd., Cambridge, UK
6.7	–	Erimbetov et al. (2020a)
9	–	Mironov et al. (2013)
9.12	37 °C	Wang et al. (2022)
9.5	20 °C	Drugabilis, Villejust, France
~10	–	Cayman Chemicals
10	25 °C	BOC Sciences, Shirley, NY, USA
11.6	–	Dilda et al. (2025)
14.4	37 °C	Meteleva et al. (2023)

*Online Chemical Database with Modelling Environment (www.ochem.eu)
**Online Structural and Analytics-based Database for Herbs of India (www.neist.res.in/osadhi)

Table 7.2 Polarity: the wide range of calculated LogP data for 20E

LogP	References
2.92	Nambaaru et al. (2024)
1.98	Fang et al. (2022)
1.85	ANPDB*
1.79	Usmanov et al. (2025)
1.36	Dinan et al. (2021b)
1.317	OSADHI**
1.31	ALOGPS***
0.61	ChemAxon (https://chemaxon.com)
0.5	Ambrosio 2019/ChemSpider (www.chemspider.com)
−0.04 ± 0.01	**Measured** (Dilda et al. 2025)
−0.270	Human metabolome database (www.hmdb.ca)
−0.53	Parr et al. (2019)/ACD Labs prediction (www.acdlabs.com)
−1.25	Lapenna (2006)

*African Natural Products Database (Betow et al. 2025)
**Online Structural and Analytics-based Database for Herbs of India (www.neist.res.in/osadhi)
***Online Chemical Database with Modelling Environment (www.ochem.eu)

Positive values express lipophilicity, and negative values hydrophilicity. LogP can be directly measured or calculated taking into consideration the molecule's different structural elements (CLogP, e.g. Lipinski et al. 2001). As can be seen from Table 7.2, according to the in silico methodology used, the calculated values obtained are very different, and only a few are close to the experimentally determined value, which indeed raises questions about the reliability of those methodologies.

7.3.2 Drugability of Ecdysteroids

Since the development of new drugs is a time-consuming and very expensive undertaking where many of the promising candidates fall by the wayside during the search for an effective, safe, and commercially viable preparation, several procedures have been designed to act as preliminary screens to focus on the candidate analogues which possess suitable physicochemical and ADMET properties. The best known of these is Lipinski's Rule of Five (to be an orally active drug, it should have no more than one violation of < 5 H-bond donors, MW < 500, ClogP < 5 and < 10 H-bond acceptors; Lipinski et al. 2001). This has been developed to produce druggability scores (ADMET Traffic Light Scores; Lobell et al. 2006) which are allocated dependent on the structure and polarity of the molecule (molecular weight, LogP, polar surface area, number of rotatable bonds, and numberof H-bond donors and acceptors) to generate a score between 0 and 10 (the allocated parameter score is associated with a traffic light colour: 0 = green, 1 = amber, and 2 = red; the scores are added to give the overall Traffic Light Score [0–10], where 0–2 is deemed green [good druggability], 3–6 orange [moderate druggability], and 7–10 red [poor druggability]), which gives 20E an overall score of 3 ('amber'; poor to moderate suitability as drug: Dinan et al. 2021b). Data for the ADMET Traffic Light Scores for hundreds of ecdysteroid analogues have been obtained and are included in the online ecdysteroid resource 'Ecdybase' (Lafont et al. 2002; Dinan and Lafont 2021). Table 7.3 presents the scores for a few common analogues of the many known ecdysteroids to indicate how the structure affects the ADMET TLS.

Over 500 ecdysteroid analogues have been identified (Lafont et al. 2002). The majority of these have been isolated from plants as phytoecdysteroids. From an examination of Table 7.3 and the database from which this selection was taken (Dinan and Lafont 2021), it can be seen that ecdysteroids as a chemical class demonstrate a wide range of predicted physicochemical properties and that some of them accord fully with the criteria desired for drug-like compounds (low TL scores), since they should possess good bioavailability and permeability. In the case of poststerone (TL score = 0), it is known (see below) that the bioavailability is significantly better (ca. 20x) than that of 20E. However, applied analogues also must possess biological activity or have the potential to be converted to an active analogue within the organism. Although a few preliminary structure–activity relationship studies have been initiated for ecdysteroid effects on mammals (Syrov 2000; Syrov et al. 2001; Saatov et al. 1999; Báthori et al. 2008), these have not used a large enough number of analogues to provide really meaningful QSAR information suitable for modelling studies. The reasons for this are clear: the biological assays are expensive, complicated, and time-consuming. Moreover, since most are based on whole-animal assessments, they require strict regulation, large numbers of experimental mammals, and relatively large amounts of each (pure) analogue in addition to the currently readily available 20-hydroxyecdysone.

Table 7.3 ADMET Traffic Light Scores for 15 commonly occurring phytoecdysteroids (Lobell et al. 2006; data taken from Dinan and Lafont 2021)

Ecdysteroid	Formula	MW	ClogP	PSA (Å^2)	Rotatable bonds	H-Acceptors	H-Donors	No. violations	TL Score
Ajugasterone C	$C_{27}H_{44}O_7$	480.64	1.65	138.44	5	7	6	1	3
Cyasterone	$C_{28}H_{44}O_8$	520.66	0.71	144.52	4	8	5	1	4
Ecdysone	$C_{27}H_{44}O_6$	464.64	2.30	118.21	5	6	5	0	1
20-Hydroxyecdysone	$C_{27}H_{44}O_7$	480.30	1.36	138.45	5	7	6	1	3
20-Hydroxyecdysone 3- acetate	$C_{29}H_{46}O_8$	522.68	2.06	144.52	7	8	5	1	3
20-Hydroxyecdysone 3-β-D-glucoside	$C_{33}H_{54}O_{12}$	642.78	−0.35	217.59	8	12	9	3	7
Inokosterone	$C_{27}H_{44}O_7$	480.64	1.30	138.44	6	7	6	1	3
Makisterone C	$C_{28}H_{48}O_7$	508.70	2.34	138.44	6	7	6	2	4
Polypodine B	$C_{27}H_{44}O_8$	496.64	0.41	158.67	5	8	7	1	4
Ponasterone A	$C_{27}H_{44}O_6$	464.64	2.56	118.21	5	6	5	0	1
Poststerone	$C_{21}H_{30}O_5$	362.46	1.03	94.83	1	5	3	0	0
Pterosterone	$C_{27}H_{44}O_7$	480.64	1.39	138.44	5	7	6	1	3
Rubrosterone	$C_{19}H_{26}O_5$	334.41	0.28	94.83	0	5	3	0	0
Taxisterone	$C_{27}H_{44}O_6$	464.64	2.51	118.21	5	6	5	0	1
Turkesterone	$C_{27}H_{44}O_8$	496.64	0.44	158.67	5	8	7	1	4

The canonical isomeric SMILES string was created for each ecdysteroid at http://endocrinedisruptome.ki.si (accessed on 15 July 2019) (Kolšek et al. 2014), which allows full stereochemical specification of the structure (including C-5 and C-9), and pasted into the Molinspiration calculator (www.molinspiration.com, accessed on 15 July 2019) to obtain values for the above parameters. Scoring was as follows: MW: <400 = 0; 400–500 = 1; >500 = 2; calculated LogP (calculated by fragment summation method): <3 = 0; 3–5 = 1; >5 = 2; polar surface area PSA (Å^2): <120 = 0; 120–140 = 1; >140 = 2; rotatable bonds (excluding C-O-H): ≤7 = 0; 8–10 = 1; ≥11 = 2; number of N + O (H-bond acceptors): ≤10 = 0; >10 = 1; number of NH + OH (H-bond donors): ≤5 = 0; >5 = 1. This generates a score between 0 and 10 for each compound; the lower the value, the more suitable the overall physicochemical properties are predicted to be for an orally administered drug

7.3.3 Protein Binding

Most drugs in the systemic system will circulate to some extent in a form non-covalently bound to protein. Human serum albumin (HSA) is the most abundant carrier protein in plasma (0.6 mM) and is known to bind many drugs, but there are currently no studies on the binding of ecdysteroids to mammalian plasma proteins. Protein-bound proteins may be protected from metabolism but may also reduce distribution to other compartments (Stielow et al. 2023).

7.3.4 Ecdysteroid ADME

Below, we successively describe the different components of ADMET with regard to ecdysteroids (mainly 20E) in mammals.

7.3.4.1 Dose

The optimal dosage is complicated to assess and is rarely thoroughly determined. It should be effective and non-toxic and produces minimal side effects as possible. Generally, the dosing regime is developed by consideration of existing experimental data deriving from biochemical assays and preclinical and clinical trials. The goal is to achieve a concentration of the API at the active site which is within the therapeutic window over an effective time period. If one knows the effective concentration at the target site (potency of the API), one can work backwards, taking into account what is known about the ADME and bioavailability (pharmacokinetics), while bearing in mind the possibilities of acute and chronic toxicities and side effects. Much of the available data will derive from model species and may not be directly transferable to humans, owing to differences in size, distribution, metabolism, and rates of elimination. Over the years, an empirical understanding has been developed to allow conversion of doses found to be effective in various model animals to those which could be appropriate for humans (Nair and Jacob 2016).

7.3.4.2 Administration/Absorption

Per os administrations of 20E to rats as water solutions/suspensions were performed and compared with intravenous injections with subsequent measurement of plasma levels of 20E over the next 24 hours (Fig. 7.2). The obtained data allow the calculation of 20E exposure expressed as the AUC (area under the curve) in each case, and the ratio of AUC_{oral} to $AUC_{intravenous}$ allows calculation of the bioavailability, which is close to 1% in this case, thus confirming the predicted low oral bioavailability of 20E.

7.3.4.3 Formulation

Different strategies have been tried in order to improve this rather disappointing oral bioavailability. Many of them have explored complexation with water-soluble substances (Table 7.4).

7.3.4.3.1 Cyclodextrins

Cyclodextrins (CDs) are cyclic molecules consisting of 6, 7, or 8 (α-, β-, γ-cyclodextrins, respectively) α1–4-linked glucose units, which may be modified further (e.g. by addition of hydroxypropyl ether groups to enhance aqueous solubility further). The molecule is in the form of a hollow truncated cone, which is hydrophilic on the outside and more hydrophobic on the inside, allowing drug molecules to insert within it by non-covalently interacting with the inner surface. Thus, the cyclodextrin can act as a carrier to take the drug into solution (reviewed in Nicolaescu et al. 2025). Cyclodextrins based on β-cyclodextrin tend to be the most commonly used because the cavity is able to form inclusion complexes with the widest range of drugs. Unfortunately, β-cyclodextrin has much lower aqueous solubility than either α- or γ-cyclodextrin. Consequently, a number of derivatives of β-cyclodextrin have been developed which enhance the solubility (Saokham et al. 2018), of which the most commonly used is 2-hydroxypropyl-β-cyclodextrin (2HPβCD).

To date, most attempts to improve ecdysteroid aqueous solubility and bioavailability involve complexation with cyclodextrins, even though it is recognised that the impact of cyclodextrins on the oral availability of BCS Class III drugs is negligible (Saokham et al. 2018; Rincon-Lopez et al. 2021). The justification used was to raise the concentration in water solutions, which is partly true since the amount of 20E increases in linear proportion with cyclodextrin concentration (Fig. 7.3), but since the maximum water solubility of β-cyclodextrin is ca. 40 mM, the amount of 20E can only be just over doubled.

The formation of CD complexes is simple; it suffices to dissolve a mixture of 20E and cyclodextrin in alcohol and then to evaporate it to dryness before dissolving the powder in water. The potential increase in solubility of 20E depends on the solubility of the cyclodextrin itself. The increase was ca. 2.5x (Wang et al. 2022) and was claimed to reach 7x with 500 mg/mL 2-(hydroxypropyl)-β-cyclodextrin (Fig. 7.3: Mironov et al. 2013). However, the maximum aqueous solubility of 2HPβCD is only about 50 mg/mL (Saokham et al. 2018). The resulting improvement of oral bioavailability was, however, modest and not proportional to the increase in water solubility. Some authors limited their studies to solubility and postulated that it would improve bioavailability. When bioavailability was investigated, it was found to be 1.3–1.4 (Erimbetov et al. 2020a, b) or 2.97 times (Wang et al. 2022) greater. However, other authors observed no increase in AUC of CD complexes compared to the same amount of free 20E (i.e. the relative bioavailability was 1; Dilda et al. 2025). This approach may be limited by the aqueous solubility of β-cyclodextrin itself (18.5 g/L = 16.3 mM; Hetal et al. 2010; Saokham et al. 2018),

Table 7.4 Strategies to improve bioavailability of 20E (adapted from Hetal et al. 2010)

Strategy	General method	Specific method	References
Enhance solubility and dissolution rate	Inclusion, complexation	Cyclodextrins	Berkenov and Datkhayev (2017), Chiruta (2020), Dilda et al. (2025), Erimbetov et al. (2020a), Kozhanova et al. (2021), Meteleva et al. (2023), Mironov et al. (2013), Temirgaziyev et al. (2019a, b), Wang et al. (2022)
		Arabinogalactan	Roziev et al. (2013)
		Arabinoglycan	Erimbetov et al. (2020a)
		Gum arabic	Erimbetov et al. (2020a)
		Alginate microspheres	Guan et al. (2011)
Formulation-based approaches	Addition of a solubilizer	Transcutol	Not yet assessed
		Kolliphor	Not yet assessed
		Labrasol	Graf et al. (2014), Anthony et al. (2014)
		Plurol oleic	Not yet assessed
		Mixtures	Not yet assessed
Modification of partition coefficient	Ester formation	20E undecanoate	Udintsev et al. (2014)
		20E triacetate (+CD)	Tuleuov et al. (2020, 2024)
	Novel formulation approaches	Liposomes	Volodin et al. (2020)
Targeted tissue delivery	Peptide conjugation	Peptide conjugation	Lesma et al. (2016)
	Sensibilization of tumour cells to radiations	Squalenoylated particles	Fumagalli et al. (2018) Vágvölgyi et al. (2020)
	Neuroprotection	ROS-induced delivery	Xing et al. (2025)
Inhibition of P-glycoprotein efflux	P-gP inhibitors	Not tested	Not yet assessed
Alternative mode of administration	Subcutaneous implants	PLGA polymers	Luo et al. (2025)
	Intranasal delivery	As CD complexes	Dilda et al. (2025)
	Nebulisation	Not tested	Not yet assessed
	Transdermal application	+ Permeation enhancers	Wang et al. (2022)
		Dermal cream	Tsuji et al. (1995)
Design of a more active analogue/derivative	Hemisynthesis	20,22-phosphate ester	Zhang et al. (2011)

CD: cyclodextrin; PLGA: poly(lactic-co-glycolic acid)

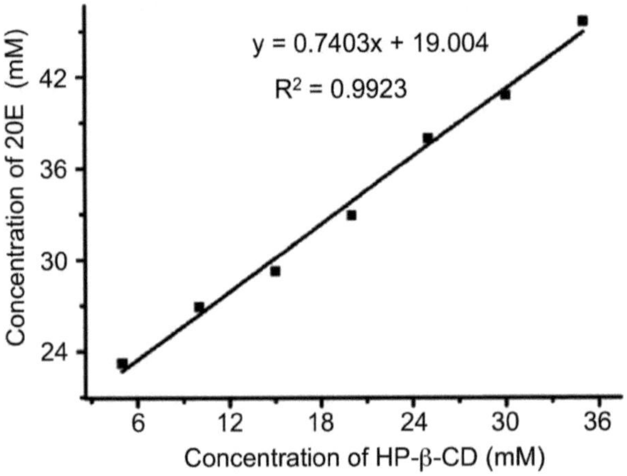

Fig. 7.3 Water solubility of 20E in the presence of increasing concentrations of 2-hydroxypropyl-β-cyclodextrin (redrawn from Wang et al. 2022)

which is lower than the solubility of 20E in water (ca. 20 mM), but it may enhance the rate of dissolution of crystalline 20E.

In addition to 20E (Nazarov et al. 2014; Berkenov and Datkhayev 2017; Temirgaziyev et al. 2019a; Kozhanova et al. 2021; Wang et al. 2022; Meteleva et al. 2023), the abilities of several other ecdysteroids to form complexes with cyclodextrins have been investigated: 2-deoxyecdysone (Kozhanova et al. 2020), 2-deoxy-20-hydroxyecdysone (Temirgaziyev et al. 2018), 3-*epi*-2-deoxyecdysone (Temirgaziyev et al. 2019b; Tuleuov et al. 2025), 20E 3-acetate (Tuleuov et al. 2020), and 20E 2,3,22-triacetate (Tuleuov et al. 2024), all of which are less polar than 20E and thus have poorer water solubilities alone. These studies have investigated the nature of the internal complex formed with the ecdysteroid by ^{1}H-nmr and looked for changes in chemical shift induced by complexation. Most authors concur on a 1:1 CD/ecdysteroid molar ratio in the complex, but the interpretations of the nature and orientation of the ecdysteroid in the complex differ. It is currently not clear if this is due to the structural differences between the ecdysteroids or the small chemical shifts observed, making it difficult to interpret the precise nature of the interaction. 2HPβCD was found to increase the transdermal bioavailability in rats by 1.92-fold (Wang et al. 2022).

In summary, complexes of 20E with β-CD do not appear to significantly improve bioavailability, in accord with 20E's BCS Class III status. 20E/CD complexes might be more effective with other application routes (see below).

7.3.4.3.2 Other Clathrates

Erimbetov et al. (2020a, b) compared the bioavailability of nanosized forms of 20E mixed with arabinoglycan (AG), gum arabic (GA), and β-cyclodextrin (CD). The study of the pharmacokinetics and bioavailability of 20E and its clathrates was

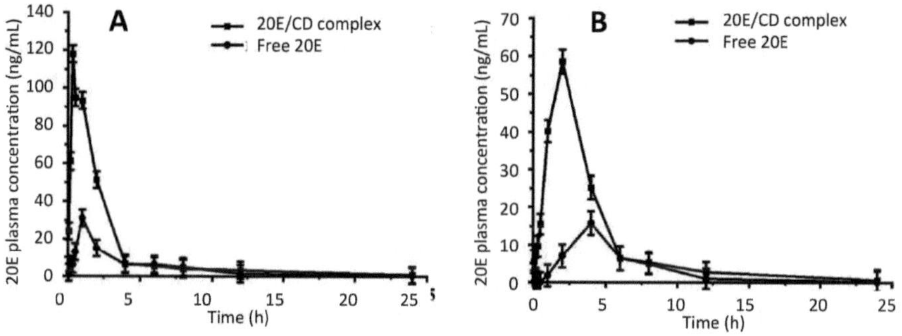

Fig. 7.4 Improved bioavailability of 20E-cyclodextrin complex as compared to free 20E in per os (**a**) and transdermal (**b**) administration (after Wang et al. 2022). Plasma concentration of 20E/2HPβCD inclusion complex and free 20E after oral (left) and transdermal (right) administration to female rats (50 mg 20E/kg in all cases). The results were expressed as the mean and standard deviation (n = 6) of the analyses

carried out on male Wistar rats. The test compound (20E) was administered intragastrically at a dose of 230 mg/kg of body weight. According to the research results, the bioavailabilities of 20E in the form of clathrates with AG, CD, and GA were 191, 165, and 156%, respectively, relative to 20E alone (100%) (Erimbetov et al. 2020a).

7.3.4.3.3 Transdermal Application

Cosmetic use of ecdysteroids has led to the development of ecdysteroid-containing creams to improve skin quality (e.g. Meybeck et al. 1997), from which 20E incorporated into liposomes can cross the external skin layer (*stratum corneum*), which provides the main barrier to skin penetration. It appears that the penetration is high enough to allow significant amounts of 20E to be found in plasma (Meybeck et al. 1997). Uptake of 20E can also be enhanced by applying cyclodextrin complexes (Fig. 7.4; Wang et al. 2022). The AUC obtained after transdermal application is about one half that of oral administration. The authors calculated a penetration rate of 0.31 ± 0.05 mg/cm^2 in 24 h. Application of ecdysteroid-containing creams to control in psoriatic patients did not result in any allergic or irritant reactions and reduced psoriatic lesions (Kroma et al. 2022). A range of other permeation enhancers are being developed to facilitate the transdermal entry of drugs (Dvořáková et al. 2022), but none of these appear to have yet been applied to ecdysteroids.

7.3.4.3.4 Subcutaneous Implants

Luo et al. (2025) incorporated 20E into poly(lactic-co-glycolic acid) polymer (PLGA) and used it to promote healing of cranial defects after subcutaneous implantation into the heads of diabetic rats. The purpose of this was twofold: to bring the

20E close to the damaged bone site and to provide a slow-release system for 20E. The 20E/PLGA implants were shown to be non-toxic to bone marrow cells and enhanced new bone formation to repair the cranial defects in the diabetic model.

7.3.4.3.5 Administration of 20E Esters

Udintsev et al. (2014) compared the intragastric bioavailability of chemically synthesised 20E 22-undecanoate with that of 20E in rats. When 20E undecanoate was administered, rather low plasma levels of this molecule were observed, but high concentrations of free 20E were found, and these were much higher than those obtained after giving 20E. This means that (1) 20E undecanoate is much better absorbed than 20E and (2) it is rapidly hydrolysed back to 20E (Fig. 7.5; note the ten-fold difference in the two vertical axes). The bioavailability based on 20E increased 9.6-fold following administration of the 22-undecanoate, and, in accordance with this, significantly more 20E and its metabolite ('Epi-20E'; not fully identified) were excreted in the urine. Acyl (acetyl and fatty acyl) derivatives of 20E and their incorporation into liposomes have also been proposed as a controlled-release application method for apolar ecdysteroid esters (Politova et al. 2001).

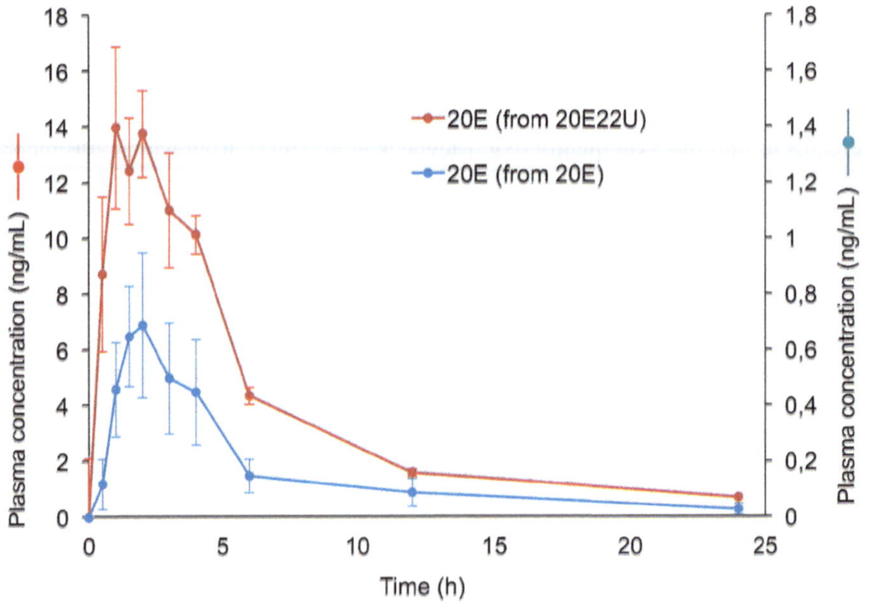

Fig. 7.5 The significant increase of 20E bioavailability when giving 20E 22-undecanoate (20E22U) as compared with 20E (single dose of 5 mg/kg of 20E or the equivalent amount of 20E in the form of 20E undecanoate was given per os to rats; drawn from the data published by Udintsev et al. 2014)

A 1:1 complex of 2,3,22-20E triacetate and β-cyclodextrin was prepared by Tuleuov et al. (2020, 2024). It was shown to possess anti-inflammatory activity in rats, whereas 20E-2,3,22-triacetate did not (unfortunately, 20E itself was not tested). The same complex proved biologically active in adult male rats in anabolic (increase in body weight) and adaptogenic ('open field' and 'free swimming') assessments at an intragastric dose of 10 mg/kg, presumably by hydrolysis to release 20E after uptake. In a totally different context, 20E triacetate was active as a slow-release form to induce successful moulting in lobsters (Gilgan & Burns 1977).

At the other end of the polarity spectrum of ecdysteroid derivatives, a cyclic diphospho ester of 20E (TAPA; 20E 20,22-phosphonic acid) has been prepared in relatively high overall yield (67%; adequate to prepare kg amounts) in six steps from 20E and found to effectively reduce glucose levels in diabetic model rats and to be practically non-toxic (Zhang et al. 2011; Wang et al. 2014). Compared to the water solubility of 20E, the solubilities of TAPA were found to be 10- and 50-fold higher for the acid and sodium salt, respectively, but a value for 20E is not given (cf. Table 7.1). Also, data for the relative bioavailabilities have not yet been published. Initial studies show that TAPA promotes cellular uptake of glucose, but does not enhance insulin secretion, as is the case for 20E and rosiglitazone, a diabetes medication acting as a positive control (Yoshida et al. 1971; Zhang et al. 2011). It has not yet been determined if TAPA is active in its own right or is hydrolysed within the body to release 20E.

7.3.4.4 Distribution

After oral administration, some 20E is absorbed in the stomach/small intestine (Dinan et al. 2021a) and is then distributed among tissues, except for the central nervous system, as it does not cross the blood–brain barrier (Kalász et al. 2017). Detailed distribution kinetics have been performed using radioactive 20E in mice by Dzhukarova et al. (1987) and Wu et al. (2011) (Table 7.5).

20E was isolated from *Achyranthes bidentata* and tritiated to a specific activity of 1.02 Ci/mmol. 20 μCi (9.6 μg)/mouse was injected into the tail vein and the mice were sacrificed at the above-stated times. The radioactivity associated with each of the organs was determined by scintillation counting and converted to μg/g (assuming that all radioactivity remains associated with 20E; see the discussion in the text for possible caveats).

Upon oral delivery, the concentration of radioactivity in the urine peaked after 30 minutes. After 6 hours, the radioactivity concentration reached a peak in the large intestine and faeces. A limitation of the analysis is that the authors measure total radioactivity (20E + possible metabolites + tritiated water formed by isotopic exchange, as observed by Dinan et al. 2021a). Wu et al. (2011) do not describe the tritiation method used, but we can speculate this was performed through isotopic exchange in ^3H_2O, as described by Russell & Price (1977) and Dzhukarova et al. (1987); so we believe that the low radioactivity remaining at 24–36 hours corresponds mainly to tritiated water. Nevertheless, the general information provided by early samples remains fully consistent, such as:

Table 7.5 The effect of [^{3}H]-20-hydroxyecdysone (20E) on the quantitative distribution in different tissues of mice over time (mean ± SEM, n = 5); values for 20E are given in μg.g^{-1} (after Wu et al. 2011)

	5 min	10 min	30 min	1 h	3 h	6 h	12 h	24 h	36 h
Blood plasma	0.061	0.057	0.052	0.047	0.032	0.023	0.015	0.011	0.010
Urine	0.482	0.921	1.534	0.281	0.102	0.096	0.087	0.046	0.025
Faeces	0.015	0.019	0.035	0.068	0.099	0.931	0.312	0.041	0.021
Bile	0.421	0.456	1.042	0.901	0.301	0.209	0.198	0.094	0.074
Liver	0.312	0.251	0.213	0.112	0.078	0.061	0.056	0.046	0.034
Heart	0.062	0.052	0.035	0.033	0.031	0.029	0.023	0.022	0.018
Spleen	0.041	0.026	0.024	0.018	0.023	0.027	0.032	0.027	0.02
Lungs	0.116	0.069	0.057	0.053	0.049	0.045	0.036	0.028	0.025
Kidney	0.137	0.123	0.098	0.071	0.055	0.043	0.037	0.032	0.023
Adrenals	0.184	0.139	0.098	0.081	0.073	0.065	0.052	0.043	0.041
Testis	0.048	0.036	0.029	0.026	0.023	0.021	0.019	0.017	0.015
Skeletal muscle	0.021	0.023	0.028	0.017	0.015	0.013	0.011	0.010	0.009
Spinal cord	0.019	0.028	0.067	0.041	0.033	0.024	0.020	0.018	0.014
Brain	0.013	0.014	0.015	0.013	0.012	0.01	0.008	0.007	0.006
Small intestine	0.254	0.139	0.094	0.083	0.076	0.061	0.05	0.033	0.025
Stomach	0.034	0.047	0.075	0.061	0.053	0.041	0.032	0.023	0.016
Large intestine	0.011	0.023	0.029	0.031	0.035	0.048	0.038	0.023	0.010
Bladder	0.013	0.024	0.038	0.026	0.023	0.018	0.015	0.013	0.010

- The high concentrations in urine (early) and faeces (later),
- The high concentrations in bile (indicative of an efficient enterohepatic cycle),
- A transitory accumulation in the liver, lungs, and adrenals.
- The low values in the brain extracts are consistent with a blood–brain barrier (Kalász et al. 2017).

The penetration of ecdysteroids across the blood–brain barrier can be achieved by using non-polar derivatives such as 20E 2,3:20,22-diacetonide (Kalász et al. 2017; Müller et al. 2017). As yet, little is known about the effects of ecdysteroids on the mammalian brain. Such non-polar ecdysteroid derivatives have also been shown to possess chemosensitising activity on drug-sensitive and multidrug-resistant cancerous cell lines (Martins et al. 2012, 2013, 2015).

7.3.4.5 Pharmacokinetics of 20E

For most studies in humans, 20E has been delivered in the form of pills (Ecdysten, 5 mg/pill) or as capsules of dietary supplements (non-certified composition) or as BIO101 (175 mg + colloidal silica and microcrystalline cellulose as excipients: Dioh et al. 2023). Only some studies used hydroalcoholic solutions of 20E (Ambrosio et al. 2021).

Fig. 7.6 Pharmacokinetics of ecdysone in man (redrawn from Simon and Koolman 1989)

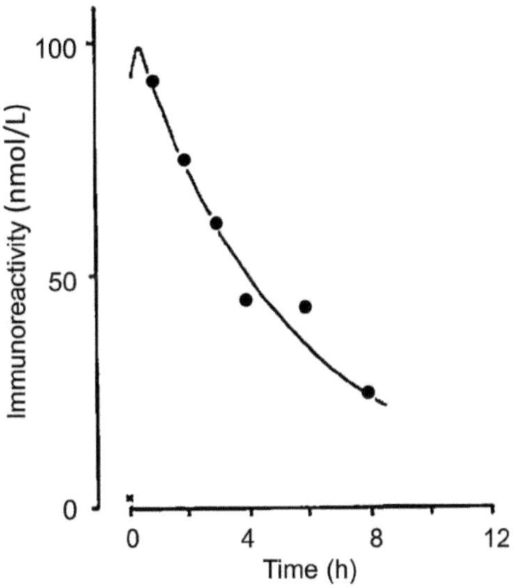

Table 7.6 Pharmacokinetic parameters deduced from plasma analyses (single ascending dose experiment, fasted individuals, n = 6/group; Dioh et al. 2023)

Parameter	Dose (mg)			
C_{max} (ng/mL)	100	350	700	1400
T_{max} (h)	141	317	399	710
$AUC_{0-\tau}$ (ng*h/mL)	2.03	3.00	3.00	3.50
AUC_{0-t}/dose (ng*h/mL*mg)	819	1927	2589	4097
$T_{1/2}$ (h)	7.67	5.50	3.68	2.96
Bioavailability	2.40	3.26	4.85	3.84

Most available human data concern the assessment of urine samples (e.g. Brandt 2003; Ambrosio et al. 2021); thus, circulating plasma levels were seldom investigated. The first data on plasma levels were provided after ecdysone oral administration (Simon & Koolman 1989). Ecdysone was given orally (0.2 mg/kg), and plasma levels were then determined by radioimmunoassay, which gave an approximate half-life of 4.1 h (Fig. 7.6). A limitation of this approach is that the immunoassay used recognises both ecdysone and at least some of its metabolites; this situation somewhat recalls the measurement of total radioactivity in rats after application of [^{3}H]20E (see Table 7.6).

More extensive information was provided recently by the Phase 1 study by Dioh et al. (2023) using a set of single ascending doses from 100 to 1400 mg (1.2 to 19.9 mg/kg) and measuring plasma levels over the next 24 h (Fig. 7.7 and Table 7.6).

The use of increasing doses shows that the absorption process is saturable, as the AUC/dose (the dose being expressed here as mg/kg b.w.) decreases as the dose increases (Fig. 7.8).

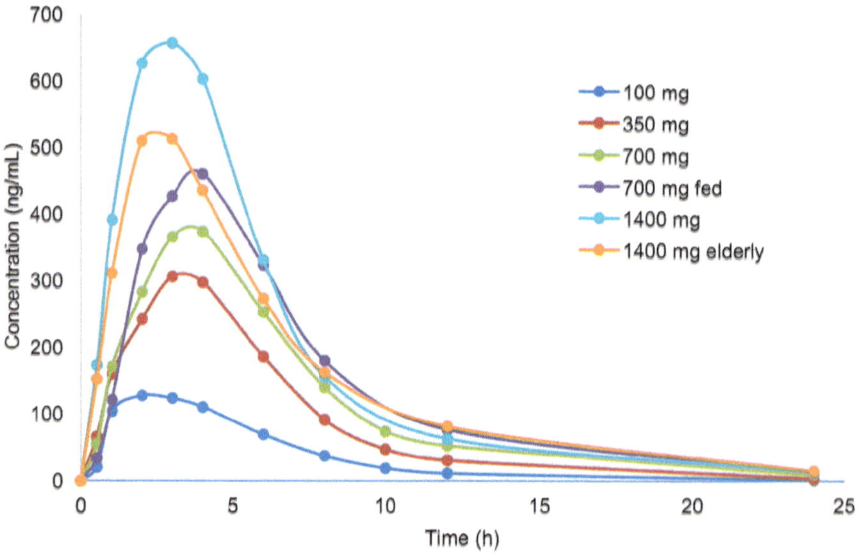

Fig. 7.7 Plasma levels of 20E following a single-dose oral administration with increasing amounts of 20E administered as capsules (Dioh et al. 2023). On this basis, it was proposed for human treatments to use two daily administrations at 12 h intervals

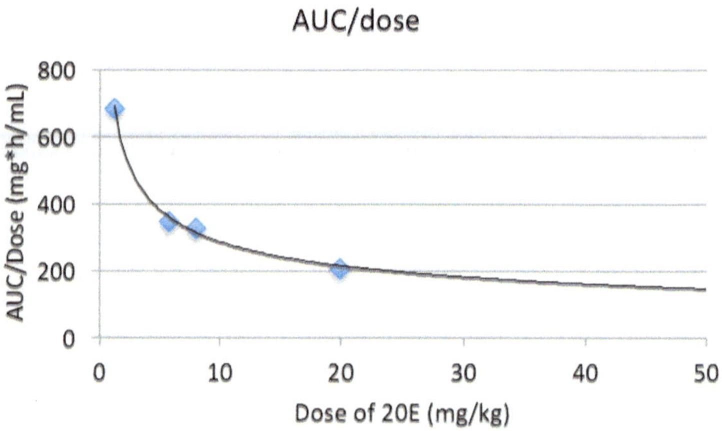

Fig. 7.8 Relationship between the administered dose and the AUC/dose in humans, which provides evidence for a saturable process (Dioh et al. 2023). The expected trend is shown beyond 20 mg/kg

This mode of representation allows an easier comparison with other animal data (Table 7.7), and (unsurprisingly) it appears that the bioavailability increases with body size, which is consistent with general correction parameters used when calculating the putative active dose in humans from data established in rodents (Reigner

Table 7.7 Comparison of AUC/dose in humans (20 mg/kg, capsules) with rats or mice (50 mg/kg, suspension)

Species	Dose (mg/kg)	AUC ng*h/mL	AUC/dose (mg*h/mL)
Mouse	50	98	1.97
Rat	50	259	5.18
Human	20	4097	205

AUC/dose values were calculated from the data of Dinan et al. (2021a) for rodents and Dioh et al. (2023) for humans

& Blesch 2002), although the ratio rodent/man is higher than usual (e.g. Nair & Jacob 2016).

Based on the quantification of 20E and its major metabolites in urine, a high individual variability was observed by Ambrosio et al. (2021). Thus, approximately 3% to 50% of the administered dose (50 mg 20E in 500 mL of an aqueous 'alcoholic' solution [presumably EtOH], but the proportion of H_2O is not defined) was recovered in urine as 20E + metabolites. This is in contrast with the lower bioavailability observed by Isenmann et al. (2024; ca. 1.7–2.6% recovered in urine after consumption of ecdysteroid-containing food). This suggests that the mode of administration (liquid or solid) may be of importance for the oral bioavailability. In particular, a higher proportion of EtOH may facilitate the ability of 20E to cross the gut (see Sect. 7.8.3.7 on Transient Permeability Enhancement).

The rather high water solubility of 20E means that it probably binds only weakly to plasma proteins, leading to a rather efficient urine elimination of the low level in the plasma. This, in addition to efficient bile secretion (as shown in rats: Dinan et al. 2021a), contributes to maintaining low plasma levels of 20E.

7.3.4.6 Metabolism

The metabolism of 20E in mammals has been investigated in mice (Lafont et al. 1988; Girault et al. 1988; Kumpun et al. 2011; Kraiem et al. 2021), rats (Gharib et al. 1993; Ramazanov et al. 1996; Dinan et al. 2021a; Balducci et al. 2021), calves (Destrez et al. 2008, 2009), and man (Tsitsimpikou et al. 2001; Brandt 2003; Bolduc 2008; summarised in Dinan et al. 2021b; Piper and Thevis 2023; Yuliandra et al. 2023), and has included misidentification of some metabolites, but has nevertheless been found to be complex. Here we provide a brief overview of the current consensus. Primary reactions involve dehydroxylation at C-14 and side-chain cleavage between C20 and C22, thereby yielding 14-deoxy-20-hydroxyecdysone, poststerone, and 14-deoxypoststerone. These primary metabolites then undergo several reactions of reduction involving, in particular, the 6-keto group (Kumpun et al. 2011). These already identified reactions are summarised in Fig. 7.9 and result in a combinatorial formation of a mixture of many metabolites, the proportions of which may greatly differ between species and even between individuals.

The 14-dehydroxylation of 20E is a typical prokaryotic (bacterial) reaction that recalls the metabolism of bile acids. It was observed long ago in mammals (Girault

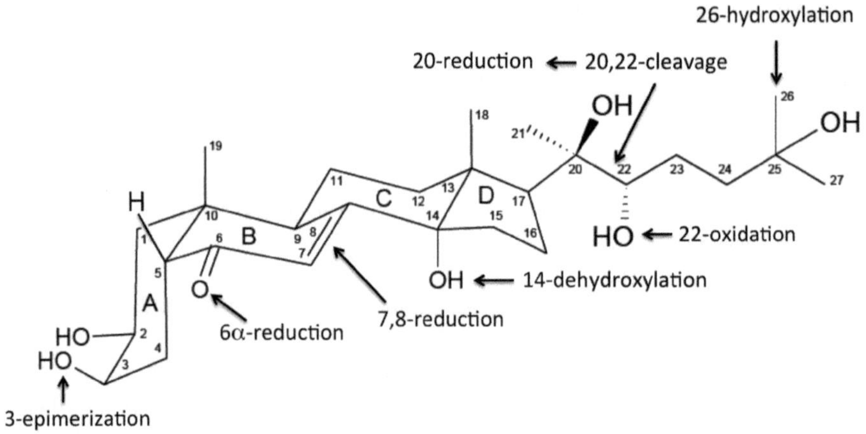

Fig. 7.9 The complex metabolic transformations of 20E in mammals and humans. A large number of metabolites can be formed, which result from one or a combination of several of these reactions so that the proportion of the different metabolites may differ with the species used, the dose administered, the duration of administration, and an individual's genetic and microbiome variation

et al. 1988; Brandt 2003) and has also been observed in insects (Hoffmann et al. 1990). Kinetic studies indicate that 20E resists the acidic conditions of the stomach after oral delivery and undergoes dehydroxylation when reaching the large intestine (Dinan et al. 2021a, b). It is expected that side-chain cleavage is also performed by the gut microbiome, perhaps with the intermediate formation of 22-oxo-20-hydroxyecdysone (by analogy with 17,20-lyase, which has an α-hydroxyketone [17α-OHprogesterone or 17α-OHpregnenolone] as substrate; Guengerich et al. 2023; Lee et al. 2024).

The total reduction of the B-ring was previously observed for ecdysone (Girault et al. 1988; Lafont et al. 1988). The tissue/cellular localisation(s) of the different reductions have not yet been identified. By analogy with the hepatic reductions of steroid hormones (e.g. the conversion of cortisol to tetrahydrocortisol), we may expect that the liver is involved through an efficient enterohepatic cycle. This again recalls the well-documented metabolism of bile acids (e.g. Cai et al. 2022).

Some additional reactions have been observed, such as 26-hydroxylation in calves (Destrez et al. 2009) and epimerisation of the 3-OH (3β-OH to 3α-OH), as observed with ecdysone in mice (Girault et al. 1988) and poststerone in rats (Balducci et al. 2021).

Some of the described reactions may become more important upon chronic administration of 20E, as evidenced by the progressive reduction of the proportion of unconverted 20E in rat faeces (Dinan et al. 2021a). It is highly possible that microbiome populations are modified or that some intestinal and/or liver enzymes are upregulated upon chronic exposure to 20E.

The physiological significance of all these metabolites is an open question; are they simply inactivation products, or do they contribute to the overall biological activity? It has already been established that side-chain cleavage products retain some biological activity (Csábi et al. 2019; Issaadi et al. 2019).

7.3.4.7 Elimination/Excretion

Excretion of 20E and its metabolites has been documented through the use of radioactive molecules in mice and rats (Dinan et al. 2021a). Two major conclusions can be drawn from those studies:

(a) Excretion in faeces by far exceeds that in urine, in accordance with a low oral bioavailability,
(b) All the administered 20E are eliminated within 72 hours following oral intake (Fig. 7.10).

The relative importance of urine as an excretory pathway for 20E may nevertheless vary among individuals (Ambrosio et al. 2021, see above). It is also much increased in rats (19%) when the administered ecdysteroid is poststerone (Balducci et al. 2021), in accordance with that molecule's higher bioavailability. The variability in humans is associated with the extent of metabolism to 14-deoxy-20-hydroxyecdysone and 14-deoxypoststerone, which have greater bioavailabilities than 20E (Ambrosio et al. 2021).

As a consequence of this rapid and total elimination along with individual variation in the metabolic profile, the proposal to detect the doping use of 20E is faced

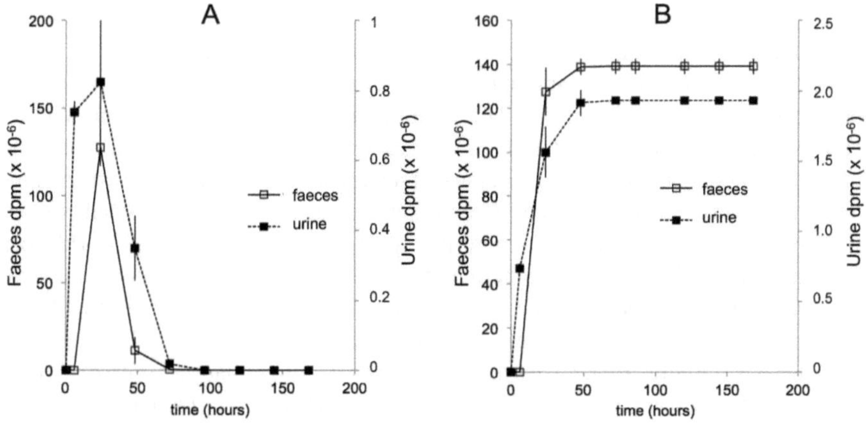

Fig. 7.10 Comparison of the temporal (**a**) and cumulative (**b**) urinary and faecal elimination of radioactivity after the oral application of [5,7,9-^{3}H]20E to 6–7 week-old male Wistar rats. After oral application of [^{3}H]20E, elimination of radioactively labelled 20E and its metabolites was completed within 48 h. The amount of radioactivity recovered from the faeces was by far the major route of excretion, since the amount in the urine corresponded to only 1.40% of the total radioactivity recovered. (Reproduced, with permission, from Dinan et al. 2021a)

with the difficulty of establishing the use of 20E if application is stopped ≥3 days before control of urine samples (Ambrosio et al. 2021). This difficulty is increased by the presence of 20E in some common food plants, such as spinach or quinoa (Dinan et al. 2020; Isenmann et al. 2024).

7.4 Sourcing and Purity

This topic is more extensively discussed elsewhere in other chapters in this book, but it is worth summarising the key points concerning the source and purity of ecdysteroidal preparations before discussing the various reports on the safety and toxicity in mammals; adverse findings can, and do, have major impacts on the development of a potential drug as an effective and commercially viable medication, so it is of existential importance for the viability of the drug to know if any adverse effects are associated with the API or any impurity. As previously discussed (Dinan et al. 2021b), the combination of low bioavailability of orally ingested 20E and the need to isolate the drug from plant sources generates a major limitation to the widespread, large-scale development of ecdysteroids as pharmaceuticals. Ecdysteroids are chemically too complex for efficient chemical synthesis from a readily available cheap precursor substance. Reported syntheses (reviewed in Kametani and Tsubuki 1980) are multistep and have reported overall yields of 1% or less and would generate a large amount of toxic waste if conducted on an industrial scale. The semisynthesis of ecdysteroids (i.e. the conversion of readily available phytoecdysteroids, such as 20E, to other desired analogues) can usually be attained in a few steps and some of them proceed with high yields (reviewed in Kovganko 1997, 1998; Kovganko et al. 2003; Coll-Toledano 2009). Thus, ecdysteroids for experimental and pharmaceutical testing require the isolation of at least 20E from plant sources in large amounts and in high purity. The initial discoveries of phytoecdysteroids (Galbraith and Horn 1966; Takemoto et al. 1967; Heinrich and Hoffmeister 1967; Kaplanis et al. 1967) were serendipitous, but already revealed the much larger concentrations and diversity of analogues present in plants when compared to insects. Since then many more plant species have been examined, so that we have a much better idea of where phytoecdysteroids occur in the world flora (Dinan et al. 2023) and which genera and species are high accumulators and what their ecdysteroid profiles are (see other chapters in this book). However, that being said, only a very few species are currently used for the commercial isolation of ecdysteroids (largely 20E): *Cyanotis arachnoidea* (Commelinaceae), *Serratula coronata* (Asteraceae), *Rhaponticum* [*Leuzea*] *carthamoides* (Asteraceae), and *Achyranthes* sp. (Amaranthaceae) (Dinan et al. 2021b). This derives not only from the need for the presence of a high concentration of ecdysteroids, but also from the presence of largely only 20E (i.e. a simple ecdysteroid profile), an amenable plant matrix which contains few compounds which co-purify with the ecdysteroids and the need for plant species which can be cultivated and grown rapidly to generate a large amount of biomass. The chosen species contain 1–5% (w/dw) ecdysteroids (cf. most ecdysteroid-containing species which contain 0.001–0.1%), so even these species require the processing of several tonnes of fresh plant material to isolate even a kilogram of purified 20E. When 20E is purified to >97% from these species, the remaining few percent are predominantly small amounts of other ecdysteroids which co-purify with 20E. In the pharmaceutical context, the supply problem is exacerbated by the low bioavailability of orally delivered 20E where only 1% of the ingested dose reaches the plasma, meaning that large doses (gram amounts) would have to be consumed by adult humans to reach the therapeutic threshold and this also enhances the possibility of undesirable side effects. In the future, it may be possible to modify plants or yeasts to produce higher concentrations of ecdysteroids, but this is currently hampered by our incomplete understanding of phytoecdysteroid biosynthesis and regulation (Dinan et al. 2021b).

It is essential that the API in a pharmaceutical preparation should be as pure as possible and that the purity is consistent from batch to batch. In the case of isolation from ecdysteroid-containing species, the compounds co-purifying with 20E tend to be ecdysteroid analogues, either endogenous to the plant or artefactually produced during processing (Tóth et al. 2021, 2023). Other components with other biological activities and toxicities may also co-purify and result in undesired side-effects, resulting in the need for extensive experimental investigation to clarify the situation. Ideally, one should demonstrate co-chromatography (preferably in two diverse systems) of the biological activity with the compound to assess if the biological activity/toxicity is truly associated with it (e.g. Dinan et al. 2001), but this is rarely done. In view of the complexities, there is an inertia to stick with the few currently used and more or less well-characterised plant sources. The registration of a new plant source for the pharmaceutical preparation of ecdysteroids is simplified if the plant is a recognised food or nutrient source, but very few of these are ecdysteroid-containing and none are hyperaccumulators (Dinan et al. 2020, 2023). Owing to the complexities of obtaining truly pure ecdysteroids from plant sources, many of the biological assessments of the effects/toxicities of ecdysteroids in animals have been performed with preparations where the precise content of a named ecdysteroid (and consequently the amount and identities of the non-ecdysteroidal components) is unknown. This is discussed in Sect. 7.7. A highly relevant example of this is the large number of preparations sold online as nutrient supplements which purport to contain significant amounts of specific ecdysteroids (20E, turkesterone), but which on subsequent analysis by independent research laboratories are revealed to contain little, or even no, ecdysteroid (e.g. Hunyadi et al. 2016).

7.5 Structure-Activity Relationships (SAR)

Only a few studies have been performed to compare the activity of several ecdysteroids in assays for effects on mammalian systems (Saatov et al. 1999; Syrov 2000; Syrov et al. 2001; Báthori and Pongracz 2005; Báthori et al. 2008; Martins et al. 2012, 2013). Most of these have not been extensive enough to be able to make more than preliminary deductions about the ecdysteroid structural features associated with biological activity (the exceptions to this, the studies by Martins et al. are discussed in Sect. 7.7.2.2). Also, the assays are complex and poorly defined; therefore, it is not known what contributions penetration, sequestration, metabolism, and rate of elimination make, in addition to the activity at the site(s) of action, which are themselves not fully identified. Also, it is not known to what extent the metabolites circulating in the plasma after oral administration or intravenous injection contribute to the detected activities at any time point over the course of exposure after administration (Dinan et al. 2021a).

7.6 Cellular Mode of Action

The cellular mode of action of ecdysteroids is an area of active research in mammals and is covered in more detail in Chap. 3. Suffice it to say here that in view of the pleiotropic effects of ecdysteroids in mammalian systems and the μM concentrations required to bring about these effects, this does not correspond to what one would expect of a classical (much more sensitive; responding to nM ligand concentrations) steroid hormone receptor system directly altering gene expression. Rather, interactions with membrane-bound receptors seem more probable, and multiple modes of action, depending on responding cell type, are to be expected.

7.7 Safety and Benefits

7.7.1 General Acute and Chronic Toxicity Studies for Ecdysteroids in Mammals

The consensus is that 20E and its metabolites (where they have been assessed) are non-toxic to mammals, but it has to be accepted that, as yet, only a few ecdysteroids have been thoroughly assessed on only a few mammalian species.

Lack of acute toxicity after per os (p.o.) or intraperitoneal (i.p.) administration to rats was demonstrated in early Japanese studies (Matsuda et al. 1970; Ogawa et al. 1974); they found an LD50 (determined over a 3-day observation period) p.o. > 9 g/kg for 20E or inokosterone, and LD50 i.p. of 6.4 g/kg (20E) and 7.8 g/kg (inokosterone). Intravenous injection of 1, 10, and 100 mg/kg in rabbits had no effect on respiration or blood pressure (Ogawa et al. 1974). More recently, 20E 2,3:20,22-diacetonide was shown to possess no severe toxicity when tested on CFLP mice at 250 mg/kg (Martins et al. 2012).

Administration of 20E or inokosterone to castrated rats was ineffective: in males, administration of 0.1, 0.5, and 5 mg/day/kg for 7 days did not alter the weight of prostate, seminal vesicles, or *levator ani* muscle, and, in females, administration of

0.1, 1.0, 10, and 100 mg/day/kg for 3 days had no effect on uterine weight when compared to non-treated controls (Ogawa et al. 1974). Thus, sex-hormonal effects were not observed. This was confirmed more recently in castrated female rats (Seidlova-Wuttke et al. 2010).

Daily intraperitoneal injection of 20E into mice at up to 1 mg/day/mouse over 30 days (total of 100 mg 20E/g) did not result in any observed toxicity (Stopka et al. 1999).

In rats receiving daily subacute doses of 200–2000 mg/day/kg of a 20E:inokosterone (1:1) mixture for 35 days, no changes were observed for blood parameters (erythrocytes, haemoglobin, cholesterol, and transaminase activity) or in the weight of various organs (pituitary, thyroid, thymus, lungs, heart, kidneys, adrenals, prostate, testes, epididymis, seminal vesicles, uterus, and ovaries), and no toxicosis was observed (Ogawa et al. 1974).

In their study on the comparative bioavailabilities of 20E and 20E 22-undecanoate, Udintsev et al. (2014) showed that these compounds were not acutely toxic to rats after intragastric administration at doses up to 2.0 and 2.8 g/kg, respectively, while temporary side effects (salivation and inertia) were observed at 1.5 and 2.0 g/kg.

More recently, during clinical studies, humans receiving pure 20E (>97%) at up to 900 mg daily for 14 days (in a Phase I clinical trial; Dioh et al. 2023) and 350 or 700 mg daily for up to 9 months (in a Phase 2b clinical trial; Fielding et al. 2025) showed no serious adverse events, and the patients tolerated the drug well.

7.7.2 *More Specific Effects of Ecdysteroids in Mammals*

We briefly summarise the more specific effects which have been accorded to ecdysteroids according to area of medical interest. We and others have reviewed the literature in this area before (Lafont and Dinan 2003; Dinan and Lafont 2006; Lafont and Dinan 2009; Cahliková et al. 2011). The aim here is to demonstrate the breadth of potential applications of ecdysteroids, to give an idea of the balance between 'positive' and 'negative' reported findings and to critically review them. Where deleterious/toxic effects of ecdysteroids in mammals are reported, they are included below. However, it is important to mention at the outset that several of these reports can be criticised on the basis of the experimental approach or the uncertain purity of the ecdysteroid preparations used.

7.7.2.1 Genotoxicity and Mutagenicity

Neves et al. (2016) observed some genotoxic effects of a 10% aq. MeOH extract of roots of *Pfaffia glomerata* (Brazilian ginseng) standardised on the content of 20E (only 1.93%) in two assays: rodent bone marrow micronuclei and *Allium cepa* assays. The significance of these findings for the genotoxicity of 20E is dubious because the toxicity could derive from other components in the rather crude extract, and no controls were performed with pure 20E. Also, *P. glomerata* itself may not contain genotoxic components, since the extraction with 10% aq. MeOH was performed over 15 days at room temperature, which could have resulted in microbial contamination. Further, this Brazilian study is in strong discrepancy with other authors who described genoprotective properties of ecdysteroids against poisoning in rats, Gubsky et al. (1993) showed a hepatoprotective effect of injected ecdysteroids against the damage induced by tetrachloromethane. Levitsky et al. (1993) obtained similar genoprotective effects against damage induced by phosphorus oxychloride (summarised in Chabanny et al. 1994).

As concerns mutagenicity, all available papers describe antimutagenic properties of 20E (Gagalac-Nicolas and Lim Sylianco 1981; Gubsky et al. 1993; Zainullin et al. 2003; Kudyasheva et al. 2012). 20E was shown to reduce the number of revertant colonies of *Salmonella typhimurium* TA 1535 and TA 1537 caused by two mutagens, sodium nitrite and quinacrine (Gagalac-Nicolas and Lim Sylianco 1981). Kudyasheva et al. (2012) showed a prophylactic/therapeutic antimutagenic effect on mice of 20E (50 mg/kg) or inokosterone (5 or 15 mg/kg) against lower dose (22.6 cG) γ-irradiation (see also Yang & Pan 2019, for radioprotective effects of 20E).

7.7.2.2 Cancer Promoting and Cancer-Protective Effects

Two articles from the same laboratory (El-Mofty et al. 1987, 1994) describe carcinogenic effects of *ecdysone* (not 20-hydroxyecdysone). Their choice of this ecdysteroid is related to its presence in the bracken fern (*Pteridium aquilinum*; Kaplanis et al. 1967), which is known to be carcinogenic. They used commercial ecdysone preparations (Fluka, Sigma) but do not seem to have verified the purities of these preparations for themselves. In their first experiment, they used Egyptian toads (*Bufo regularis*; not a mammal!) receiving only 0.1 mg/kg per os 3 days a week for 4 months: 10 of the 58 treated animals developed various tumours. It is not detailed how many control animals were used and whether they received saline or nothing. In an older study, performed by others (Ogawa et al. 1974), application of 200, 400, or 600 mg/kg of 20E or inokosterone into the lymph sinus of bullfrogs (*Rana catesbeiana*) had not resulted in any significant effects. In El-Mofty et al.'s second experiment, the authors used mice (100 treated and 50 controls and commercial ecdysone from Sigma). Treated mice received 36 μg/kg of ecdysone twice a week for 22 months, whereas controls received 0.3 mL saline. Among the ecdysone-treated mice, 23 developed, in particular, breast and lung tumours. These results are somewhat puzzling, given the very low doses administered to both the bullfrogs and the mice.

By contrast, most other laboratories (see, e.g. Martins et al. 2013, 2015; Shuvalov et al. 2020; Romaniuk-Drapala et al. 2021) have described no cancer-promoting effects for ecdysteroids, but none of them has tested ecdysone itself, which is a rare phytoecdysteroid. Lagova and Valueva (1981) found that 20E (daily subcutaneous injections of up to 300 mg/kg for 5 days into mice and rats) had no effect on the growth of several tumour types, but it stimulated the growth of mammary gland carcinomas.

Cancer arises from loss of regulation of cell growth, which can occur by combinations of several subtle changes in regulatory processes which ultimately result in unrestrained cell growth. Rather than being one disease with common mechanistic causes, cancer is a broad family of diseases where causes (and treatments) vary significantly. Thus, different cancers in the same organ can have quite different aetiologies and require different approaches to their successful control. The study by Romaniuk-Drapala et al. (2021) demonstrates the complexities of the possible responses of cancerous cells to ecdysteroids. The authors compared the responses of 3 breast cancer cell lines which differed in their hormone responsiveness in various permutations (+/− oestrogen receptor and progesterone receptor; +/− human epidermal growth factor receptor 2) and a non-tumorigenic breast cell line to 3 ecdysteroids (20E, ajuC, polB). Firstly, none of the ecdysteroids was found to affect the growth and viability of the non-tumorigenic MCF10A cell line and polB did not affect the viability and cell-cycle distribution of the 3 breast cancer cell lines. On the other hand, 20E and ajuC reduced the viability of MDA-MB-23 (ER/PR -; HER2 -), while increasing the viability of MCF-7 cells (ER/PR +; HER2 -; MKI67 - [a marker of cell proliferation]). 20E decreased the viability of T-47D cells (ER/PR +; HER2 +/-; MKI67 +), whereas ajuC increased the viability in these cells. The 20E response in MDA-MB-23 and T-47D cells was associated with proapoptotic activity as revealed by several biomarkers. Thus, not only do different breast cell lines respond differently to ecdysteroids, but also the nature of the response in cancerous cells depends on the structure of the ecdysteroids (which differ only by the presence/absence of hydroxyl groups at C-5β, C-11α, and C-25). These findings substantiate another similar study comparing the responses of various breast cancer cell lines and non-cancerous cells (Shuvalov et al. 2020), where 20E was also found to synergise the induction of cell death by doxorubicin in all the tested breast cancer cell lines. The same group (Shuvalov et al. 2023) have investigated the antineoplastic activity of 20E (0.1–100 μM) on 3 human non-small cell lung cancer cell lines (A549, H1299, and H460). They showed that 20E reduced reactive oxygen species (ROS) by up to 20%, which at least in part was a consequence of induced expression of several antioxidative genes. 20E also slightly reduced growth of the lung cancer cells, but did not enhance the death rate or apoptosis. 20E treatment down-regulated respiration, glycolysis, one-carbon metabolism, and ATP levels, which is significant because enhanced glycolysis is a characteristic of the cancerous state and one-carbon metabolism is significant in nucleotide synthesis, methylation reactions, and generally in providing cancer cells with anabolic capacity, so 20E broadly impacts the cancer cells' ability to grow rapidly. 20E also synergised the activity of several metabolic inhibitors and reduced the expression of several cancer stem-like cell markers. Further, the authors demonstrated that most of the enzymes of glycolysis, serine biosynthesis, the mitochondrial folate cycle, and the master regulators c-Myc and ATF4 (CREB-2) which are down-regulated by 20E in the cancer cell lines are highly expressed in human lung cancer patients and strongly associated with shortened survival rates. In contrast, in the non-transformed mouse cell lines C2C12 (myoblasts) and NIH/3 T3 (embryonic fibroblasts), 20E enhanced glycolysis, respiration and ATP production, and protein synthesis. This induction of energy metabolism is mediated through the PI3K/AKT/mTOR pathway (Shuvalov et al. 2024).

The contribution that ecdysteroids might make to the alleviation of multidrug resistance in human cancers has been broached in a series of papers by Martins et al. (2012, 2013, 2015). Multidrug resistance can be associated with the presence of ABCB1 (P-gp1) efflux pump, and the authors assessed the ability of a wide range of ecdysteroid analogues to affect the proliferation of cancer cells (ABCB1+ or ABCB1-) in the presence or absence of doxorubicin. In the first study (Martins et al. 2012), 58 natural and hemi-synthetic ecdysteroids were assessed and it was found that, by themselves, they showed no or very few antiproliferative or cytotoxic effects. However, when 20E was applied to the ABCB1+ cells in conjunction with doxorubicin (at various ecdysteroid/doxorubicin ratios), strong synergistic or antagonistic effects were observed. Ecdysteroid acetonides (especially the di- and tri-acetonides (e.g. 20E 2,3:20,22-diacetonide) were particularly synergistic (i.e. more strongly antiproliferative), whereas most natural ecdysteroids were antagonistic (proliferative). In view of the synergistic activity of the acetonides, 32 non-polar analogues of 20E 2,3:20,22-diacetonide were synthesised and assessed on mouse L5178 T-cell lymphoma cells (with or without transformation to express ABCB1; Martins et al. 2013). All the new derivatives were found to be synergistic with 28-diethyl-2,3-*O*-methylidene-20E being more active than 20E 2,3:20,22-diacetonide, 20E 2,3-monoacetonide, or 20E 20,22-monoacetonide. In the third phase (Martins et al. 2015), the study was extended to 6 other cell lines (MCF7 [human mammary gland] and KB-3-1 [human cervix], and their ABCB1 expressing sub-cell lines, and the androgen-dependent LNCaP and androgen-independent PC3 prostate cancer cell lines) as well as the previously used L5178 (mouse lymphoma: ABCB1- and ABCB1+ lines) and assessed synergism with doxorubicin, paclitaxel, vincristine, and cisplatin, while focussing on 3 ecdysteroids (20E, 20E 20,22-monoacetonide and 20E 2,3:20,22-diacetonide). The three ecdysteroids, by themselves, had very low toxicity (>80 μM) to all 8 cell types. 20E20:22-acetonide and 20E2,3:20,22-diacetonide were able to exert a sensitising effect to doxorubicin, paclitaxel, and vincristine in all the multidrug resistant cell lines, and to cisplatin in some of the cell lines. Further, the synergistic effect of 20E diacetonide could be enhanced by enclosing it in liposomes, presumably by enhancing solubility and entry to the cells.

7.7.2.3 Hepato-Toxicity/Protection

There are no reports for 20E toxicity to the liver. On the contrary, hepatoprotective actions in rats of plant extracts rich in phytoecdysteroids or 20E and other purified ecdysteroids have been documented in several experimental intoxication protocols with heliotrine (an alkaloid used as a model for hepatitis: Syrov et al. 1986; Badal'yants et al. 1996; Syrov et al. 2004), carbon tetrachloride (Syrov et al. 1992; Xia et al. 2013b), or indomethacin (Ramazonov et al. 2016). 20E also protected the liver from non-alcoholic fatty liver disease (NAFLD) development induced by hypercaloric/hyperglucidic diets (in rats: Li et al. 2010; in gerbils: Agoun et al. 2021). A recent paper (Kang and Li 2022) has shown that makisterone A (24-methyl 20E) attenuates experimental cholestasis (dysregulation of bile synthesis and release in the liver) by activating the nuclear receptor FXR. Makisterone A was identified as a potential regulator of FXR by the screening in silico of a database of compounds found in herbal medicines. This was then supported by biochemical assays to show that MakA activates FXR transcriptional activity in HepG2 cells at μM concentrations and reduces liver damage in a mouse model of cholangitis (inflammation/swelling of the bile duct).

7.7.2.4 Renal Effects: Damage-Inducing or Protective?

Two articles from the same laboratory (Lu et al. 2018a,b) have observed renal damage-inducing effects of ecdysone (but not 20E) similar to those caused by excess aldosterone. They described glomerular injury and proteinuria in mice after daily subcutaneous treatment with 6 mg/kg of ecdysone (Sigma-Aldrich) for 2 weeks. The authors did not verify the purity of the purchased ecdysone, nor did they demonstrate that the mineralocorticoid-like activity co-chromatographs with ecdysone. The authors suggest that ecdysone and aldosterone are structurally very similar, but this is based on a 2D structural comparison which is of limited biochemical significance. They related the experimental results to the effect of excess mineralocorticoids (aldosterone) and supported the similarity by in silico docking experiments showing ecdysone binding to the mineralocorticoid receptor, but this was not substantiated by receptor ligand displacement assays. They also observed the damaging effects of ecdysone (0.1 μM) on cultures of immortalised mouse podocytes. The authors do not explain why they focussed on ecdysone and not 20E, and in their conclusions they alert those using ecdysteroid nutritional supplements, even though none of them contains this rare phytoecdysteroid.

By contrast, several articles describe kidney-protective effects against experimental toxic renal damage, but for 20E or turkesterone, not for ecdysone. These inlude intoxication with uranyl acetate (Syrov et al. 2001), TGF-β1-induced renal cellular fibrosis (Hung et al. 2012), diabetic nephropathy (Gui et al. 2014), and daunorubicin-induced nephrotoxicity (Khudair et al. 2025).

7.7.2.5 Reproduction

Kosar et al. 1997 is the only article published so far on embryotoxicity of ecdysteroids. The authors tested the effects of 20-hydroxyecdysone and polypodine B (>96% purity) on the development of chicken embryos (again not a mammal). The compounds were administered in ethanol solution (3 μL of 10^{-5} to 10^{-3} M) placed in the vicinity of embryos, and some toxicity was observed with the highest administered dose (i.e. 1.5 μg) of both compounds. The susceptibility to toxicity showed a decreasing trend (not statistically significant) as the development of the embryo proceeded.

As concerns the lack of interference with sex steroid hormones, several authors indeed failed to observe any direct binding to the nuclear receptors of androgens and oestrogens in vitro. Docking studies in silico have provided somewhat contradictory results (Lapenna et al. 2015; Mo et al. 2021; Zaghloul et al. 2023). There is some pharmacological evidence for an (indirect) interaction of 20E with a membrane-bound isoform of oestrogen receptor (Lafont et al. 2021), and this topic is still under investigation. On the other hand, Parr et al. (2014) proposed that 20E binds an ERbeta receptor isoform (nuclear or membrane form?); see Chap. 3 in this volume.

Interference with the reproductive functions of male and female mice has been studied by Stopka et al. (1999). Mice were injected intraperitoneally with 0.1, 0.1, or 1 mg 20E daily for 30 days. They observed an increase in the duration of the oestrous cycle and a small reduction in sperm count and seminal vesicle weight. Enhanced sexual activity was observed in adult male rats after per os application of 20E (5 or 10 mg/kg: Mirzaev and Syrov 1992; Mirzaev et al. 2000).

The administration of 20E (Ecdysten, 5 mg, twice a day for 30 days) to men with an infertility diagnosis (disturbed spermatogenesis as a complication of some urological diseases) increased copulative function and improved sperm quality (Mirzaev et al. 2000). The same treatment administered to patients in the stage of recovery from myocardial infarction also improved their sexual function (Mirzaev et al. 2000). By contrast, Bandara et al. (1989) described in vitro spermicidal activity (immediate immobilisation) of 20E on human sperm, but when using a very high concentration (20 mg/mL = 40 mM).

7.7.2.6 Effects on the Nervous System

Once again no toxic effects have been reported and, on the contrary, several neuroprotective effects have been described, including antinociceptive effects (Sarpong et al. 2016; Archana et al. 2022; Frachet et al. 2025), related to 20E's anti-inflammatory and antioxidant activities. It is suggested that 20E could find application in the relief of pain associated with diabetic and chemotherapy-induced neuropathies.

Many efforts have been devoted to developing treatments for Alzheimer's disease, but these are faced with the difficulty of crossing the blood–brain barrier. In studies in vitro, Chakraborty and Basu (2017) showed that 20E can inhibit amyloid-β peptide formation and aggregation in the formation of amyloid plaques. Gholipour et al. (2025) showed the beneficial effects of oral 20E (10 mg/kg/day) together with high-intensity interval training in a rat model of the disease. Recently, Xing et al. (2025) have developed a very promising brain-targeting nanodelivery system (CMCS-CPBA-Ang2/20E) based on carboxymethyl chitosan (CMCS) functionalised with 4-carboxyphenylboronic acid (CPBA) and further modified with angiotensin-2 peptide to provide targeting to lipoprotein receptor-related protein (LRP) on target human cerebral endothelial cells. This construct provides a hydrophobic nanocarrier which can encapsulate 20E and which can cross the blood–brain barrier and locally release 20E by a ROS-responsive mechanism and reduce off-target side effects.

Three calonysterone-related analogues (one definitely artefactual and the other two possibly so; Tóth et al. 2023) have been assessed in a model system (hCNEC/D3 human brain microvascular endothelial cells) to protect the blood–brain barrier against production of ROS brought about by *t*-butyl hydroperoxide or proinflammatory cytokines (TNF-α and IL-1β) and all 3 compounds were found to have significant effects in protecting the cells at 1 μM. It would be interesting to learn if this protective effect is limited to such unusual analogues (which do not possess the typical ecdysteroid chromophore) or is more widely associated with more common/natural ecdysteroid structures.

7.7.2.7 Diabetes

20E and inokosterone did not alter normal glycaemia in mice, but they suppressed hyperglycaemia evoked by glucagon in the case of diabetes induced by alloxan treatment (which eliminates pancreatic beta cells producing insulin) (Yoshida et al. 1971). Similar conclusions were found more recently in diabetic gerbils (Mallek et al. 2018; Fig. 7.11) and for ecdysteroid-containing aqueous leachates of quinoa (*Chenopodium quinoa*) seeds (0.89% 20E, 1.0% total ecdysteroids; Graf et al. 2014).

Injected 20E (10 mg/kg) reduced diabetic cardiomyopathy in streptozocin-induced rats by lowering hyperglycaemia, hyperlipidaemia, and inflammation (Xia et al. 2013a). Further, 20E promotes proliferation and differentiation of rat

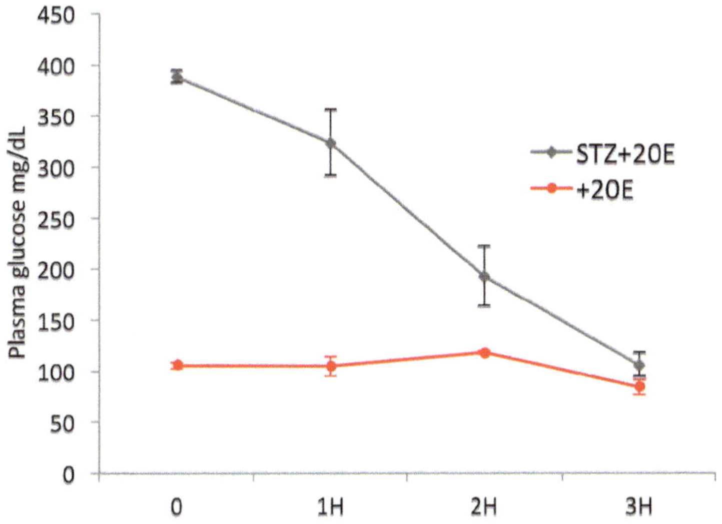

Fig. 7.11 Effects of 20-hydroxyecdysone (5 mg/kg, i.p.) on plasma glucose in STZ-treated and control male adult gerbils. Values are given as mean ± SEM for 9 animals in each group (reproduced from Mallek et al. 2018)

osteoblasts induced by high glucose levels by inhibiting the p38 MAPK signalling pathway (Zhao et al. 2021).

7.7.2.8 Wound and Bone Healing

An anti-gastric ulcer activity of 20E has been shown (Syrov et al. 1989); 20E can improve the high expression of EGF, EGFR, and PGE2 in gastric tissue so as to promote gastric ulcer healing processes (Zhou et al. 2010).

20E is also efficient against allergic reactions (Bouyahya et al. 2020). It was recently used to treat atopic dermatitis (Khan 2025).

7.8 Some Further Potential Approaches to Improving the Bioavailability of Ecdysteroids

Many methods have been proposed for the improvement of the bioavailability of drugs (reviewed in Hetal et al. 2010; Wu et al. 2025). Some of these are not applicable to 20E because they are intended for compounds with low water solubility or which are ionisable, neither of which applies to 20E. Below we summarise some of the potential methods which, at least theoretically, might be applied to ecdysteroids:

7.8.1 Chemical Structure

An analysis of >1100 drug candidates (Veber et al. 2002) found that greater rigidity (low number of rotatable bonds: <10) and low polar surface area (<140 Å or H-bond count: ≤12) are important predictors of good oral bioavailability. Further, a lower MW of 300–350 may be associated with higher bioavailability (Lipinski et al. 2001) than the 500 used in Lipinski's Rule of 5. It may seem a tall order to find appropriate ecdysteroids, but there are a few ecdysteroids which fulfil these criteria (e.g. rubrosterone: Table 7.3). Further, it is worth pointing out that smaller non-steroidal analogues (diacylhydrazines: Dinan and Hormann 2005) of ecdysteroids have been identified as ecdysteroid agonists in insects and several of them have been developed as commercial insecticides, so the identification of chemically simpler classes possessing the same biological activity is not impossible, although the development of such agents would require far better understanding of the mode of action and SAR in mammals than we currently have.

7.8.2 *Route of Application*

Absorbed molecules from the mouth and rectum avoid gastric and hepatic first-pass metabolism to a large extent. Further, small-molecule enzyme metabolic activities in the mouth and colon are relatively low, and pH values are close to neutral (saliva: pH 5.5–7; rectal fluid: pH 7–8; Backsay et al. 2025). Permeabilities are generally much higher than for transdermal applications. Rectal application of drugs by means of suppositories is possible but not popular with many patients. Backsay et al. (2025) review the oral mucosal dosage of drugs for local and systemic therapies. The volume of the adult mouth is ca. 110 mL which allows significant volumes of drug solution to be held to allow absorption. This could limit the pharmacological dose of poorly water-soluble drugs, but could allow up to 1 g 20E to be applied in one dose. Alternatively, the drug can be incorporated into a gum and chewed to provide a slow-release formulation (medicated chewing-gum).

7.8.3 *Formulation*

7.8.3.1 Inclusion Complexes in Addition to Cyclodextrins

Liposomes (phospholipid vesicles) and niosomes (non-ionic surfactant vesicles) can be used to encapsulate hydrophilic drugs and enhance permeability and bioavailability. It has been proposed that the liposomes accumulate at the brush-border membranes of intestinal enterocytes and increase the drug gradient across the intestinal epithelium, thereby increasing the systemic concentration (Hetal et al. 2010).

7.8.3.2 Enhanced Dissolution/Solubility

Amorphous and nanoparticulate preparations can improve solubility. Smaller particle size results in a greater rate of dissolution because of the greater surface area; the final soluble concentration remains the same, but the rate at which that concentration is reached is enhanced. 20E (and other ecdysteroids) forms hydrated crystals, which require more energy to dissolve, so anhydrous, amorphous preparations should possess greater rates of dissolution.

7.8.3.3 Inhibition of Hepatic First-Pass and Gastric Metabolism

Co-administration of metabolic inhibitor: ritonavir (originally developed as an HIV/AIDS inhibitor) is a potent inhibitor of cytochrome P450-containing metabolic enzymes, particularly CYP3A4, which is involved in the metabolism of many xenobiotics, such that co-administration with another drug can inhibit intestinal and first-pass hepatic metabolism of that drug (Hetal et al. 2010).

7.8.3.4 Administration of a Biostere/pro-Drug to Reduce Pre-Systemic Metabolism

A biostere is an analogue where, for example, H-atoms have been replaced by fluorine; the much stronger H-F bond can result in considerably reduced rates of metabolism (Hetal et al. 2010; Wu et al. 2025). Fluorinated ecdysteroids have already been chemically synthesised (Tomás et al. 1992; Csábi et al. 2016), and these can be expected to retain biological activity because the small size of the F-atom will not change the overall volume of the molecule by much. Deuterated biosteres may also lower susceptibility to metabolism (Wu et al. 2025). Prodrugs are analogues, which in themselves are not (or less) biologically active, but which convey the ability to avoid/reduce gastric or hepatic metabolism and can be metabolised in the systemic system to release the active drug (Hetal et al. 2010).

7.8.3.5 Enhancing Gastric Retention

In 'floating dosage' formulations, the drug is incorporated into a swellable matrix (polysaccharide, polycarbonate, polyacrylate, polystyrene, etc.) which expands on contact with the gastric contents to produce a gel-like matrix which is retained in the stomach and acts as a slow-release, sustained source of the drug. A variant on this approach is to use mucoadhesive polymers (either synthetic or seminatural) or gums which have an affinity for the lining of the stomach and hence are retained there longer (Hetal et al. 2010). This could be useful for ecdysteroids, as 20E does not appear to be affected by the low pH of the stomach.

7.8.3.6 Inhibition of Efflux

Intestinal epithelial cells have the capacity not only to take up drugs from the lumen but also, to a greater or lesser extent, to return them to the lumen, thus lowering bioavailability. Efflux by epithelial cells mediated by P-glycoprotein (a membrane transporter protein) can be inhibited by verapamil and a range of natural flavonoids (Hetal et al. 2010; Wu et al. 2025). Certain surfactants also affect P-glycoprotein efflux and increase co-administered drug permeability (Hetal et al. 2010).

7.8.3.7 Transient Permeability Enhancement (TPE)

TPE is applicable to macromolecular drugs and small-molecule drugs which are poorly absorbed in the gastrointestinal tract. The TPE preparation usually consists of the API together with a mixture of excipients leading to a suspension of the hydrophilic drug in a hydrophobic medium. TPE appears to act paracellularly by temporarily (60–90 min) opening up the tight junctions between gastric enterocytes, thereby enhancing permeability along the length of the intestine (Hetal et al. 2010).

7.8.4 *Peptide Conjugates*

Recently, coupling of 20E with peptides through the C-6 position has been proposed to produce derivatives with selective properties depending on the nature of the peptide (Lesma et al. 2016). These compounds have been explored as putative multi-drug resistance inhibitors in cancer therapy. It is possible that they could allow for specific tissue targeting of 20E. Attaching a peptide to the C-6 position is attractive, as a 6-carboxymethyl derivative attached to albumin retains the biological activity of 20E since this conjugate is capable of activating the Mas1 receptor in mouse C2C12 cells (Lafont et al. 2021; see also Chap. 3 in this book).

7.8.5 *20E/PLGA Complexes as Implants for Local 20E Delivery*

Poly(lactic-co-glycolic acid) (PLGA), a biodegradable polymer formed by the random polymerisation of lactic acid and glycolic acid monomers, is widely used in pharmaceutical and biomedical engineering owing to its excellent biocompatibility. PLGA also functions as a localised drug delivery carrier, enabling precise drug release and increasing local drug concentration. It was used by Luo et al. (2025) as a scaffold to promote repair of bone defects by 20E in a diabetic rat model.

7.9 Discussion and Future Prospects

Too few ecdysteroid analogues are currently available in larger amounts and at sufficient purity and reasonable cost. Currently, only 20E fulfils these criteria. A vast array of chemically diverse ecdysteroid analogues are known (Lafont et al. 2002), but at best most of these are only available in mg amounts. The prospects for full synthesis of ecdysteroid analogues are, and will remain, poor. All the currently known high-accumulating plant species contain predominantly 20E, but only about 5–6000 of the probable 3–400,000 terrestrial vascular plant species have so far been investigated in any way for their phytoecdysteroid content (Dinan et al. 2023). Even the few most extensively studied phytoecdysteroid-containing species are still yielding new analogues on closer investigation (e.g. *Cyanotis arachnoidea*; Tóth et al. 2021). Extensive chemical procedures for ecdysteroids have been developed, such that hemisynthesis of significant amounts of a more extensive range of analogues from 20E would be possible. Further, other techniques such as fungal conversions could be used to generate specific analogues (e.g. rubrosterone: Li et al. 2015). The ready availability of a wider range of ecdysteroid analogues would greatly facilitate studies into the mode of action of ecdysteroids in mammals, QSAR of receptor binding sites, pharmacokinetic studies, improved bioavailability, and ultimately further clinical trials.

Investigation of the distribution, metabolism, and elimination of exogenous ecdysteroids in mammals requires the ability to follow the fate of the molecules in the mammal. Traditionally, this would be done by the use of radiolabelled (^{3}H or ^{14}C) molecules. In addition to the adverse safety aspects of this approach for studies in mammals, previous studies on ecdysteroid metabolism have revealed two other major limitations to the use of radioligands: the importance of the location of labelling and the need for stable labelling. Since side-chain cleavage (conversion of 20E to poststerone) is a significant metabolic route, it would be desirable to have radiolabelling in both the steroid rings and the side-chain portions of the molecule, so that the fates of both parts can be followed. Owing to the unavailability of stably labelled radioactive ecdysteroids, 20E may be tritiated by exchange labelling from $^{3}H2O$. Unfortunately, this form of labelling is not stable in biological systems, and the ^{3}H can exchange back to yield tritiated water, leading to a mixture of radiolabelled starting material, metabolites, and water. The radiolabelled water can be removed by lyophilisation, but this results in quantification problems, and the labelling may not be exchanged at the same rate from each position of the ecdysteroid molecule. More recent methods for tracking the fate of biomolecules of interest exist, and two worthy of mention are stable deuterated labelling and mass defect HRMS. Stable deuterium labelling requires the chemical synthesis of appropriate intermediates which can be reacted with deuterium-labelled reducing agents to yield analogues of the ecdysteroid with molecular masses which are x amu higher (where x = the number of H-atoms in the molecule replaced by D-atoms) than the natural ecdysteroid, and these can be detected and quantified by mass spectrometry (e.g. Piper and Thevis, 2023). The main advantages of D-labelling are the enhanced safety and the extended shelf-life of the derivative, as it is no longer subject to radioactive decay and possesses the shelf-life of the natural compound. The above-mentioned caveat for radiolabelling is still applicable, since it would be advantageous to have labelling in both the steroidal nucleus and the side chain. Mass defect HRMS (see Zhang et al. 2009 and Sleno 2012 for introductions to this topic) can be used to follow metabolism in biological systems by exploiting the difference between a molecule's exact mass and its nominal mass, focussing on the parent compound and its metabolites (which will have similar mass defects) and filtering out the molecules which do not have similar mass defects. This approach has the enormous advantage that one can follow the administered drug directly, but it does require a sophisticated HRMS machine and dedicated software.

The poor oral bioavailability requires the administration of high doses to obtain effects. Improvement of the bioavailability would reduce the need for high dosage levels and, simultaneously, lower the probability of side effects occurring. There is therefore a need for improved formulations, the use of alternative ecdysteroids with better bioavailability but which retain biological activity (e.g. less polar molecules have better bioavailability, as observed for poststerone [Balducci et al. 2021], possibly complexed with a cyclodextrin) or the design of prodrug molecules with better bioavailability. Other routes of application (nasal, buccal, and rectal), which are associated with higher permeation and avoid first-pass metabolism, should be assessed (see Madden 2022 and Dilda et al. 2025 for the potential of nasal application).

Much progress has been made in the identification of the metabolites in the complex metabolism of 20E in model mammalian and human systems, but the associated biochemistry is currently little understood. The enzymology and regulation of the metabolism of exogenous ecdysteroids are virtually unknown. It is not known with certainty whether these reactions occur in the microbiome and/or gut wall/liver and whether ecdysteroids act as substrates for enzymes of steroid hormone, bile acid, or xenobiotic metabolism (or a combination thereof). Further, it is not clear what the contribution of these metabolites is to the biological activity of 20E. Such knowledge will probably open up new routes to the enhancement of the bioavailability and efficacy of ecdysteroids as pharmaceutical agents.

Individual variation in the metabolic response to drugs is an increasing area of study, and this has to be borne in mind as research on ecdysteroids as potential medicines progresses. The microbiome differs significantly among individuals (see Dioh et al. 2023 as concerns initial evidence on side-chain cleavage of 20E and its metabolites). This could be related to the genetics or food habits of the individuals, as it is clear that there is a complex interplay between genetics, diet, and composition of the gut microbiome. This variability may have consequences for observed biological effects of ecdysteroids in humans. Experiments should be performed with axenic model animals to better understand the contribution of the gut microbiome to the metabolism and efficacy of exogenously applied ecdysteroids.

Up to now, there are no unequivocal findings of toxicity for 20E and its metabolites, but many studies report a lack of toxicity. The situation is less clear for the structure-related ecdysone, where there are a few reports of negative effects, but these are not large in number, and this needs to be investigated further, even though ecdysone is a very rare phytoecdysteroid.

There is a need for pharmacokinetic studies with other ecdysteroid analogues to assess their bioavailabilities and metabolism.

There is a need for extended QSAR for the various effects observed in mammals. The potential is there because of the large range of ecdysteroid analogues (>500) known from plants, fungi, and animals (Lafont et al. 2002), which can be further extended by semisynthetic derivatives, but where this has been done so far, too few analogues have been assessed to provide significant conclusions. There are clear reasons why this has not yet been undertaken, ranging from the cost and commercial availability of only a few of the many known ecdysteroid analogues, through to the complexity of undefined bioassays where the various contributions of distribution, sequestration, metabolism, and elimination cannot be separated from potency at the site of action, to the cost and legal complexities of conducting experiments on model mammals. Clarification of the mode(s) of action of ecdysteroids in mammals would facilitate the development of cellular-based assays where the competing processes could be properly assessed and quantified, and smaller amounts of a larger number of highly purified ecdysteroid analogues could be used to generate meaningful and predictive QPAR/QSAR.

The availability of many mammalian cell lines (including human ones) provides the opportunity to investigate aspects of the pharmacokinetics where one process predominates and the others are minor or not present at all. One such is the Caco-2 (human colonic adenocarcinoma cancer) cell line which forms tight confluent polarised monolayers formed of columnar cells with microvilli on the apical membranes and tight junctions between neighbouring cells and can be used as a model for permeability across the gut wall (van Breemen and Li 2005; Sun et al. 2008). In the standard version of the assay (Hubatsch et al. 2007), the monolayers are grown over a period of 3 weeks on supports in 12-well plates so that the monolayer, once developed, forms a barrier between the upper and lower part of the well, such that compounds can be added to the upper or lower chamber and the ability of the compound to cross the monolayer and/or be metabolised can be measured over time. Although this is not a perfect model for intestinal epithelial cells (e.g. the expression of cytochrome P450 enzymes, particularly CYP3A4, is reduced; van Breemen and Li 2005), it provides a clearer picture of the factors influencing permeability for a drug or class of drugs in isolation from all the other components affecting bioavailability. The long time required to develop the confluent monolayers (21 days) and the rather narrow window over which they can be used for assays (day 21–day 28) are a limitation to throughput, but it has been found that the monolayers may be used up to 3 times during the window (Pires et al. 2021).

The development of modern drugs is associated with the accumulation of massive amounts of theoretical and experimental data, and it is necessary to accurately curate the information and maintain a reliable overview. This is especially important as nowadays the data can be generated by many people in different sections of the organisation or subcontracted companies over an extended period of time, with no one person being responsible for the collation. It is important to have an effective data strategy in place from the start, as this avoids many later problems. Paper records have been replaced by Electronic Laboratory Notebooks (ELN) and Laboratory Information Management Systems (LIMS), but these are now being superseded by Biopharmaceutical Lifecycle Management (BPLM) systems, where data are captured at the time and location of execution and used to create a contextualised data backbone (Marcazzo 2022).

The advancements in computational tools for prediction of physicochemical properties, quantitative physicochemical activity relationships (QPAR), molecular models, QSAR, etc. will inevitably continue apace. The attraction of such computer-based approaches is obvious as they are easier and less time-consuming than obtaining the equivalent amount of experimental data. However, as is clear from the large variation in the various predicted Clog P values for 20E (Table 7.2), the resulting values should only be taken as indications and, when the parameter is of real significance to the study, should be determined/verified by other means, including experimentally.

Artificial Intelligence (AI) is starting to have an impact on all aspects of drug discovery and development (Jumper et al. 2021; Abramson et al. 2024; Buterez et al. 2024, 2025; Krokidis et al. 2025), including the prediction of physicochemical properties, bioactivity, toxicity, target protein structure, pharmaceutical product development, pharmaceutical manufacturing, quality control and assurance, clinical trial design, and product management (Paul et al. 2021). AI is also being used specifically for natural product drug discovery (Mullowney et al. 2023; Othman et al. 2025; Xu et al. 2026), since ca. 25% of all drugs are, or are derived from, natural products. AI can be used advantageously to search through large volumes of pre-existing data to find those relevant to a particular topic, which can be done in seconds, as opposed to days or weeks if done manually. However, again, caution is required, so that at least part of the search is done manually and then it is verified if there is correspondence between the manual and AI searches. The algorithms used in AI are opaque and generate many errors, so all important conclusions should be verified by other means before proceeding to further research based on these conclusions. Newer iterations of the algorithms are resulting in more accurate and more sophisticated predictions (cf. AlphaFold [Jumper et al. 2021] and AlphaFold3 [Abramson et al. 2024]), but still suffer from some surprising and counter-intuitive limitations (Abramson et al. 2024).

7.10 Summary

It is clear that ecdysteroids have considerable potential as pharmaceutical leads, owing to the wide range of diverse 'positive' effects that they have on diseased and healthy animals and people and the low toxicity of the compounds in mammals under most circumstances. However, two major constraints exist to the full development of these compounds as viable and commercially effective medicaments: the sourcing of adequately large (tonne) amounts of the pure compounds and the poor oral bioavailability of 20E. At least one of these barriers needs to be overcome, but of the two, it would, from medical, industrial, and environmental points of view, be more beneficial if the bioavailability could be significantly enhanced, since this would also mean reduction in the dose needed to bring about the therapeutic effect and, hence, reduce the pressure to isolate such large amounts of pure ecdysteroid. The chemical complexity of ecdysteroids precludes the full chemical synthesis of the compounds, but hope perhaps exists for the identification of non-steroidal analogues, which are chemically simpler to synthesise and which retain biological activity, but this will require far greater understanding of the modes and sites of action of ecdysteroids in mammals. In this respect, the vast array of natural and hemisynthetic ecdysteroid analogues which already exist can facilitate the development of QSARs and pharmacophore models. We have outlined various strategies which might be used to improve the bioavailability through chemical modification, formulation, route of application, and altered metabolism, which may help others to realise the full medical potential of these fascinating molecules, which fulfil so many diverse roles in the biological world such as steroid hormones in arthropods and other invertebrates, defensive compounds in plants and pycnogonids, and now as pharmaceutical agents in mammals.

Declaration by the Authors Artificial Intelligence (AI) has not been used in the preparation or writing of this chapter.

References

Abramson J, Adler J, Dunger J, Evans R, Green T, Pritzel A, Ronnenberger O, Willmore L, Ballard AJ, Bambrick J, Bodenstein SW, Evans DA, Hung CC, O'Neill M, Reiman D, Tunyasuvanakool K, Wu Z, Zemgulyte A, Arvaniti E, Beattie C, Bertolli O, Bridgland A, Cherepanov A, Congreve M, Cowen-Rivers AI, Cowie A, Figurnov M, Fuchs FB, Gladman H, Jain R, Khan YA, Low CMR, Perlin K, Potapenko A, Savvy P, Singh S, Stecula A, Thillaisundaram A, Tong C, Yakneen S, Zhong ED, Zielinski M, Zidek A, Bapst V, Kohli P, Jaderberg M, Hassabis D, Jumper JM (2024) Accurate structure prediction of biomolecular interactions with AlphaFold 3. Nature 630, 493–500 https://doi.org/10.1038/s41586-024-07487-w

Agoun H, Semiane N, Mallek A, Bellahreche Z, Hammadi S, Madjreb M, Abdlalli M, Khalkhal A, Dahmani Y (2021) High-carbohydrate diet-induced metabolic disorders in *Gerbillus tarabuli* (a new model of non-alcoholic fatty liver disease). Protective effects of 20-hydroxyecdysone. Archives of Physiology and Biochemistry 127 (2): 127–135 https://doi.org/10.1080/13813455.2019.1621350.

Ambrosio G. (2019) Ecdysterone use and misuse in sports: Diagnostic markers in human specimens and determination in supplements. PhD Thesis, University of Berlin, BRD.

Ambrosio G, Yuliandra T, Wuest B, Mazzarino M, de la Torre X, Botrè F, Diel P, Isenmann E, Parr MK (2021) Urinary elimination of ecdysterone and its metabolites following a single-dose administration in Humans. Metabolites 11: 366 https://doi.org/10.3390/metabo11060366.

Anthony T, Mirek E, DeOliveira C, Bargoud A, Marella D, Philipson-Weiner L, Graf B, She P, Kuhn P, Wilson G, Raskin I (2014) Evaluating the effect of 20-hydroxyecdysone on leucine signaling in the skeletal muscle and liver of male rats. FASEB J 28(1): Supplement 820.7. https://doi.org/10.1096/fasebj.28.1_supplement.820.7

Archana J, Annapurna A, Devayani P (2022) Neuroprotective role of *Tinospora cordifolia* extract in streptozotocin induced neuropathic pain. Brazilian Journal of Pharmaceutical Sciences 58: article e18501 https://doi.org/10.1590/s2175-97902020000118501.

Backsay I, Arany P, Feher P, Josza L, Vasvari G, Nemes D, Peto A, Kosa D, Haimhoffer A, Ujhelyi Z, Sinka D (2025) Bioavailability enhancement and formulation technologies of oral mucosal dosage forms: a review. Pharmaceutics 17: article 148 https://doi.org/10.3390/pharmaceutics17020148.

Badal'yants KL, Nabiev AN, Khushbaktova ZA, Syrov, VN (1996) Mechanism of hepatoprotective action of ecdystene in acute heliotrine intoxication. Doklady Akademii Nauk Respubliki Uzbekistan (10): 46–48.

Balducci C, Dinan L, Guibout L, Foucault AS, Carboonne C, Durand JD, Caradeuc C, Bertho G, Girault JP, Lafont R (2021) The complex metabolism of poststerone in male rats. Journal of Steroid Biochemistry and Molecular Biology 212: article 105897 https://doi.org/10.1016/j.jsbmb.2021.105897.

Bandara BMR, Jayasinghe I, Karunaratne V (1989) Ecdysterone from stem of *Diploclisia glaucescens*. Phytochemistry 28: 1073–1075 https://doi.org/10.1016/0031-9422(89)80185-2.

Báthori M, Pongracz Z (2005) Phytoecdysteroids - from isolation to their effects on humans. *Current Medicinal Chemistry* 12: 153–172 https://doi.org/10.2174/0929867053363450.

Báthori M, Tóth N, Hunyadi A, Márki A, Zádor E (2008) Phytoecdysteroids and anabolic-androgenic steroids—structure and effects on humans. Current Medicinal Chemistry 15: 75–91 https://doi.org/10.2174/092986708783330674.

Berkenov K, Datkhayev UM (2017) Preparation and analysis of the NMR spectra of the pharmaceutical substance OOSE-11, -12, -13. Asian Journal of Pharmaceutical and Clinical Research 10(4): 292–296.

Betow JY, Qaseem A, Bekono BD, Feng Y, Moumbock AFA, Simoben CV, Babiaka SB, Tanyi SA, Shu VA, Ndi AT, Onguéné PA, Metuge CS, Akame S, Eni DB, Namba-Nzanguim CT, Tjegbe MJM, Dikande MNL, Akeh PT, Balgah S, Meh CD, Ibezim A, Tabi IF, Malange YI, Ndingkokhar B, Njume LE, Telukunta KK, Amrani S, Ebob OT, Sippl W, Günther S, Ntie-Kang F (2025) African Natural Products Database (ANPDB): A resource for exploring the therapeutic potential of natural compounds from Africa, Nucleic Acids Research 2025: article gkaf1186, https://doi.org/10.1093/nar/gkaf1186.

Bolduc TM (2008) Human urinary excretion profiles after exposure to ecdysterone. MSc Thesis, University of Utah, USA.

Bouyahya A, El Omari N, Elmenyiy N, Guaouguaou FE, Balahbib A, El-Shazly M, Chamkhi I (2020) Ethnomedicinal use, phytochemistry, pharmacology, and toxicology of *Ajuga iva* (L.) Schreb. *J. Ethnopharmacol* 258: article 112875 https://doi.org/10.1016/j.jep.2020.112875.

Brandt W (2003) Pharmakokinetik und Metabolismus des 20-Hydroxyecdysons im Menschen. PhD Thesis, Philipps-Universität Marburg, BRD, 17192/2004.0136.

Buterez D, Janet JP, Kiddle SJ, Oglic D, Lio P (2024) Transfer learning with graph neural networks for improved molecular property prediction in the multi-fidelity setting. Nature Communications 15: article 1517 https://doi.org/10.1038/s41467-024-45566-8.

Buterez D, Janet JP, Oglic D, Lio P (2025) An end-to-end attention-based approach for learning on graphs. Nature Communications 16: article 5244 https://doi.org/10.1038/s41467-025-60252-z.

Cahliková L, MacáKová K, Chlebek J, Host'álková A, Kulkhánková A, Opletal L (2011) Ecdysterone and its activity on some degenerative diseases. Nat Prod Commun 6(5): 707–718

Cai J, Rimal B, Jiang C, Chiang JYL, Patterson AD (2022) Bile acid metabolism and signaling, the microbiota, and metabolic disease. Pharmacology & Therapeutics 237: article 108238 https://doi.org/10.1016/j.pharmthera.2022.108238.

Chabanny VN, Levitsky EL, Gubskii II, Kholodova ID, Vistunova IE, Budmaska MI (1994) Genoprotective effect of drugs based on ecdysteroids in poisoning of rats with tetrachloromethane and chlorophos. Ukr Biokhim Zh 66(5): 67–77 PMID: 7747349 [in Russian].

Chakraborty S, Basu S (2017) Dual inhibition of BACE1 and Aβ aggregation by β-ecdysone: application of a phytoecdysteroid scaffold in Alzheimer's disease therapeutics. International Journal of Biological Macromolecules 95: 281–287 https://doi.org/10.1016/j.ijbiomac.2016.11.061.

Chiruta V (2020) 20-Hydroxyecdysone's cyclodextrin complex - a novel anabolic phytosteroid (APS)? *International Journal of Healthcare Sciences* 8(1): 250–254 https://doi.org/10.6084/m9.figshare.12361139.

Coll-Toledano J (2009) Synthetic ecdysteroidal compounds. In Smagghe G (Ed) Ecdysone: structures and functions. Springer Science + Business Media BV, pp. 99–123 https://doi.org/10.1007/978-1-4020-9112-4_4.

Csábi J, Martins A, Sinka I, Csorba A, Molnár J, Zupkó I, Tóth G, Tillekeratne LMV, Hunyadi A (2016) Synthesis and in vitro evaluation of the antitumor potential and chemo-sensitizing activity of fluorinated ecdysteroid derivatives. *Medicinal Chemistry Communications* 7: 2282–2289 https://doi.org/10.1039/C6MD00431H.

Csábi J, Rafai T, Hunyadi A, Zádor E (2019) Poststerone increases muscle fibre partly similar to its metabolically parent compound, 20-hydroxyecdysone. Fitoterapia 134: 459–464 https://doi.org/10.1016/j.fitote.2019.03.017.

Daina A, Michielin O, Zoete V (2017) SwissADME: a free web tool to evaluate pharmacokinetics, drug-likeness and medicinal chemistry friendliness of small molecules. Scientific Reports 7: article 42717; https://doi.org/10.1038/srep42717.

Danish KA, Lubhan S (2016) Various techniques of bioavailability enhancement: a review. *Journal of Drug Delivery & Therapeutics* 6(3): 34–41 https://doi.org/10.22270/jddt.v6i3.1228.

Destrez B, Pinel G, Bichon E, Monteau F, Lafont R, Le Bizec B (2008) Detection of 20-hydroxyecdysone in calf urine by comparative liquid-chromatography/high-resolution mass spectrometry and liquid chromatography/tandem mass spectrometry measurements: application to the control of the potential misuse of ecdysteroids in cattle. Rapid Communications in Mass Spectrometry 22(24): 4073–4080 https://doi.org/10.1002/rcm.3816.

Destrez B, Pinel G, Monteau F, Lafont R, Le Bizec B (2009) Detection and identification of 20-hydroxyecdysone by liquid chromatography-high resolution or tandem mass spectrometry measurements and establishment of their kinetics of elimination after 20-hydroxyecdysone administration. Analytica Chimica Acta 637(1–2): 178–184 https://doi.org/10.1016/j.aca.2008.10.015.

Dilda P, Latil M, Cukerman L, Lafont R, Veillet S (2025) Compositions of phytoecdysones for intranasal administration. Patent application, PCT/EP2024/087433, WO 2025/132808 A1 [in French].

Dinan L. 2001. Phytoecdysteroids: biological aspects. Phytochemistry 57(3): 325–39. https://doi.org/10.1016/s0031-9422(01)00078-4.

Dinan L, Hormann RE (2005) Chapter 3.4: Ecdysteroid agonists and antagonists. In: Gilbert LI, Iatrou K, Gill SS (Eds) Comprehensive Molecular Insect Science. Volume 3: Endocrinology, Elsevier, Oxford, UK. pp 197–242.

Dinan L, Whiting P, Bourne P, Coll J (2001) 8-*O*-Acetylharpagide is not an ecdysteroid agonist. Insect Biochemistry and Molecular Biology 31: 1077–1082 https://doi.org/10.1016/s0965-1748(01)00055-8.

Dinan L, Mamadalieva NZ, Lafont R (2020) Dietary phytoecdysteroids. In: Xiao J, Sarker SD, Asakawa Y (Eds), Handbook of Dietary Phytochemicals, Springer Nature Singapore Pte Ltd., pp. 1–54, https://doi.org/10.1007/978-981-13-1745-3_35-1.

Dinan L, Lafont R (2006) Effects and applications of arthropod steroid hormones (ecdysteroids) in mammals. Journal of Endocrinology 191: 1–8 https://doi.org/10.1677/joe.1.06900.

Dinan L, Lafont R (2021) Ecdysteroid physico-chemical data. Ecdybase (compiled by Lafont R, Harmatha J, Marion-Poll F, Dinan L and Wilson ID), https://ecdybase.org.

Dinan L, Balducci C, Guibout L, Foucault AS, Bakrim A, Kumpun S, Girault JP, Tourette C, Dioh W, Dilda PJ, Veillet S, Lafont R (2021a) Ecdysteroid metabolism in mammals: The fate of ingested 20-hydroxyecdysone in mice and rats. *J Steroid Biochem Molec Biol* 212: article 105896 https://doi.org/10.1016/j.jsbmb.2021.105896.

Dinan L, Dioh W, Veillet S, Lafont R (2021b) 20-Hydroxyecdysone, from plant extracts to clinical use: Therapeutic potential for the treatment of neuromuscular, cardio-metabolic and respiratory diseases. *Biomedicines* 9: article 492 https://doi.org/10.3390/biomedicines9050492.

Dinan L, Lafont F, Lafont R (2023) The distribution of phytoecdysteroids among terrestrial vascular plants: a comparison of two databases and discussion of the implications for plant/insect interactions and plant protection. Plants 12: article 776. https://doi.org/10.3390/plants12040776.

Dioh W, Tourette C, Del Signore S, Daudigny L, Dupont P, Dilda PJ, Lafont R, Veillet S (2023) A phase I study for safety and pharmacokinetics of BIO101 (20-hydroxyecdysone) in healthy young and older adult adults. Journal of Cachexia, Sarcopenia and Muscle 14: 1259–1273 https://doi.org/10.1002/jcsm.13195.

Dvořáková, K.; Štěpánek, P.; Kroupová, J.; Zbytovská, J (2022) N-Alkylmorpholines: potent dermal and transdermal skin permeation enhancers. Pharmaceutics 14: article 64 https://doi.org/10.3390/pharmaceutics14010064.

Dzhukarova MK, Sakhibov AD, Kasymov B, Syrov VN, Takanaev AA and Saatov Z (1987) Molecular biological problems of the creation of drugs and study of the mechanism of their action: pharmacokinetics of ecdysterone in experiments. Khimiko-farmatsevticheskii Zhurnal

21(10): 1163–1167 [in Russian] [English version: Chemical-Pharmaceutical Journal 21: 689–692, 1988].

El-Mofty M, Sadek I, Soliman A, Mohamed A, Sakre S (1987) α-Ecdysone: a new bracken fern factor responsible for neoplasm induction in the Egyptian toad (*Bufo regularis*). Nutr Cancer 9: 103–107 https://doi.org/10.1080/01635588709513917.

El-Mofty MM, Sakre SA, Rizk AM, Moussa EA (1994) Induction of breast and lung neoplastic lesions in mice by alpha-ecdysone. *Oncology Reports* 1: 435–438 https://doi.org/10.3892/or.1.2.435.

Erimbetov KT, Fedorova AV, Goncharova AY, Bondarenko EV (2020a) Creation of the nano-sized form of 20-hydroxyecdysone and research its biological availability. Problemy Biologii Produktivnykh Zhivotnykh (Problems of Productive Animals Biology) (3): 106–113 [in Russian].

Erimbetov KT, Fedorova AV, Goncharova AY, Bondarenko EV, Khomyakova TI, Zemlyanoy RA (2020b) The toxicity of the nanoscale form of 20-hydroxyecdysone in a single intragastric administration to rabbits. Medline.Ru Experimental Toxicology 21 (46): 593–602 https://doi.org/10.36946/0869-7922-59-63 [in Russian].

Fang X, Szoltysik R, Tang J, Bajkacz S (2022) Efficient extraction and sensitive HPLC-MS/MS quantification of selected ecdysteroids in plants. Journal of Food Compounds and Analysis 110: art. 104580. https://doi.org/10.1016/j.jfca.2022.105580.

Fielding RA, Dao MM, Cannon K, Desvarieux M, Miller SS, Gimness MP, Brandon DM, Villareal DT, Bruyère O, Bautmans I, Rickner K, Perry R, Kritchevsky SB, Musi N, Chehade JM, Kirsten JL, Gielen E, Pickrell P, Dilda P, Lafont R, Margalev C, Rolland Y, Del Signore S, Mariani J, Agus S, Tourette C, Dioh W, van Maanen R, Veillet S (2025) BIO101 in sarcopenic seniors at risk of mobility disability: Results of a double-blind randomized interventional phase 2b trial. Journal of Cachexia, Sarcopenia and Muscle 16(2): article e13750 https://doi.org/10.1002/jcsm.13750.

Frachet S, Danigo A, Latil M, Dilda PJ, Bessaguet F, Richard L, Sturtz F, Magy L, Demiot C (2025) Neuroprotective effect of Mas activation by BIO101 in vincristine-induced small fiber-neuropathy. *Journal of the Peripheral Nervous System* 30: article e70055 https://doi.org/10.1111/jns.70055.

Fuart Gatnik M, Worth A (2010) Review of software tools for toxicity prediction. European Commission Joint Research Centre Report EUR24489 https://doi.org/10.2788/60101, pp 22.

Fumagalli G, Giorgi G, Vágvölgyi M, Colombo E, Christodoulou MS, Collico V, Prosperi D, Dosio F, Hunyadi AA, Montopoli M, Hyeraci M, Silvani A, Lesma G, Dalla Via L, Passarella D (2018) Heteronanoparticles by self-assembly of ecdysteroid and doxorubicin conjugates to overcome cancer resistance. ACS Medicinal Chemistry Letters 9: 468–471 https://doi.org/10.1021/acsmedchemlett.8b00078.

Gagalac-Nicolas MG, Lim Sylianco CY (1981) Antimutagenic effects of *Cyanotis vaga* Lour. (Schultes) extracts. Bull Phil Biochem Soc 4(1&2): 8–14.

Galbraith MN, Horn DHS (1966) An insect-moulting hormone from a plant. Journal of the Chemical Society, Chemical Communications 905–906 https://doi.org/10.1039/C19660000905.

Gharib B, Nugon-Baudon L, Lafont R, De Reggi M (1993) Ecdysteroids, a new pathological marker in man. In: Biologie prospective: Compte-rendus du 8è Colloque de Pont-à-Mousson, John Libbey Eurotext 203–206.

Gholipour P, Komaki A, Parsa H, Ralezani M (2025) Ecdysterone and high-intensity interval training mitigate Alzheimer's pathology in rats: impacts on depression, synaptic plasticity, and neuroinflammation. Molecular Neurobiology 62 (11): 13933–13957 https://doi.org/10.1007/s12035-025-05168-x.

Gilgan MW, Burns BG (1977) Molt induction in lobsters (*Homarus americanus*) by intramuscular injection of ecdysterone triacetate. Experientia 33(8): 1114–1115 https://doi.org/10.1007/BF01946003.

Girault JP, Lafont R, Kerb U (1988) Ecdysone catabolism in the white mouse. Drug Metabolism and Disposition 16: 716–720

Graf BL, Poulev A, Kuhn P, Grace MH, Lila MA, Raskin I (2014) Quinoa seeds leach phytoecdysteroids and other compounds with anti-diabetic properties. Food Chemistry 163:178–185 https://doi.org/10.1016/j.foodchem.2014.04.088.

Guan H, Chi D, Yu J, Li H (2011) Encapsulated ecdysone by internal gelation of alginate microspheres for controlling its release and photostability. Chemical Engineering Journal 168: 94–101 https://doi.org/10.1016/j.cej.2010.12.040.

Gubsky YI, Levitsky EL, Kholodova YuD, Goriushko AG, Primak RG, Vistunova IE, Savchenko LG (1993) Mechanisms of genoprotective action of a phytoecdysteroid drug (BTK-8L) in chromatin damage by tetrachloromethane. Ukr Biokhim Zhurnal 65(6): 75–83 [in Russian]

Guengerich FP, McCarty KD, Tateishi Y, Liu L (2023) Steroid 17α-hydroxylase/17, 20-lyase (cytochrome P450 17A1). Methods in Enzymology 2023; 689:39–63. https://doi.org/10.1016/bs.mie.2023.04.001.

Gui G, Zhang Q, Xia X, Liang G, Wang Z, Liu R (2014) Protective effects of 20-hydroxyecdysone on diabetic nephropathy of rat. Modern Preventive Medicine 41 (10): 1844–1847.

Heinrich G, Hoffmeister H (1967) Ecdyson als Begleitsubstanz des Ecdysterons in *Polypodium vulgare*. Experientia 23: 995 [in German] https://doi.org/10.1007/BF02136403.

Hetal T, Bindesh P, Sneha T (2010) A review on techniques for oral bioavailability enhancement of drugs. Int J Pharm Sci Reviews and Research 4(3): article 033.

Hoffmann KH, Thiry E, Lafont R (1990) 14-Deoxyecdysteroids in an insect (*Gryllus bimaculatus*). Z Naturforsch 45c: 703–708 https://doi.org/10.1515/znc-1990-0622.

Hung TJ, Chen WM, Liu SF, Liao TN, Lee TC, Chuang LY, Guh JY, Hung CY, Chen PY, Hsieh PF, Yang YL (2012) 20-Hydroxyecdysone attenuates TGF-β1-induced renal cellular fibrosis in proximal tubule cells. J Diabetes Complic 26(6):463–469 https://doi.org/10.1016/j.jdiacomp.2012.06.014.

Hubatsch I, Ragnarsson EG, Artursson P (2007) Determination of drug permeability and prediction of drug absorption in Caco-2 monolayers. Nature Protocols 2(9):2111–9. https://doi.org/10.1038/nprot.2007.303.

Hunyadi A, Herke I, Lengyel K, Báthori M, Kele Z, Simon A, Tóth G, Szendrei K (2016) Ecdysteroid-containing food supplements from *Cyanotis arachnoidea* on the European market: evidence for spinach product counterfeiting. Scientific Reports 6: article 37322. https://doi.org/10.1038/srep37322.

Isenmann E, Yulandra T, Touvleliou K, Broekmann M, de la Torre X, Botrè F, Diel P, Parr MK (2024) Quinoa as phytopharmaceutical? Urinary elimination of ecdysterone after consumption of quinoa alone and in combination with spinach. Arch Pharm (Weinheim) 357(6): article e2300689 https://doi.org/10.1002/ardp.202300689.

Issaadi HM, Csábi J, Hsieh TJ, Gáti T, Tóth G, Hunyadi A (2019) Side-chain cleaved phytoecdysteroid metabolites as activators of protein kinase B. Bioorganic Chemistry 82: 405–413 https://doi.org/10.1016/j.bioorg.2018.10.049.

Jumper, J., Evans, R., Pritzel A, Green T, Figurnov M, Ronneberger O, Tunyasuvunakool K, Bates R, Žídek A, Potapenko A, Bridgland A, Meyer C, Kohl SAA, Ballard AJ, Cowie A, Romera-Paredes B, Nikolov S, Jain R, Adler J, Back T, Petersen S, Reiman D, Clancy E, Zielinski M, Steinegger M, Pacholska M, Berghammer T, Bodenstein S, Silver D, Vinyals O, Senior AW, Kavukcuoglu K, Kohli P, Hassabis D (2021) Highly accurate protein structure prediction with AlphaFold. Nature 596, 583–589 https://doi.org/10.1038/s41586-021-03819-2.

Kalász H, Hunyadi A, Tekes K, Dolesal R, Karvaly G (2017) HPLC analysis and blood-brain penetration of 20-hydroxyecdysone diacetonide. Acta Chromatographica 29(3): 3755–383 https://doi.org/10.1556/1326.2017.29.3.07.

Kametani T, Tsubuki M (1980) Strategies for the synthesis of ecdysteroids. In: Koolman J (Ed) Ecdysone: from chemistry to mode of action, Thieme Verlag, Stuttgart, BRD, pp. 74–96.

Kang P, Li S (2022) Makisterone A attenuates experimental cholestasis by activating the farnesoid x receptor. Biochem Biophys Res Comm 623: 162–169 https://doi.org/10.1016/j.bbrc.2022.07.030.

Kaplanis JN, Thompson MJ, Robbins WE, Bryce M (1967) Insect hormones: alpha-ecdysone and 20-hydroxyecdysone in bracken fern. Science 157:1436–1438 https://doi.org/10.1126/science.157.3795.1436.

Khan SA (2025) Network pharmacology and *in silico* analysis reveal Kochiae fructus as a potential therapeutic against atopic dermatitis through immunomodulatory pathway interactions. PLoS One 20(4): article e0320818 https://doi.org/10.1371/journal.pone.0320818.

Khudair AR, Oleiwi MA, Khudair IR (2025) Nephroprotective effect of turkesterone against daunorubicin-induced acute kidney injury in rats. Iraqi Journal of Pharmaceutical Sciences 34(3): 249–256 https://doi.org/10.31351/vol34iss3pp249-256.

Kolšek K, Mavri J, Sollner Dolenc M, Gobec S, Turk S (2014) Endocrine disruptome—An open source prediction tool for assessing endocrine disruption potential through nuclear receptor binding. Journal of Chemical Information and Modeling 54: 1254–1267 https://doi.org/10.1021/ci400649p.

Koolman J (1989) Ecdysone: from chemistry to mode of action. Georg Thieme Verlag, Stuttgart, BRD, pp. 482.

Kosar K, Opletal L, Vokác K, Harmatha J, Sovová M, Cerovsky J, Krátky F, Dvorák J (1997) Embryotoxicity of 20-hydroxyecdysone and polypodine B from *Leuzea carthamoides*. Pharmazie 52(5): 406–407

Kovganko NV (1997) Recent advances in the chemical synthesis of ecdysteroids and compounds related to them. Chemistry of Natural Compounds 33(3): 506–526 https://doi.org/10.1007/BF02254793.

Kovganko NV (1998) Chemical transformations of ecdysteroids: latest advances. Chemistry of Natural Compounds 34(2): 111–127 https://doi.org/10.1007/BF02249125.

Kovganko NV, Kashkan ZN, Chernov YG, Ananich SK, Sokolov SN, Survilo V (2003) Synthesis of ecdysteroids and related compounds. Chemistry of Natural Compounds 39(5): 411–437 https://doi.org/10.1023/B:CONC.0000011114.96205.98.

Kozhanova AM, Tuleuov BI, Kudabayeva PK, Temirgaziyev BS, Seilkhanov TM, Seidakhmetova RB, Salkeyeva LK, Adekenov SM (2020) Synthesis, NMR spectroscopic study of α-, β- and γ-cyclodextrin inclusion complexes of 2-deoxyecdysone and their anti-inflammatory activity. Macroheterocycles 13(3): 292–297 [in Russian] https://doi.org/10.6060/mhc200602t.

Kozhanova AM, Temirgaziyev BS, Zhanarbek A, Tuleuov BI, Seilkhanov TM, Adekenov SM (2021) Synthesis of a hydrophilic derivative of ecdysterone and development of its water-soluble form. Bulletin of the University of Karaganda—Chemistry 104 (4): 138–148 https://doi.org/10.31489/2021Ch4/138-148.

Kraiem S, Al-Jaber MY, Al-Mohammed H, Al-Menhali AS, Al-Thani N, Helaleh M, Samsam W, Touil S, Beotra A, Georgakopoulas C, Bouabdallah S, Mohamed-Ali V, Al Maadheed M (2021) Analytical strategy for the detection of ecdysterone and its metabolites in vivo in uPA(+/+)-SCID mice with humanized liver, human urine samples, and estimation of prevalence of its use in anti-doping samples. Drug Test Analysis 13(7):1341–1353. https://doi.org/10.1002/dta.3032.

Krokidis MG, Koumadorakis DE, Lazaros K, Ivantsik O, Exarchos TP, Vrahatis AG, Kotsiantis S, Vlamos P (2025) AlphaFold3: An overview of applications and performance insights. International Journal of Molecular Sciences 26: article 3671. https://doi.org/10.3390/ijms26083671.

Kroma A, Pawlaczyk M, Feliczak-Guzik A, Urbanska M; Jenerowicz D, Seraszek-Jaros A, Kikowska M, Gornowicz-Porowska J (2022) Phytoecdysteroids from *Serratula coronata* L. for psoriatic skincare. Molecules 27 (11): article 3471 https://doi.org/10.3390/molecules27113471.

Kudyasheva AG, Shevchenko OG, Zagorskaya NG, Bashlykova LA, Raskosha OV, Ermakova OV, Shishkina LN (2012) The study of anti-ray action of ecdysteroid containing preparations at chronic irradiation at lower doses. Theoretical and Applied Ecology (1): 24–30 [in Russian].

Kumpun S, Girault JP, Dinan L, Blais C, Maria A, Dauphin-Villemant C, Yingyongnarongkul B, Suksamrarn A, Lafont R (2011) The metabolism of 20-hydroxyecdysone in mice: relevance to pharmacological effects and gene switch applications of ecdysteroids. J Steroid Biochem Mol Biol 126(1–2): 1–9 https://doi.org/10.1016/j.jsbmb.2011.03.016.

Lafont R, Dinan L (2003) Practical uses for ecdysteroids in mammals including humans: an update. Journal of Insect Science 3: article 3 https://doi.org/10.1093/jis/3.1.7.

Lafont R, Dinan L (2009) Chapter 23: Innovative and future applications for ecdysteroids. In: Smagghe G (Ed) Ecdysone: structures and functions. Springer Science + Business BV 551–578 https://doi.org/10.1007/978-1-4020-9112-4_23.

Lafont R, Girault JP, Kerb U (1988) Excretion and metabolism of injected ecdysone in the white mouse. Biochemical Pharmacology 37: 1174–1177 https://doi.org/10.1016/0006-2952(88)90529-1.

Lafont R., Harmatha J., Marion-Poll F., Dinan L., Wilson ID (2002) The Ecdysone Handbook, 3, on-line (continuously updated), https://ecdybase.org.

Lafont R, Serova M, Didry-Barca B, Raynal S, Guibout L, Dinan L, Veillet S, Latil M, Dioh W, Dilda PJ (2021) 20-Hydroxyecdysone activates the protective arm of the RAAS via the Mas receptor. J Mol Endocrinol 68 (2): 77–87 https://doi.org/10.1530/JME-21-0033.

Lagova ND, Valueva IM (1981) Effect of ecdysterone isolated from *Rhaponticum carthamoides* on the growth of experimental tumours. Eksp Onkol 3: 69–71 [in Russian].

Lapenna S. (2006) A study of the ligand-receptor interactions of ecdysteroids. PhD Thesis, University of Exeter, Exeter, UK.

Lapenna S, Gemen R, Wollgast J, Worth A, Maragkoudakis P, Caldeira S (2015) Assessing herbal products with health claims. Critical Reviews in Food Science and Nutrition 55: 1918–1928 https://doi.org/10.1080/10408398.2012.726661.

Le J, Vivian EM (2025) Drug bioavailability. MSD Manual Professional Version. Merck &Co., Rahway, NJ, USA, pp3

Lee JHZ, Coleman T, Mclean MA, Podgorski MN, Hayball EF, Stone ISJ, Bruning JB, Whelan F, Voss JJ, Sligar SG, Bell SG (2024) Selective α-hydroxyketone formation and subsequent C-C bond cleavage by cytochrome P450 monooxygenase enzymes. ACS Catalysis 14(11): 8958–8971. https://doi.org/10.1021/acscatal.4c01766.

Lesma G, Luragh A, Rainoldi G, Mattuzzo E, Bortolozzi R, Viola G, Silvani A (2016) Multicomponent approach to bioactive peptide-ecdysteroid conjugates: creating diversity at C6 by means of the Ugi reaction. Synthesis 48: 3907–3916 https://doi.org/10.1055/s-0035-1562497.

Levitsky EL, Kholodova YD, Gubsky YI, Goryushko AG, Primak RG, Vistunova IE, Sachenko LG (1993) Mechanism of genoprotective action of steroidal preparation BTK-8L based on phytoecdysteroids at chromatin damage by chlorvos. Ukrainian Biochem Journal 65(6): 84–91 [in Russian].

Li X, Li W, Li P (2010) Protective effects of ecdysterone on nonalcoholic fatty liver disease in rats. Lishizhen Medicine and Materia Medica Research (3): 743–745 [in Chinese].

Li Y, Cai L, Dong JW, Duan WH, Zhou H, Ding ZT (2015) Innovative approach to the accumulation of rubrosterone by fermentation of *Asparagus filicinus* with *Fusarium oxysporum*. Journal of Agricultural and Food Chemistry 63 (29): 6596–6602 https://doi.org/10.1021/acs.jafc.5b02570.

Lipinski CA, Lombardo F, Dominy BW, Feeney PJ (2001) Experimental and computational approaches to estimate solubility and permeability in drug discovery and development settings. Advanced Drug Delivery Reviews 46: 5–26 https://doi.org/10.1016/s0169-409x(00)00129-0.

Lobell M, Hendrix M, Hinzen B, Keldenich J, Meier H, Schlmeck C, Schohe-Loop R, Wunberg T, Hillisch A (2006) *In silico* ADMET traffic lights as a tool for the prioritization of HTS hits. ChemMedChem 1(11): 1229–1236 https://doi.org/10.1002/cmdc.200600168.

Lu M, Wang P, Ge Y, Dworkin L, Brem A, Liu Z, Gong R (2018a) Activation of mineralocorticoid receptor by ecdysone, an adaptogenic and anabolic ecdysteroid, promotes glomerular injury and proteinuria involving overactive GSK3β pathway. Scientific Reports 8: article 12225 https://doi.org/10.1038/s41598-018-29483-7.

Lu M, Wang P, Zhou S, Flickinger B, Malhotra D, Ge Y, Tatar M, Dworkin L, Liu Z, Gong R (2018b) Ecdysone elicits chronic renal impairment via mineralocorticoid-like pathogenic activities. Cellular Physiology and Biochemistry 49: 1633–1645 https://doi.org/10.1159/000493499.

Luo Y, Wu Z, Zhang Y, Qiao Y, Wei Y, Yan X, Ma X, Huang X, Zhong X, Ye Z, Lu X, Liao H (2025) β-ecdysone/PLGA composite scaffolds promote skull defect healing in diabetic rat. Frontiers in Bioengineering and Biotechnology 12: article 1536102 https://doi.org/10.3389/fbioe.2024.1536102.

Madden S (2022) Nasal delivery: improving nasal drug delivery with permeation enhancing technology. Drug Development & Diversity 22(2): 22–29.

Mallek A, Movassat J, Ameddah S, Liu J, Semiane N, Khalkhal A, Dahmani Y (2018) Experimental diabetes induced by streptozotocin in the desert gerbil, *Gerbillus gerbillus*, and the effects of short-term 20-hydroxyecdysone administration. Biomedicine & Pharmacotherapy 102: 354–361 https://doi.org/10.1016/j.biopha.2018.03.070.

Mamadalieva NZ, Hussain H, Mollica A, Zengin G, Mamadalieva RZ, Elhady SS, Fadil SA, Ashour ML, Youssef FS (2022) Ecdysteroids as potent enzyme inhibitors and verification of their activity using *in vitro* and *in silico* docking studies. Life (Basel) 12 (6): article 824 https://doi.org/10.3390/life12060824.

Marcazzo C (2022) Data strategies: as we shift towards biologics, we also need to shift toward smarter data management. Drug Development & Diversity 22(2): 41–43.

Marsault E (2011) BCM400_ADME. Pharmacocinétique Absorption, Distribution, Métabolisme, Excrétion. Institut de Pharmacologie de Sherbrooke, downloaded on March 9, 2020.

Martins A, Tóth N, Ványolós A, Béni Z, Zupkó J, Báthori M, Hunyadi A (2012) Significant activity of ecdysteroids on the resistance to doxorubicin in mammalian cancer cells expressing the human ABCB1 transporter. J Med Chem 55: 5034–5043 https://doi.org/10.1021/jm300424n.

Martins A, Csábi J, Balazs A, Kitka D, Amaral L, Molnár J, Simon A, Tóth G, Hunyadi A (2013) Synthesis and structure-activity relationships of novel ecdysteroid dioxolanes as MDR modulators in cancer. Molecules 18: article 15255 https://doi.org/10.3390/molecules181215255.

Martins A, Sipos P, Dér K, Csábi J, Miklos W, Berger W, Zalatnai A, Amaeral L, Molnár J, Szabó-Révész P, Hunyadi A (2015) Ecdysteroids sensitize MDR and Non-MDR cancer cell lines to doxorubicin, paclitaxel, and vincristine, but tend to protect them from Cisplatin. Biomed Res Int. 2015: article 895360 https://doi.org/10.1155/2015/895360.

Matsuda H, Kawaba T, Yamamoto Y (1970) Pharmacological studies of insect metamorphosing steroids from *Achyranthis* radix I: ponasterone A, ecdysterone and inokosterone in the rat. Folia Pharmacological Japonica 66(5): 551–563 https://doi.org/10.1254/jjp.20.142.

Meteleva ES, Polyakov NE, Tuleuov BI, Temirgaziyev BS, Dushkin AV, Adekenov SM (2023) Preparation and physicochemical properties of supramolecular complexes of ecdysterone. Chemistry of Plant Raw Materials 2023(4): 209–217 https://doi.org/10.14258/jcprm.20230412999.

Meybeck A, Bonte F, Redziniak G (1997) Use of an ecdysteroid for the preparation of cosmetic or dermatological composition intended, in particular, for strengthening the water barrier function of the skin or for the preparation of a skin cell culture medium, as well as to the compositions. US Patent 5,609,873.

Mironov VA, Nazarov AK, Volkov VL (2013) Increasing solubility of ecdysterone using 2-hydroxypropyl-β-cyclodextrin. Market Dietary Supplements 75(1): 35–37 [in Russian].

Mirzaev YR, Syrov VN (1992) Effect of phytoecdysteroids on the sexual activity of male rats. Doklady Akademii Nauk Respubliki Uzbekistan (3): 47–49 [in Russian].

Mirzaev YR, Syrov VN, Krushev SA, Iskanderova SD (2000) Study of the effects of ecdysten on the sexual function under experimental and clinical conditions. Eksp Klin Farmakol 63: 35–37 [in Russian]

Mo JH, Xie HK, Zhou YM, Ng SB, Li SX, Wang L (2021) Inokosterone is a potential drug target of estrogen receptor 1 in rheumatoid arthritis patients: analysis from active ingredients of *Cyathula officinalis*. Chin J Integr Med 27 (suppl 3): 767–773, https://doi.org/10.1007/s11655-021-3492-5.

Müller J, Martins A, Csábi J, Fenyvesi F, Könczöl Á, Hunyadi A, Balogh GT (2017) BBB penetration-targeting physicochemical lead selection: ecdysteroids as chemo-sensitizers against CNS tumors. European Journal of Pharmaceutical Science 96: 571–577 https://doi.org/10.1016/j.ejps.2016.10.034.

Mullowney MW, Duncan KR, Elsayed SS, Garg N, van der Hooft JJJ, Martin NI, Meijer D, Terlouw BR, Biermann F, Blin K, Durairaj J, Gorostiola González M, Helfrich EJN, Huber F, Leopold-Messer S, Rajan K, de Rond T, van Santen JA, Sorokina M, Balunas MJ, Beniddir MA, van Bergeijk DA, Carroll LM, Clark CM, Clevert DA, Dejong CA, Du C, Ferrinho S, Grisoni F, Hofstetter A, Jespers W, Kalinina OV, Kautsar SA, Kim H, Leao TF, Masschelein J, Rees ER, Reher R, Reker D, Schwaller P, Segler M, Skinnider MA, Walker AS, Willighagen EL, Zdrazil B, Ziemert N, Goss RJM, Guyomard P, Volkamer A, Gerwick WH, Kim HU, Müller R, van Wezel GP, van Westen GJP, Hirsch AKH, Linington RG, Robinson SL, Medema MH (2023) Artificial intelligence for natural product drug discovery. Nature Reviews in Drug Discovery 22(11):895–916. https://doi.org/10.1038/s41573-023-00774-7.

Nair AB, Jacob S (2016) A simple practice guide for dose conversion between mammals and human. Journal of Basic and Clinical Pharmacy 7: 27–31 https://doi.org/10.4103/0976-0105.177703.

Nambaaru S, Sahu PN, Katta S, Ganapavarapu VRS, Kancharla B, Sen A (2024) Phytochemical analysis and anti-inflammatory potential of *Floscopa splendes* Lour. and exploring ecdysterone as promising therapeutic agent: An HR-LCMS screening and molecular docking analysis. ChemistrySelect 9: article e202403146. https://doi.org/10.1002/slct.202403146.

Nazarov AK., Antokhin AM, Aksenov AV, Nazarov GV, Mitrofanov DA, Goncharov VM, Berizovskaya EI (2014) Increasing the solubility of ecdysterone. Pharmacy (5): 35–38 [in Russian].

Neves CS, Gomes SSL, dos Santos TR, de Almeida MM, de Souza YO, Garcia RMG, Otoni WC, Chedier LM, Raposo NRB, Viccini LF, de Campos JMS (2016) 'Brazilian ginseng' (*Pfaffia glomerata* Spreng. Pedersen, Amaranthaceae) methanolic extract: cytogenotoxicity in animal and plant assays. South African Journal of Botany 106: 174–180 https://doi.org/10.1016/j.sjab2016.07.003.

Nicolaescu OE, Belu I, Mocanu AG, Manda VC, Rau G, Pirvu AS, Ionescu C, Ciulu-Costinescu F, Popescu M, Ciocilteu MV (2025) Cyclodextrins: enhancing drug delivery, solubility and bioavailability for modern therapeutics. Pharmaceutics 17: article 288 https://doi.org/10.3390/pharmaceutics17030288.

Ogawa S, Nishimoto N, Matsuda H (1974) Pharmacology of ecdysones in vertebrates. In Burdette WJ Ed., Invertebrate Endocrinology and Hormonal Heterophylly, Springer-Verlag, pp. 341–344 https://doi.org/10.1007/978-3-642-65769-6_27.

Othman ZK, Ahmed MM, Kasimieh O, Musa SS, Branda F, Cue EG, Ocampo JMA, Lucero-Prisno DE, Vimolmangkang S (2025) Artificial intelligence for natural product drug discovery and development: current landscape, applications, and future directions. Intelligence-Based Medicine 12: article 100316 https://doi.org/10.1016/j.ibmed.2025.100316.

Parr MK, Ambrosio G, Wuest B, Mazzarino M, de la Torre X, Sibilia F, Joseph JF, Diel P, Botré F (2019) Targeting the administration of ecdysterone in doping control samples. Forensic Toxicology https://doi.org/10.1007/s11419-019-00504-y.

Parr MK, Zhao P, Haupt O, Tchoukouegno Ngueu S, Hengevoss J, Fritzemeier KH, Piechotta M, Schlörer N, Muhn P, Zheng WY, Xie MY, Diel P (2014) Estrogen receptor beta is involved in skeletal muscle hypertrophy induced by the phytoecdysteroid ecdysterone. Mol Nutr Food Res 58: 1861–1872 https://doi.org/10.1002/mnfr.201300806.

Paul D, Sanap G, Shenoy S, Kalyane D, Kalia K, Tekade RK (2021) Artificial intelligence in drug discovery and development. Drug Discovery Today 26(1): 80–93. https://doi.org/10.1016/j.drudis.2020.10.010.

Piper T, Thevis M (2023) Investigation into the human metabolism of ecdysterone. Drug Test Analysis 15: 1503–1520 https://doi.org/10.1002/dta.3582.

Pires CL, Praça C, Martins PAT, Batista de Carvalho ALM, Ferreira L, Marques MPM. Moreno MJ (2021) Re-Use of Caco-2 monolayers in permeability assays—validation regarding cell monolayer integrity. Pharmaceutics 13: article 1563. https://doi.org/10.3390/pharmaceutics13101563.

Politova NK, Kovler LA, Volodin VV, Lushka VG, Pshunetleva EAP (2001) Chemical modification of 20-hydroxyecdysone and the study of membranotropic properties of its conjugates. Chemistry of Plant Raw Material 5(2): 69–81 [in Russian].

Ramazanov NS, Saatov S, Syrov VN (1996) Study of ecdysterone metabolites isolated from rat urine. Chemistry of Natural Compounds 32, 545–549 https://doi.org/10.1007/BF01372610.

Ramazonov NS, Bobayev ID, Yusupova UY, Mahmudova MM, Usmanov DA, Tursunova NB, Syrov VN (2016) Allocation phytoecdysteroids from *Stachys hissarica* and studying their hepatoprotective activity. Sciences of Europe 4(4): 85–88.

Reigner BG, Blesch KS (2002) Estimating the starting dose for entry into humans: principles and practice. European Journal of Clinical Pharmacology 57(12):835–45. https://doi.org/10.1007/s00228-001-0405-6.

Rincon-Lopez J, Almanza-Arjona YC, Riascos AP, Rojas-Aguirre Y (2021) Technological evolution of cyclodextrins in the pharmaceutical field. Journal of Drug Delivery Science and Technology 61: article 102156 https://doi.org/10.1016/j.jddst.2020.102156.

Romaniuk-Drapala A, Lisiak N, Toton, Matysiak A, Nawrot J, Nowak G, Kaczmarek M, Rybscynska M, Rubis B (2021) Proapoptotic and proautophagic activity of 20-hydroxyecdysone in breast cancer cells *in vitro*. Chemico-Biological Interactions 342: article 109479 https://doi.org/10.1016/j.cbi.2021.109479.

Roziev RA, Goncharova AJ, Erimbetoc KT, Podgordnichenko VK (2013) Clathrate complex of arabinogalactan or arabic gum with 20-hydroxyecdysones, method for producing it (versions), pharmaceutical composition and medicinal product. Russian Patent RU 2572334 C2.

Russell GB, Price GM (1977) Metabolism of β-ecdysone during the larval and white puparial stage of the blowfly, *Calliphora erythrocephala*. Insect Biochemistry 7: 197–202 https://doi.org/10.1016/0020-1790(77)90014-2.

Saatov Z, Agzamhodzhaeva DA, Syrov VN (1999) Distribution of phytoecdysteroids in plants of Uzbekistan and the possibility of using drugs based on them in nephrological practice. Chemistry of Natural Compounds 35(2): 186–191 https://doi.org/10.1007/BF02234932.

Saokham P, Muankaew C, Jansook P, Loftsson T (2018) Solubility of cyclodextrins and drug/cyclodextrin complexes. Molecules 23: article 1161 https://doi.org/10.3390/molecules23051161.

Sarpong FM, Armah FA, Amponsah IK, Atchoglo PK (2016) Antinociceptive ecdysteroids and other constituents of *Palisota hirsuta* K. Schum (Commelinaceae). J Applied Pharmaceutical Science 6(10): 147–153 https://doi.org/10.7324/JAPS.2016.601020.

Seidlova-Wuttke D, Ehrhardt C, Wuttke W (2010) Metabolic effects of 20-OH-ecdysone in ovariectomized rats. J Steroid Biochem Mol Biol 119: 121–126 https://doi.org/10.1016/j.jsbmb.2010.01.006.

Shafiq N, Zameer R, Attik N, Moveed A, Farooq A, Imtiaz F, Parveen S, Rashid M, Noor N (2024) Integration of virtual screening of phytoecdysteroids as androgen receptor inhibitors by 3D-QSAR model, CoMFA, molecular docking and ADMET analysis: an extensive and interactive machine learning. Journal of Steroid Biochemistry and Molecular Biology 237: article 106427 https://doi.org/10.1016/j.jsbmb.2023.106427.

Shuvalov O, Fedorova O, Tananykina E, Gnennaya Y, Daks A, Petukhov A, Barlev NA (2020) An arthropod hormone, ecdysterone, inhibits the growth of breast cancer via different mechanisms. Frontiers in Pharmacology 11: article 561587 https://doi.org/10.3389/fphar.2020.561537.

Shuvalov O. Kirdeeva Y. Fefilova E. Netsvetay S. Zorin M. Vlasova Y. Fedorova O. Daks A. Parfenyev S. Barlev N (2023) 20-Hydroxyecdysone confers antioxidant and antineoplastic properties in human non-small cell lung cancer cells. Metabolites 13: article 656. https://doi.org/10.3390/metabo13050656.

Shuvalov O. Kirdeeva Y. Fefilova E. Daks A. Fedorova O. Parfenyev S. Nazarov A. Vlasova Y. Krasnov GS. Barlev NA (2024) 20-Hydroxyecdysone boosts energy production and biosynthetic processes in non-transformed mouse cells. Antioxidants 13: article 1349. https://doi.org/10.3390/antiox13111349.

Simon P, Koolman J (1989) Ecdysteroids in vertebrates: pharmacological aspects. In: (Koolman J, Ed) Ecdysone: from chemistry to mode of action, Georg Thieme Verlag, Stuttgart, BRD, pp. 254–259,

Sleno L (2012) The use of mass defect in modern mass spectrometry. Journal of Mass Spectrometry 47(2): 226–236 https://doi.org/10.1002/jms.2953.

Smagghe G (2009) Ecdysone: structures and functions. Springer Science + Media BV, pp 583 ISBN: 978-1-4020-9111-7.

Stopka P, Stancl J, Sláma K (1999) Effect of insect hormone, 20-hydroxyecdysone on growth and reproduction in mice. Acta Soc Zoologica Bohemoslovaka 63: 367–378.

Strickley RG (2004) Solubilizing excipients in oral and injectable formulations. Pharmaceutical Research 21 (2): 201–230 https://doi.org/10.1023/b:pham.0000016235.32639.23.

Stielow M, Witczynska A, Kubryn N, Fijalkowski L, Nowaczyk J, Nowaczyk A (2023) The bioavailability of drugs - the current state of knowledge. Molecules 28, article 8038 https://doi.org/10.3390/molecules28248038.

Sun H, Chow EC, Liu S, Du Y, Pang KS (2008) The Caco-2 cell monolayer: usefulness and limitations. Expert Opin Drug Metab Toxicol 4(4):395–411. https://doi.org/10.1517/17425255.4.4.395.

Syrov VN (2000) Comparative experimental investigations of the anabolic activity of ecdysteroids and steranabols. Pharmaceutical Chemistry Journal 34: 193–197 https://doi.org/10.1007/BF02524596.

Syrov VN, Nabiev AN, Sultanov MB (1986) The effect of phytoecdysteroids on the bile secretion function of the liver in normal rats and in animals with experimental hepatitis. Farmakologiya i Toksikologiya 49(3): 100–103 [in Russian]

Syrov VN, Khushbaktova ZA, Mirzaev YR, Baltaev UA (1989) Antiulcer activity of phytoecdysteroids. Khimiko-Farmatsevticheskii Zhurnal 23(4): 441–445 [in Russian].

Syrov VN, Khushbaktova ZA, Nabiev AN (1992) An experimental study of the hepatoprotective properties of phytoecdysteroids and nerobol in carbon tetrachloride - induced liver injury. Eksp Klin Farmakologiya 55(3): 61–65 [in Russian]

Syrov VN, Saatov V, Sagdullaev SS, Mamatkhanov AU (2001) Study of the structure—anabolic activity relationship for the phytoecdysteroids extracted from some plants of central Asia. Pharmaceutical Chemistry Journal 35: 667–671. https://doi.org/10.1023/A:1015344614064.

Syrov VN, Khushbaktova ZA, Komarin AS, Abidov AB, Pechenitsina TV, Aripkhodzhaeva FA (2004) Experimental and clinical evaluation of the efficacy of ecdysten in the treatment of hepatitis. Eksperimental'naia klinicheskaia Farmakologiia 67(5): 56–59 [in Russian].

Takemoto T, Ogawa S, Nishimoto N (1967) Studies on the constituents of Achyranthes radix. I. Yakugaku Zasshi 87: 1463–1468.

Temirgaziyev BS, Seilkhanov TM, Tyanakh S, Kozhanova AM, Seilkhanov T, Minayeva YeV, Salkeyeva LK, Tuleuov BI, Adekenov SM (2018) Obtaining and investigation of supramolecular inclusion complex of 2-deoxy-20-hydroxyecdysone with γ-cyclodextrine by NMR spectroscopic method. Chemical Journal of Kazakhstan (Khimicheskiy Zhurnal Kazakhstana) (2): 36–44.

Temirgaziyev BS, Kucakova K, Baizhigit YA, Jurasek M, Dzubak P, Hajduch B, Drasar P, Tuleuov BI, Adekenov SM (2019a) Bioavailability and structural study of 20-hydroxyecdysone complexes with cyclodextrins. Steroids 147: 37–41 https://doi.org/10.1016/j.steroids.2018.11.007.

Temirgaziyev BS, Tuleuov BI, Romanova Ma, Seidakhmetova RB, Seilkhanov TM, Seilkhanov TM, Salkeeva IK, Adekenov SM (2019b) Supramolecular complexes of 3-*epi*-2-deoxyecdysone with cyclodextrins and their antiinflammatory activity. Russian Journal of General Chemistry 89(3): 424–428 https://doi.org/10.1134/S1070363219030095.

Tomás J, Camps F, Coll J, Mélé E, Pascual N (1992) Synthesis of 25-fluoroponasterone A. a fluorinated analogue of 20-hydroxyecdysone. Tetrahedron 48(44): 9809–9818 https://doi.org/10.1016/S0040-4020(01)81197-2.

Tóth G, Herke I, Gáti T, Vágvölgyi M, Berkecz R, Parfenova LV, Ueno M, Yokoi T, Nakagawa Y, Hunyadi A (2021) A commercial extract of *Cyanotis arachnoidea* roots as a source of unusual ecdysteroid derivatives with insect hormone receptor binding activity. Journal of Natural Products 84(7): 1870–1881, https://doi.org/10.1021/acs.jnatprod.0c01274.

Tóth G, Santa-Maria AR, Herke I, Gáti T, Galvis-Montes D, Walter FR, Deli MA, Hunyadi A (2023) Highly oxidized ecdysteroids from a commercial *Cyanotis arachnoidea* root extract as potent blood-brain barrier protective agents. Journal of Natural Products 86: 1074–1080 https://doi.org/10.1021/acs.jnatprod.2c00948.

Tsitsimpikou C, Tsamis GD, Siskos PA, Spyridaki MH, Georgakopoulos CG (2001) Study of excretion of ecdysterone in human urine. Rapid Communications in Mass Spectrometry 15(19): 1796–1801 https://doi.org/10.1002/rcm.433.

Tsuji K, Hirose T, Okada M, Shibatani J, Hirai Y, Muramatsu N, Inaoka Y, Fukuda T, Yagi M (1995) Skin cosmetics or bath preparations containing steroid derivatives as skin metabolism-activating and antiwrinkling agents. Patent JP 07316057 A 1995-12-05.

Tuleuov B, Temirgaziyev BS, Kozhanova AM, Seydakhmetova RB, Turdybekov KM, Seilkhanov T, Seilkhanov O, Drasar P, Adekenov S (2020) Supramolecular complex of 20-hydroxyecdysone-3-acetate with β-cyclodextrin and its biological activity. Russian Journal of General Chemistry 90: 2258–2263 https://doi.org/10.1134/S1070363220120075.

Tuleuov BI, Kozhanova AM, Temirgaziyev BS, Almagambetov K, Shaidanov MZ, Seidakhmetova RB, Drasar P, Adekenov SM (2024) Studying the specific (adaptogenic and anabolic) activity of the supramolecular complex of 20-hydroxyecdysone triacetate with β-cyclodextrin. Macedonian Journal of Medical Sciences 12 (2): 356–361 https://oamjms.eu/index.php/mjms/article/view/11906.

Tuleuov B, Abulyaissova L, Temirgaziyev B, Kassymova M, Almagambetov K, Melsov M, Kozhanova A, Draqsar P, Adekenov S (2025) Synthesis, anti-inflammatory and analgesic activity of the supramolecular complex of 3-*epi*-2-deoxyecdysone and prospects for its medical application. Fitoterapia 187: article 106952 https://doi.org/10.1016/j.fitote.2025.106952.

Udintsev AV, Ihalaynen AA, Maximov VA (2014) The comparative experimental access of toxicity and pharmacokinetic parameters of pharmacological substances based on phytosteroid ecdysterone. Medline-ru 15(2): 250–262 [in Russian].

Usmanov D, Szwieck S, Martin GC, Rasulev B (2025) From activity to structure: training AI to design a drug – an ecdysteroid-based case study. ACS meeting August 19, Poster Board #607.

Vágvölgyi M, Bélteky P, Bogdán D, Nové M, Spengler G, Latif AD, Zupkó I, Gáti T, Tóth G, Kónya Z, Hunyadi A (2020) Squalenoylated nanoparticle prodrugs of adjuvant antitumor 11α-hydroxyecdysteroid 2,3-acetonides act as cytoprotective agents against doxorubicin and paclitaxel. Frontiers in Pharmacology 11: article 552088, https://doi.org/10.3389/fphar.2020.552088.

van Breemen RB, Li Y (2005) Caco-2 cell permeability assays to measure drug absorption. Expert Opin Drug Metab Toxicol 1(2):175–85. https://doi.org/10.1517/17425255.1.2.175.

Veber DF, Johnson SR, Cheng HY, Smith BR, Ward KW, Kopple KD (2002) Molecular properties that influence the oral bioavailability of drug candidates. Journal of Medicinal Chemistry 45(12): 2615–23. https://doi.org/10.1021/jm020017n.

Volodin VV, Quang NH, Kaplun AP, Volodina SO (2020) Preparation and physicochemical properties of the ecdysteroid-containing nanodispersions based on the triterpenoids from birch bark. Nanotechnologies in Russia 15: 37–44 https://doi.org/10.1134/S1995078020010073.

Wang L, Li S, Cai H, Liu X, Feng T, Zhao X (2022) Preparation and characterization of ecdysterone/hydroxypropyl-β-cyclodextrin inclusion complex with enhanced oral and transdermal bioavailability. *J Microencapsul* 39 (2): 145–155 https://doi.org/10.1080/02652048.2022.2056251.

Wang X-l, Li H-w, Zhang J, Tan J-h (2014) Preliminary study on effects and mechanism of new ecdysterone derivative TAPA on reducing blood glucose. China Pharmacy 25(45): 4252–4255 [in Chinese] https://doi.org/10.6039/j.issn.1001-0408.2014.45.09.

Wu K, Kwon SH, Zhou X, Fuller C, Wang X, Vadgama J, Wu Y (2025) Overcoming challenges in small-molecule drug bioavailability: a review of key factors and approaches. International Journal of Molecular Sciences 25, article 13121 https://doi.org/10.3390/ijms252313121.

Wu M, Zhao S, Ren L, Wang R, Bai X, Han H, Li B, Chen H (2011) Research on relationship between tissue quantitative distribution of ^{3}H-*Achyranthes bidentata* ecdysterone and channel-tropism of herbal drugs in mice. Zhongguo Zhong Yao Za Zhi (China Journal of Chinese Materia Medica) 36(21): 3018–3022 [in Chinese]

Xia X, Zhang Q, Liang G, Lu S, Yang Y, Tian Y (2013a) Role of 20-hydroxyecdysone in protecting rats against diabetic cardiomyopathy. Zhonghua Laonian Xin Nao Xuegangbing Zazhi 15(4): 412–415 [in Chinese] https://api.semanticscholar.CorpusID:76280400.

Xia X, Zhang Q, Wang Z, Gui G, Liang G, Liu R (2013b) Protective effect of 20-hydroxy-ecdysone on diabetic hepatopathy of rats. Xiandai Yufang Yixue 40(21): 4031–4034 [in Chinese].

Xing H, Sun X, Xing L, Li Q, Puyang M, Liu L, Chen D (2025) ROS-responsive ecdysterone nanodelivery system ameliorates Alzheimer's disease via synergistic modulation of oxidative stress and amyloid-β deposition. *Chemical Engineering Journal* 2025: 167717 https://doi.org/10.1016/j.cej.2025.167717.

Xu T, Xu Y, Zhang J, Zhou Y, Feng H, Zhang A, Zhang Y (2026) Advances in AI for predicting pharmacological properties of natural medicines. Life Sciences 387. article 124180, https://doi.org/10.1016/j.lfs.2025.124180.

Yang L, Pan J (2019) Therapeutic effect of ecdysterone combine paeonol oral cavity direct administered on radiation-induced oral mucositis in rats. International Journal of Molecular Sciences 20(15): article3800. https://doi.org/10.3390/ijms20153800.

Yoshida T, Otaka T, Uchiyama M, Ogawa S (1971) Effect of ecdysterone on hyperglycemia in experimental animals. Biochemical Pharmacology 20: 3263–3268 https://doi.org/10.1016/0006-2952(71)90431-x.

Yuliandra T, Touvleliou K, de la Torre X, Botrè F, Loke S, Isenmann E, Valder S, Diel P, Parr MK (2023) Urinary excretion of ecdysterone and its metabolites following spinach consumption. Mol Nutr Food Res. 67(14):e2200518. https://doi.org/10.1002/mnfr.202200518.

Zaghloul E, Handousa H, Singab ANB, Elmazar MM, Ayoub IM, Swilam N (2023) Phytoecdysteroids and anabolic effect of *Atriplex dimorphostegia* UPLC-PDA-MS/MS profiling, *in silico* and *in vivo* models. Plants 12: article 206 https://doi.org/10.3390/plants12010206.

Zainullin VG, Mishurov VP, Punegov VV, Starabor NA, Bashlykova LA, Babkina NY (2003) Biological efficiency of two feed additives containing ecdysteroids of *Serratula coronata* L. Rastitel'nye Resursy 39(2), 95–103 [in Russian].

Zhang D, Zhang M, Ding B, Wang XL, Qiu ZY, Qin Y (2011) Synthesis of a novel phosphate analog of 20-hydroxylecdysone with potent hypoglycemic activity. J Asian Nat Prod Res 13(4): 297–303 https://doi.org/10.1080/10286020.2011.554830.

Zhang H, Zhang D, Ray K, Zhu M (2009) Mass defect filter technique and its applications to drug metabolite identification by high-resolution mass spectrometry. Journal of Mass Spectrometry 44(7): 999–1016. https://doi.org/10.1002/jms.1610.

Zhao Q, Wang Zx, He JT (2021) Effect of β-ecdysterone on the proliferation, differentiation and apoptosis of rat osteoblasts induced by high glucose. The Chinese Journal of Clinical Pharmacology (4): 443–446 [in Chinese].

Zhou Y, Wu X, Liao J, Wu C, Zhang Y, Zhang Z (2010) Effect of ecdysterone on the healing of gastric ulcer in model rats. China Pharm 21: 2332–2335 [in Chinese].

Chapter 8
Machine Learning and Artificial Intelligence in Phytoecdysteroid Discovery and Analysis

Durbek Abdikhoshimovich Usmanov, Ugiloy Yusufovna Yusupova, and Bakhtiyor Rasulev

Abstract Phytoecdysteroids (PE) constitute a unique class of naturally occurring steroids exhibiting pronounced anabolic activity and broad pharmacological potential. While quantitative structure–activity relationship (QSAR) approaches have previously been applied to rationalize the biological effects of individual ecdysteroid series, a unified and interpretable modeling framework suitable for knowledge transfer and molecular design remains limited. In this chapter, we present an expanded QSAR-based framework for the analysis of anabolic activity in PE, integrating molecular structure curation, quantum-chemical optimization, descriptor-based feature selection, statistical modeling, and chemical-space interpretation. Building on our previously published open-access study, this work introduces a conceptual modeling framework and a descriptor-space analysis using density distributions to elucidate the physicochemical patterns underlying anabolic potency. The developed framework highlights the roles of molecular symmetry, polarizability, lipophilicity, and topological complexity in governing anabolic activity, while providing chemically interpretable design rules for future ecdysteroid analogs. This chapter aims to serve as a reference workflow for QSAR-driven analysis of natural product scaffolds and as a bridge between statistical modeling and chemically meaningful insight.

Keywords Ecdysteroid · Anabolic activity · Machine learning, genetic algorithm · QSAR · Descriptors

D. A. Usmanov (✉) · B. Rasulev
Department of Coatings and Polymeric Materials, Materials and Nanotechnology Program, North Dakota State University,
Fargo, ND, USA
e-mail: durbek.usmanov@ndus.edu

U. Y. Yusupova
Laboratory of Glycosides Chemistry, Institute of the Chemistry of Plant Substances, Academy of Science of the Republic of Uzbekistan, Tashkent, Uzbekistan

V. N. Syrov et al. (eds.), *Phytoecdysteroids II*,
https://doi.org/10.1007/978-981-95-8423-9_8

8.1 Introduction

PE represent a class of naturally occurring bioactive compounds that exhibit a broad range of pharmacological activities (Parr et al. 2015; Isenmann et al. 2019; Yusupova et al. 2019; Yusupova et al. 2020; Yusupova et al. 2021; Yusupova et al. 2022; Yusupova et al. 2025). The most notable feature of these substances is their pronounced ability to stimulate protein synthesis and promote anabolic processes (Parr et al. 2015). Anabolism plays a central role in tissue and organ formation by supporting cell growth, differentiation, and overall biomass increase through the biosynthesis of complex macromolecules. In their native biological context, ecdysteroids function as hormones regulating cell proliferation, growth, and developmental cycles in insects and other invertebrates (Sláma et al. 1993; Sláma et al. 1996). Recent experimental evidence suggests that the anabolic effects of ecdysterone, a plant-derived steroid widely associated with performance enhancement, may be mediated through interactions with estrogen receptors (ERs) (Yao et al. 1993; Ono et al. 2008; Parr et al. 2015).

In animal studies, ecdysterone has demonstrated anabolic efficacy comparable to, or exceeding, that of certain prohibited anabolic agents such as metandienone, while lacking their associated adverse effects (Parr et al. 2015). However, despite these promising findings, comprehensive mechanistic investigations and large-scale clinical studies in humans remain limited. At the same time, PE are increasingly marketed as dietary supplements, particularly within athletic communities, where they are promoted for enhancing strength, supporting muscle hypertrophy during resistance training, reducing fatigue, and improving post-exercise recovery. A growing body of literature reports diverse, predominantly beneficial physiological effects of ecdysteroids in mammals, based on studies across multiple animal models, including rats, mice, Japanese quail, and cattle. This expanding use has raised concerns about their potential misuse in competitive sports, prompting heightened attention from anti-doping authorities (Syrov 2000).

8.2 QSAR Modeling of Natural Compounds: Opportunities and Challenges

QSAR modeling has become a powerful computational tool for identifying structure–activity relationships (Usmanov et al. 2020; Usmanov et al. 2025) and guiding molecular design in cases where experimental studies are costly, time-consuming, or limited by material availability (Turabekova et al. 2008; Cvetnic et al. 2019). For natural compounds, such as PE, QSAR approaches allow the extraction of maximum information from relatively small, structurally related datasets while maintaining chemical interpretability (Wurtz et al. 2000). However, the analysis of natural compounds using QSAR presents unique challenges (Juretic et al. 2014; Jagiello et al. 2016). Datasets of natural compounds are often limited in size, structurally constrained by a common scaffold, and characterized by minor variations in functional groups. In such cases, model reliability, descriptor interpretability, and applicability domain assessment become more important than simply predictive performance. Moreover, traditional QSAR studies often focus on the numerical quality of the model while providing limited insight into the underlying descriptor space and its chemical meaning.

8.3 From Predictive Models to Interpretable Structures

Recent advances in cheminformatics and molecular descriptor theory have enabled the development of statistically robust QSAR models for ecdysteroids and related steroid systems. Nevertheless, many existing studies still focus on individual models or equations, failing to integrate the modeling steps into a holistic conceptual workflow that can be easily transferred to new datasets or chemical classes. To address this issue, there is a growing need to shift from isolated QSAR equations to structure-based modeling, where data processing, descriptor generation, feature selection, model validation, and chemical interpretation are considered as interconnected components of a single analytical pipeline. This approach not only enhances transparency and reproducibility but also facilitates the transformation of statistical results into chemically meaningful knowledge.

8.4 Scope and Structure of This Chapter

This chapter presents a comprehensive QSAR framework (Fig. 8.1) for analyzing the anabolic activity of PE, building upon our previously published open-access research (Usmanov et al. 2025) but significantly expanding its conceptual scope and depth of interpretation. In addition to model development and validation, we present a descriptor-space analysis based on density distributions to visualize how key molecular descriptors differentiate between compounds with high and low activity. The chapter is structured as follows. Section 8.2 presents the conceptual QSAR framework and computational workflow. Section 8.3 describes the dataset, descriptor generation, and model development strategy. Section 8.4 presents model validation and applicability domain analysis. Section 8.5 focuses on the interpretation of the descriptor space using density-based visualization, linking statistical descriptors to chemically meaningful design rules. Finally, Section 8.6 discusses future perspectives and the broader applicability of the proposed concept to drug discovery based on natural compounds.

The proposed QSAR modeling framework, instead of being a linear workflow, is designed as an integrated system that combines data processing, statistical modeling, and chemical interpretation into a single analytical strategy. This structure ensures a close connection between model development, validation, and descriptor-level analysis, allowing for the transformation of statistical results into chemically meaningful conclusions. It also enables the extraction of chemically meaningful design rules and supports the rational modification of ecdysteroid structures for the development of future analogs.

8.5 Conceptual Structure of QSAR

The QSAR framework used in this chapter is designed to provide a transparent and interpretable link between molecular structure and the anabolic activity of ecdysteroids (Fig. 8.1). Instead of focusing on a single predictive model, this framework integrates data preparation, molecular representation, statistical learning, validation, and chemical interpretation into a single analytical pipeline. This structure

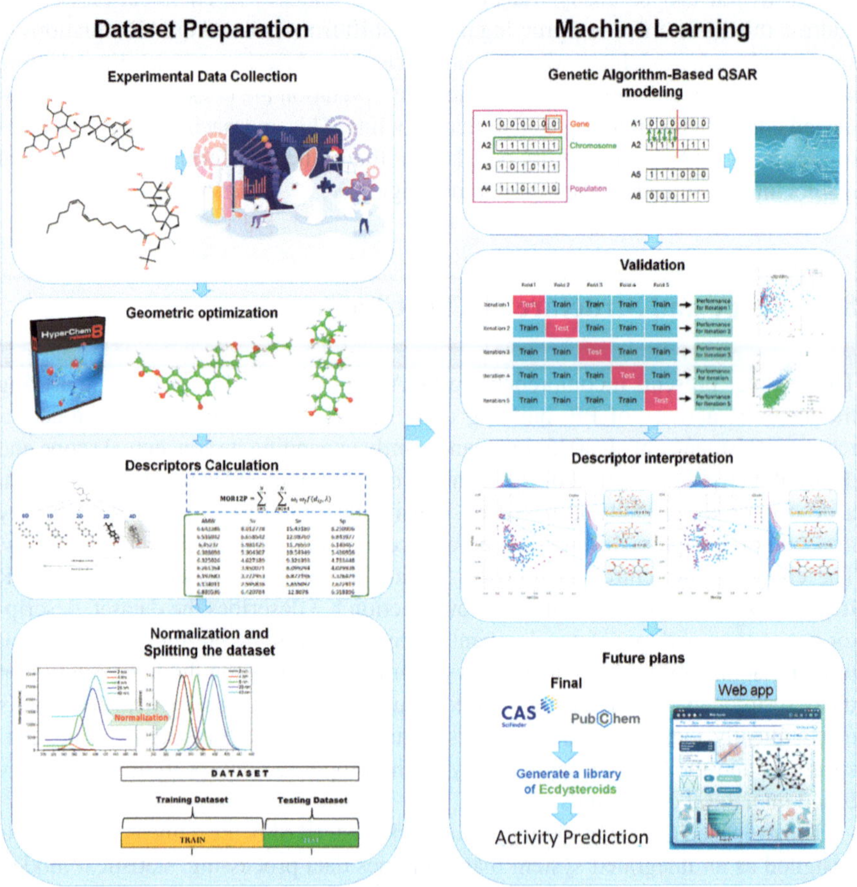

Fig. 8.1 QSAR framework for modeling the anabolic activity of ecdysteroids

ensures reproducibility, minimizes model bias, and facilitates the translation of numerical results into chemically meaningful conclusions.

The workflow begins with data preparation, where experimentally measured anabolic activity data are combined with carefully selected molecular structures. Given the sensitivity of QSAR models to structural accuracy, all compounds undergo geometric optimization before descriptor calculation. This step ensures that the three-dimensional descriptors reflect physically sound conformations and that the calculated properties are internally consistent across the entire dataset.

Following structure optimization, molecular descriptors reflecting topological, geometric, and physicochemical characteristics are calculated to encode the chemical information of each ecdysteroid. Descriptor preprocessing, including normalization and removal of redundant or uninformative variables, is performed to stabilize model training and prevent the dominance of scale-dependent features. The dataset is then split into a training and an external test subset for independent evaluation of predictive performance.

The machine learning stage focuses on feature selection using a genetic algorithm combined with multiple linear regression. The genetic algorithm efficiently explores the multidimensional descriptor space and identifies compact subsets of variables that provide optimal explanatory power while minimizing multicollinearity. This strategy prioritizes model interpretability and reliability over excessive complexity, especially for small, structurally related datasets, such as a series of natural compounds.

Model reliability is assessed using validation procedures, including cross-validation, external testing, and applicability domain analysis. These steps ensure that the resulting QSAR models are statistically sound, robust against spurious correlations, and applicable only within the defined chemical space of the studied ecdysteroids.

Finally, the framework emphasizes interpreting the descriptor space as a central component rather than an auxiliary step. Density-based visualization and structural analysis are used to explore how key descriptors differentiate compounds with high and low anabolic activity. This interpretive level enables the extraction of chemically meaningful design rules and supports the rational modification of ecdysteroid structures to develop future analogs.

8.6 Data Set and Computational Methodology

8.6.1 Data Set and Biological Data

The data set used in this study includes 23 PE compounds (Fig. 8.2) for which experimental data on anabolic activity (AA) were obtained from previously published in vivo studies. Biological activity was assessed in rat models by measuring radioactivity uptake (counts/g) in target tissues after oral administration of each

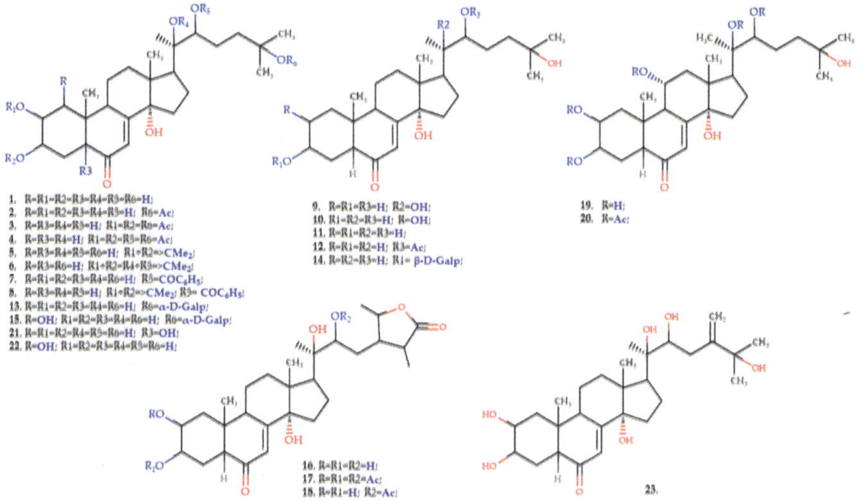

Fig. 8.2 The chemical structures of the ecdysteroids that were used in the study

compound at a dose of 5 mg/kg. To ensure consistency and simplify statistical modeling, the experimental activity values were converted to logarithmic molar units (log(AA)) and used as the dependent variable in the QSAR analysis. Molecular structures and corresponding experimental activity values are shown in Fig. 8.2 and Table 8.1, respectively.

8.6.2 Molecular Modeling and Descriptor Calculation

All molecular structures were built and geometrically optimized prior to descriptor calculation to ensure physically reasonable conformations. Geometry optimization was performed using the semi-empirical quantum-chemical method RM1, which provides an effective balance between computational cost and structural accuracy for organic molecules (Rocha et al. 2006). The optimized structures served as the basis for subsequent descriptor calculations.

To support chemical interpretation and the study of structure–activity relationships, a set of quantum-chemical and physicochemical descriptors was initially calculated, including frontier molecular orbital energies (HOMO and LUMO), dipole moment components, total energy, lipophilicity (LogP), refractivity, polarizability, and atomic charges. These descriptors were evaluated along with topological and constitutional parameters during the model development process.

Table 8.1 List of ecdysteroid compounds with experimentally determined anabolic activity values

No.	Ecdysteroids' list	Activity (Radioactivity, cpm/g)
1	20-hydroxyecdysone	$228.633 \pm 8.683^{*}$
2	Viticosterone E	$209.016 \pm 1.414^{*}$
3	20-hydroxyecdysone 2,3,22-triacetate	$211.172 \pm 2.562^{*}$
4	20-hydroxyecdysone 2,3,20,22-tetraacetate	$203.142 \pm 2.488^{*}$
5***	20-hydroxyecdysone 2,3-monoacetonide	$169.810 \pm 2.862^{*}$
6	20-hydroxyecdysone 2,3-20,22-diacetonide	$150.520 \pm 3.574^{**}$
7	20-hydroxyecdysone 22-benzoate	$188.850 \pm 2.890^{*}$
8	20-hydroxyecdysone 22-benzoate 2,3-monoacetonide	$163.280 \pm 5.280^{*}$
9	2-deoxy-20-hydroxyecdysone	$188.533 \pm 4.872^{*}$
10***	Ecdysone	$158.200 \pm 3.629^{*}$
11	2-deoxyecdysone	$153.220 \pm 4.830^{**}$
12	2-deoxyecdysone 22-acetate	$152.720 \pm 4.624^{*}$
13	SileneosideA	$236.450 \pm 7.167^{*}$
14	SileneosideE	$177.482 \pm 4.896^{*}$
15***	SileneosideC	184.120 ± 2.440
16	Cyasterone	$241.750 \pm 2.522^{*}$
17	Cyasterone 2,3,22-triacetate	$218.362 \pm 5.270^{*}$
18	Cyasterone 22-acetate	$266.166 \pm 2.363^{*}$
19	Turkesterone	$264.512 \pm 6.012^{*}$
20***	Turkesteronetetraacetate	$255.250 \pm 3.798^{*}$
21	PolypodineB	$196.890 \pm 8.250^{*}$
22	IntegristeroneA	$175.520 \pm 5.018^{*}$
23	24(28)-dehydromakisteronA	$219.400 \pm 2.758^{*}$

Note: ***prediction set of compounds; differences reliable for $^{*}p < 0.001$; $^{**}p < 0.01$

8.6.3 Chemoinformatic Analysis and QSAR Modeling

The QSAR model development was performed using a genetic algorithm–multiple linear regression (GA–MLR), which enables efficient exploration of the multidimensional descriptor space while prioritizing compact, interpretable models. Molecular descriptors reflecting constitutional, topological, and fragment information were calculated based on optimized structures, and the initial pool of descriptors was formed by removing constant and uninformative variables.

After data processing, a final set of 384 descriptors was retained for model building. Candidate models were selected based on statistical quality and robustness criteria, including high coefficients of determination for the training and external validation sets, low standard deviation, favorable Fisher statistics, and a limited number of descriptors to reduce overfitting. Particular attention was paid to minimizing multicollinearity among the selected descriptors to preserve chemical interpretability. Model validation was performed using leave-one-out cross-validation and external test sets to assess predictive performance and stability. To further ensure model reliability, an analysis of descriptor intercorrelation was conducted, and descriptors exhibiting high mutual correlation or redundancy were removed. This strategy ensured that the final QSAR models reflected independent and chemically meaningful factors influencing anabolic activity(Najafi et al. 2013; Zhuravskyi et al. 2024).

8.7 Model Development and Validation

To develop predictive and interpretable QSAR models, the dataset of 23 ecdysteroid compounds was partitioned into training and external validation subsets using an 80:20 ratio. Nineteen compounds were assigned to the training set, while four structurally representative compounds were reserved for external validation. Importantly, this partitioning was not performed randomly; instead, it was guided by structural diversity to ensure that each major structural motif present in the dataset was represented in both subsets. This strategy reduces the risk of biased performance estimates and supports a more realistic assessment of model generalizability.

Application of the GA–MLR modeling strategy resulted in the identification of several statistically significant QSAR models incorporating different combinations of molecular descriptors. The statistical characteristics of all developed models are summarized in Table 8.2. Among these, models containing two, three, and four descriptors demonstrated particularly strong performance and were selected for further analysis.

The most parsimonious model is the two-descriptor QSAR equation:

$$\log\left[\mathrm{AA}\right] = 0.808\left(\pm 0.175\right) SIC0 - 0.582\left(\pm 0.193\right) G3p + 6.491\left(\pm 0.142\right)$$

This model exhibits high coefficients of determination and cross-validated predictive ability for both the training and external validation sets. Graphical evaluation of model performance, including predicted versus experimental values, applicability domain analysis, and robustness testing, is presented in Fig. 8.3.

In this model, SIC0 represents the structural information content index of order zero and reflects neighborhood symmetry within the molecular graph, whereas G3p is a directional WHIM descriptor weighted by atomic polarizability. The regression coefficients indicate that increased structural information content positively contributes to anabolic activity, while increased three-dimensional symmetry weighted by polarizability has a negative effect. The statistical significance of both descriptors supports their relevance in describing the structure–activity relationship of ecdysteroids.

In addition to the two-descriptor model, a three-descriptor model was obtained:

$$\begin{aligned}\log\left[\mathrm{AA}\right] = {} & 0.412\left(\pm 0.217\right) GATS5e - 0.454\left(\pm 0.219\right) G3p \\ & -0.830\left(\pm 0.210\right) ALOGP2 + 6.957\left(\pm 0.182\right)\end{aligned}$$

Table 8.2 Statistical characteristics of the one-, two-, three-, and four-variable models

	Training set, n = 19				Prediction set, n = 4	
Model, no./# of descriptors	R^2	s	F	Q^2	R^2	s
– (1 descr-s)	0.67	0.19	33.87	0.58	0.53	0.27
1 (2 descr-s)	0.89	0.11	64.66	0.84	0.89	0.13
2 (3 descr-s)	0.88	0.12	37.68	0.83	0.97	0.11
3 (4 descr-s)	0.95	0.08	61.98	0.91	0.89	0.15

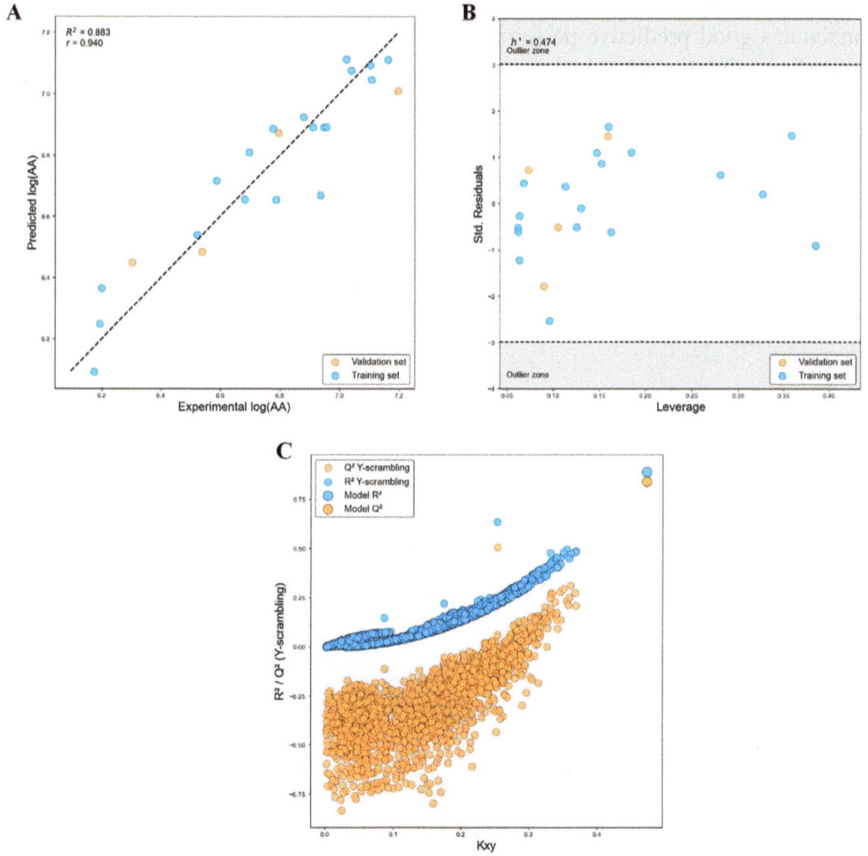

Fig. 8.3 Graphical evaluation of the statistical performance of the best two-descriptor QSAR model: (**a**) Correlation plot of predicted versus experimental log(AA) values for training (blue) and validation (orange) sets; (**b**) Williams plot showing standardized residuals versus leverage values—all compounds lie within the model's applicability domain (h * = 0.474); (**c**) Y-randomization (Y-scrambling) test confirming robustness of the model

This model demonstrates excellent fit and predictive performance, particularly for the external validation set. Its graphical representation is provided in the Supplementary Information (Fig. 8.4). The inclusion of electronegativity-weighted autocorrelation and lipophilicity terms highlights the combined influence of electronic distribution and hydrophobic effects on anabolic activity.

A four-descriptor model was also developed:

$$\log[\mathrm{AA}] = 0.643(\pm 0.367)Mp - 0.595(\pm 0.923)TIE$$
$$-0.480(\pm 0.169)G3p - 1.188(\pm 0.629)ALOGP + 7.337(\pm 0.116)$$

This model yields the highest coefficient of determination for the training set and maintains good predictive performance for the external validation set. The corresponding validation plots are shown in Fig. 8.5.

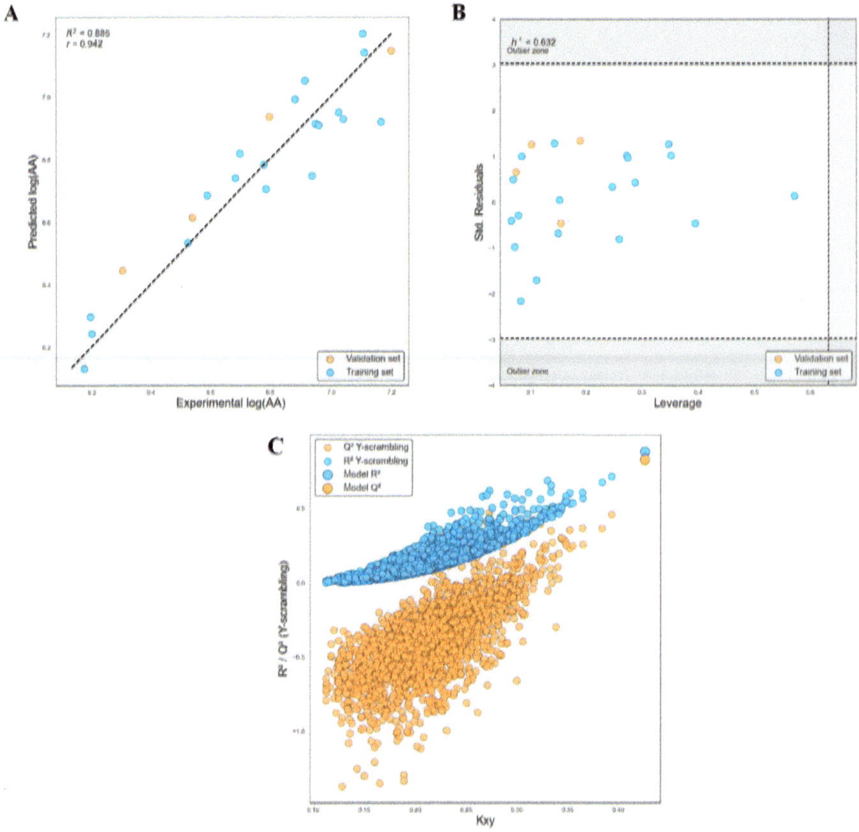

Fig. 8.4 Graphical evaluation of the statistical performance of the best two-descriptor QSAR model: (**a**) Correlation plot of predicted versus experimental log(AA) values for training (blue) and validation (orange) sets; (**b**) Williams plot showing standardized residuals versus leverage values—all compounds lie within the model's applicability domain (h * = 0.632); (**c**) Y-randomization (Y-scrambling) test confirming robustness of the model

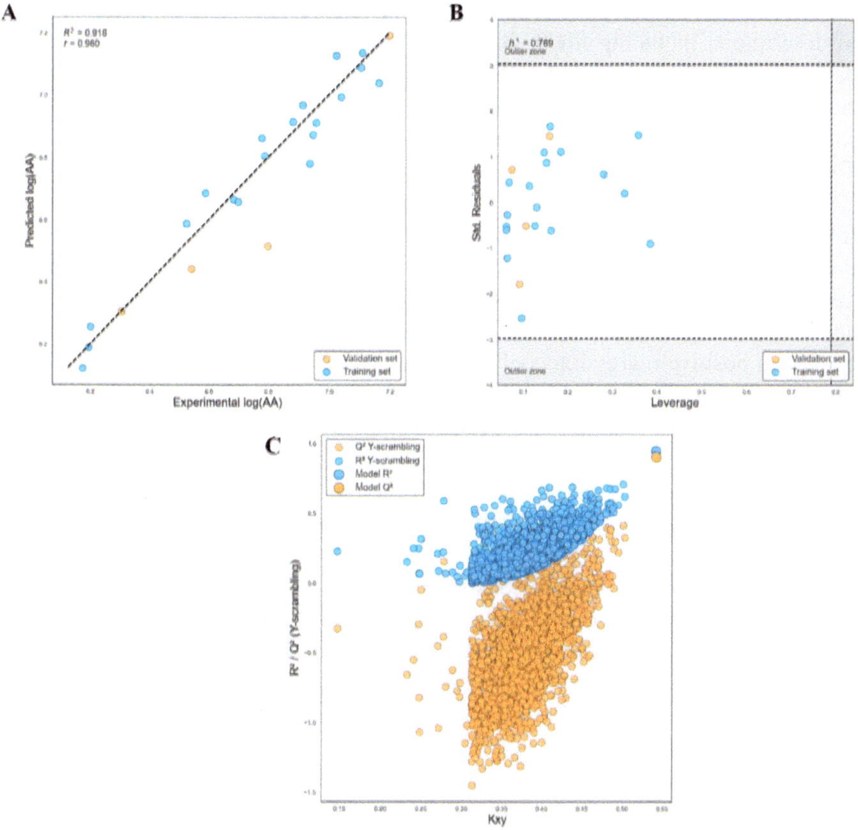

Fig. 8.5 Graphical evaluation of the statistical performance of the best two-descriptor QSAR model: (**a**) Correlation plot of predicted versus experimental log(AA) values for training (blue) and validation (orange) sets; (**b**) Williams plot showing standardized residuals versus leverage values—all compounds lie within the model's applicability domain (h * = 0.789); (**c**) Y-randomization (Y-scrambling) test confirming robustness of the model

Overall, all developed models exhibit strong statistical quality and comparable predictive accuracy. The three-descriptor model provides the highest external predictive performance (R2ext = 0.97), whereas the two-descriptor model demonstrates superior robustness in y-randomization testing. Given its balance of simplicity, interpretability, and stability, the two-descriptor model was selected as the most reliable and generalizable QSAR representation of anabolic activity in ecdysteroids.

8.8 Descriptor Interpretation

In the context of the developed QSAR models, detailed analysis of the selected molecular descriptors provides mechanistic insight into the structural and physicochemical factors governing anabolic activity in ecdysteroids. As demonstrated in the Results section, the most statistically robust models incorporate a consistent set of descriptors, including SIC0, G3p, Mp, TIE, GATS5e, and lipophilicity-related parameters (ALOGP and ALOGP2). Collectively, these descriptors emphasize the central role of molecular symmetry, electronic distribution, three-dimensional shape, and hydrophobicity in modulating biological response (Roy et al. 1984; Todeschini et al. 1994; Todeschini and Gramatica 1997; Najafi et al. 2013).

The descriptor SIC0 belongs to the class of information content indices derived from molecular graph theory and reflects zero-order neighborhood symmetry within the molecule. These descriptors quantify structural complexity by evaluating the distribution of equivalence classes among atoms, accounting for neighbor degree and edge multiplicity in the molecular graph. In the selected QSAR model, SIC0 exhibits a positive regression coefficient, indicating that increased structural information content correlates with enhanced anabolic activity. This observation is consistent with the fact that the most active compounds in the dataset (compounds 19, 21, and 22) display the highest SIC0 values, whereas compound 6, characterized by minimal activity, shows the lowest SIC0. This trend suggests that greater topological diversity and non-uniform substitution patterns favor anabolic potency in ecdysteroids.

The G3p descriptor appears consistently across all developed models and contributes substantially to activity prediction based on the magnitude of its regression coefficient. G3p is a directional WHIM (Weighted Holistic Invariant Molecular) descriptor weighted by atomic polarizability and captures three-dimensional molecular geometry, including size, shape, symmetry, and polarizability distribution. WHIM descriptors are calculated from statistical indices of atomic projections along principal molecular axes and are designed to encode essential 3D structural information relevant to molecular recognition. In particular, directional WHIM symmetry indices describe the degree of atomic symmetry along each principal axis and are sensitive to both molecular shape and electronic anisotropy.

In the present study, G3p shows a negative contribution to anabolic activity, indicating that increased three-dimensional symmetry and a uniform distribution of polarizability diminish biological response. Thus, lower G3p values, reflecting reduced symmetry or anisotropic polarizability, are associated with enhanced activity. This finding complements the positive role of SIC0, highlighting that asymmetry and structural heterogeneity are beneficial features for anabolic potency.

Importantly, although descriptors such as SIC0 and G3p are mathematically derived, they allow for chemically intuitive interpretation. Increased SIC0 values may arise from branching, selective substitution, or localized structural variation near pharmacologically relevant regions. Conversely, reduced G3p values imply disruption of global symmetry or uneven distribution of polarizable groups. In practice, such effects can be achieved through semisynthetic modification of natural ecdysteroids, such as selective acylation, esterification, or hydroxyl group derivatization. These modifications represent chemically feasible strategies for tuning molecular properties in future analog design.

The descriptor Mp, a constitutional index, represents the mean atomic polarizability normalized to a carbon atom. Its inclusion alongside G3p underscores the importance of polarizability in governing anabolic activity, as both descriptors encode complementary aspects of electronic responsiveness within the molecular framework.

The TIE descriptor belongs to the class of electrotopological state (E-state) indices and captures information related to molecular connectivity and local electronic environments. By integrating topological and electronic effects, TIE helps describe how atomic neighborhoods and electronic distributions collectively influence biological activity.

Finally, the presence of GATS5e and lipophilicity descriptors (ALOGP and ALOGP2) highlights the relevance of long-range electronic correlations and hydrophobic character. GATS5e reflects the spatial distribution of Sanderson electronegativity across five-bond separations, while ALOGP-related terms account for membrane permeability and receptor accessibility, key factors in the biological performance of steroidal compounds.

Overall, the descriptor analysis confirms that anabolic activity in ecdysteroids is governed by a delicate balance between molecular symmetry, electronic distribution, polarizability, and hydrophobicity. The consistency between statistical significance, regression trends, and chemically interpretable features reinforces the robustness of the final QSAR model and provides a rational foundation for the design of new ecdysteroid analogs with improved anabolic profiles.

8.9 Descriptor–Activity Density Analysis

The joint density plots provide a direct visualization of the relationship between key QSAR descriptors and experimental anabolic activity (log(AA)) for the studied ecdysteroids. Unlike correlation plots, these density-based representations reveal not only the direction of descriptor influence but also the regions of descriptor space where biologically relevant compounds are concentrated.

8.9.1 SIC0–Activity Relationship

The joint distribution of SIC0 and anabolic activity (Fig. 8.6a) demonstrates a clear positive association between structural information content and biological response. The highest density region is centered at SIC0 values of approximately 0.203–0.212, corresponding to log(AA) values predominantly in the range 6.8–7.2. This region encompasses the most active compounds in the dataset, including those with log(AA) exceeding 7.0.

At lower SIC0 values (below ~0.195), the density shifts toward reduced anabolic activity, typically below log(AA) ≈ 6.6. This observation indicates that ecdysteroids with lower neighborhood symmetry and reduced structural information content are less effective in stimulating anabolic processes. From a chemical perspective, SIC0

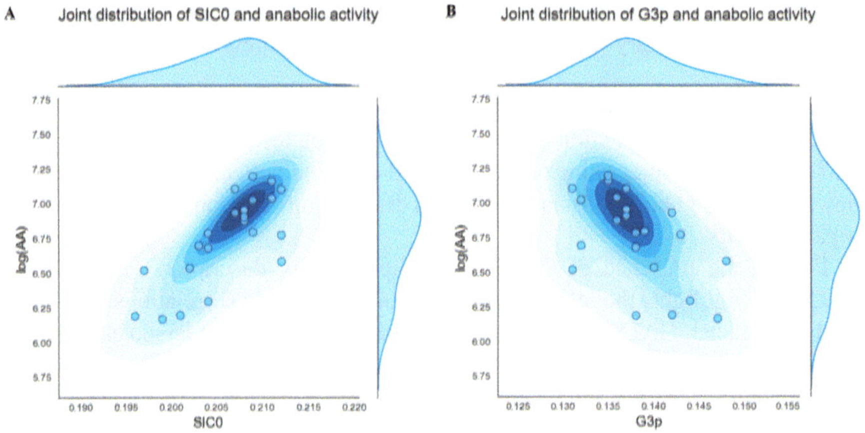

Fig. 8.6 Joint density distributions illustrating the relationships between key QSAR descriptors (**a**) SIC0 and (**b**) G3p and anabolic activity (log AA) of ecdysteroids. High-activity compounds cluster within a narrow and chemically interpretable descriptor space characterized by increased structural information content and reduced three-dimensional symmetry

reflects the degree of topological diversity and symmetry in the molecular graph. Higher SIC0 values arise from increased structural complexity and balanced substitution patterns within the steroid framework. Thus, the density plot confirms that enhanced anabolic activity is favored by molecules exhibiting greater topological richness rather than overly simplified or highly uniform structures.

8.9.2 G3p–Activity Relationship

In contrast, the joint density plot of G3p versus anabolic activity (Fig. 8.6b) reveals an inverse relationship. The highest activity region (log(AA) $\approx$ 6.9–7.2) is associated with lower G3p values, primarily in the range of 0.133–0.140. As G3p increases beyond approximately 0.145, the density progressively shifts toward lower activity values, with log(AA) frequently falling below 6.7.

G3p is a directional WHIM descriptor weighted by atomic polarizability and encodes three-dimensional molecular symmetry and polarizability distribution. Elevated G3p values correspond to more symmetric and uniformly polarizable molecular shapes, whereas lower values indicate increased anisotropy and uneven polarizability distribution. The observed density trend therefore suggests that excessive 3D symmetry and homogeneous polarizability negatively affect anabolic potency.

8.9.3 *Integrated Interpretation*

Taken together, the density plots indicate that anabolic activity in ecdysteroids is maximized within a specific and chemically meaningful descriptor space: high SIC0 combined with low G3p. This combination reflects molecules that are topologically complex yet three-dimensionally asymmetric, with non-uniform distributions of electronic and polarizability properties. Importantly, the density peaks align closely with the activity values of the most potent compounds in the dataset (log(AA) $\approx$ 7.0–7.2), providing independent visual confirmation of the trends inferred from regression coefficients in the QSAR models.

Based on the density analysis, the following descriptor ranges emerge as optimal targets for the rational design of new ecdysteroid analogs with enhanced anabolic activity:

SIC0

- → Target range: 0.203–0.215
  → Higher values are associated with increased anabolic activity.

G3p

- → Target range: 0.130–0.140
  → Lower values favor higher activity; values above ~0.145 should be avoided.

In practical terms, these criteria suggest that new molecules should maintain or increase topological complexity (to elevate SIC0) while introducing structural or electronic asymmetry that reduces global three-dimensional symmetry and uniformity of polarizability (to lower G3p). Such modifications can be achieved through selective substitution, asymmetric functionalization, or targeted derivatization of hydroxyl groups in the ecdysteroid scaffold.

The density-based descriptor analysis reinforces the robustness and interpretability of the final QSAR model. By defining a narrow and chemically interpretable region of descriptor space associated with high anabolic activity, this approach provides actionable guidelines for the design and prioritization of new ecdysteroid analogs, bridging statistical modeling with rational molecular design.

8.10 Outlook and Future Perspectives

Natural compounds display remarkable structural diversity, which underpins their broad range of biological activities (Delazar et al. 2011; Hsieh et al. 2016; Usmanov et al. 2019; Tahan et al. 2020; Esposito et al. 2020; Ali et al. 2024; Althaher et al. 2024; Djelti et al. 2024; Mahmoodi et al. 2024; Zhang et al. 2024; Li et al. 2025). The results of the present QSAR study provide a rational framework for guiding the future design and optimization of ecdysteroid-based anabolic agents. The descriptor-driven insights obtained from the final model suggest clear and chemically actionable strategies for enhancing biological activity through targeted structural modifications. One promising direction involves semisynthetic modification of natural ecdysteroids. As indicated by the descriptor analysis, increasing structural information content (SIC0) while reducing excessive three-dimensional symmetry (G3p) may be achieved through selective functionalization of hydroxyl groups, asymmetric acylation, or introduction of polarizable substituents at non-equivalent positions of the steroid scaffold. Such modifications are experimentally feasible and can be systematically explored to fine-tune electronic distribution and molecular topology.

Another important perspective is the integration of QSAR modeling with receptor-based approaches. Although the present study focuses on ligand-based modeling, the identified descriptor trends can be further validated by molecular docking and molecular dynamics (MD) simulations against relevant biological targets, such as estrogen receptor subtypes implicated in ecdysteroid-mediated anabolic effects. Combining QSAR-derived design rules with binding mode analysis would strengthen mechanistic interpretation and improve confidence in predicted activity trends.

Finally, the developed QSAR framework may be extended toward virtual screening of larger ecdysteroid libraries, including both naturally occurring analogs and in silico-generated derivatives. The clearly defined applicability domain and descriptor target ranges established in this work enable efficient prioritization of candidates with a high likelihood of anabolic potency, thereby reducing experimental cost and accelerating lead identification.

8.11 Comparison with Previous QSAR Studies

Compared to previously reported QSAR studies on steroids and phytoecdysteroids (Usmanov et al. 2025), the present work demonstrates several distinctive features. Earlier QSAR investigations have often relied on limited descriptor sets or focused primarily on lipophilicity and steric parameters, frequently lacking rigorous applicability domain assessment or mechanistic interpretation of descriptor contributions. In contrast, the current study integrates statistically robust model validation with density-based descriptor interpretation, providing both predictive reliability and chemical insight.

Notably, the consistent involvement of symmetry- and polarizability-related descriptors (SIC0 and G3p) distinguishes this model from earlier approaches, where such features were rarely emphasized. While lipophilicity descriptors (e.g., ALOGP) have been commonly reported in steroid QSAR models, their role in the present study appears complementary rather than dominant, suggesting that electronic distribution and structural topology play a more critical role in anabolic activity modulation.

Furthermore, the combination of classical validation techniques (external validation, Williams plot, and Y-scrambling) with descriptor-space visualization places this study among more rigorous contemporary QSAR frameworks. This integrated approach enhances interpretability and aligns with current best practices in cheminformatics, thereby extending the applicability and relevance of QSAR modeling within the broader context of steroid and natural product research.

8.12 Summary

In this chapter, a statistically sound and chemically interpretable QSAR framework was developed to elucidate the structural determinants of anabolic activity in ecdysteroids. The final model demonstrates strong predictive performance, robustness against chance correlation, and a well-defined applicability domain. Importantly, descriptor analysis revealed that anabolic potency is governed by a balance between structural information content, three-dimensional symmetry, polarizability, and electronic distribution.

The integration of density-based visualization with regression analysis enabled direct identification of descriptor regions associated with high biological activity, translating abstract mathematical descriptors into practical design guidelines. These findings provide clear criteria for the rational design and prioritization of new ecdysteroid analogs with enhanced anabolic potential. Overall, this work highlights the value of combining rigorous QSAR validation with chemically meaningful interpretation to support future drug design and optimization efforts.

References

Abbas Delazar, Mohammad-Reza Delnavazi, Narges Yassa, Shiva Parkhideh, Nima Delazar, Lutfun Nahar, Satyajit D. Sarker (2011) Essential oil composition and isolation of freeradical-scavenging phenolic glycosides from the aerial parts of Ajuga chamaepitys growing in Iran. Brazilian J Pharmacog 22(2):299–305. https://doi.org/10.1590/S0102-695X2011005000215.

Abbas Tahan, Mohammad Jafari, Damoun Razmjoue and Javadi Sa (2020) Relationship among some ecological factors and chemical composition of Ajuga chamaecistus Ging. plant species. Acta Ecol Sinica 41(6):268–276. https://doi.org/10.1016/j.chnaes.2019.08.001.

Ali T, Jan I, Bashir R, Mir SA, Ali S and Bader GN (2024) Attenuation of paracetamol-induced hepatotoxicity in Ajuga bracteosa extract treated mice. Heliyon 10(13). https://doi.org/10.1016/j.heliyon.2024.e33998.

Althaher AR, Saqallah FG and Qawass H (2024) Antiproliferative Effects of Methanolic Extract: In Vitro and In Silico Insights. Chemistryselect 9(41). https://doi.org/10.1002/slct.202403836.

Cvetnic M, Juretic Perisic D, Kovacic M, Ukic S, Bolanca T, Rasulev B, Kusic H and Loncaric Bozic A (2019) Toxicity of aromatic pollutants and photooxidative intermediates in water: A QSAR study. 169:918–927. https://doi.org/10.1016/j.ecoenv.2018.10.100.

Djelti F, Berkane A, Alharbi HM, Guendouzi A, Ghellai I, Nabil B, Belkhiri L, Brahim H, Tarik C, Belarbi M, Cherif Fatima Y and Guendouzi A (2024) Identification of Ajuga Iva Extracts as Potential Candidates for Antidysmenorrhea Targeting Human COX2 and PGE2S-1 through In Vitro and In Silico Drug Repurposing Approach. ChemistrySelect 9(21):e202400353. https://doi.org/10.1002/slct.202400353.

Esposito T, Sansone F, Auriemma G, Franceschelli S, Pecoraro M, Picerno P, Aquino RP and Mencherini T (2020) Study on Ajuga reptans Extract: A Natural Antioxidant in Microencapsulated Powder Form as an Active Ingredient for Nutraceutical or Pharmaceutical Purposes. Pharmaceutics 12(7). https://doi.org/10.3390/pharmaceutics12070671.

Hsieh CW, Ko WC, Ho WJ, Chang CK, Chen GJ and Tsai JC (2016) Antioxidant and hepatoprotective effects of Ajuga nipponensis extract by ultrasonic-assisted extraction. Asian Pac J Trop Med 9(5):409–414. https://doi.org/10.1016/j.apjtm.2016.03.029.

Isenmann E, Ambrosio G, Joseph JF, Mazzarino M, de la Torre X, Zimmer P, Kazlauskas R, Goebel C, Botrè F, Diel P and Parr MK (2019) Ecdysteroids as non-conventional anabolic agent: performance enhancement by ecdysterone supplementation in humans. Archives of Toxicology 93(7):1807–1816. https://doi.org/10.1007/s00204-019-02490-x.

Jagiello K, Grzonkowska M, Swirog M, Ahmed L, Rasulev B, Avramopoulos A, Papadopoulos MG, Leszczynski J and Puzyn T (2016) Advantages and limitations of classic and 3D QSAR approaches in nano-QSAR studies based on biological activity of fullerene derivatives. J Nanopart Res 18(9). https://doi.org/10.1007/s11051-016-3564-1.

Juretic D, Kusic H, Dionysiou DD, Rasulev B and Loncaric Bozic A (2014) Modeling of photooxidative degradation of aromatics in water matrix; combination of mechanistic and structural-relationship approach. Chem Eng J 257:229–241. https://doi.org/10.1016/j.cej.2014.07.053.

Li YH, Ba YH, Xiao JF, Chen S, Zhu TQ, Liu KX, He HP, Jiang HY and Yang R (2025) New 14,15-dinor-clerodane and -clerodane diterpenoids from Wall. ex benth. Nat Prod Res. https://doi.org/10.1080/14786419.2025.2473573.

Mahmoodi F, Bikdeloo M, Khadivi A and Akramian M (2024) Morphological and biochemical variation of Ging. ex Benth. in different habitats of Markazi province in the center of Iran. Bmc Plant Biol 24(1). https://doi.org/10.1186/s12870-024-05125-1.

Najafi A, Sobhanardakani S and Marjani M (2013) Exploring QSAR for Antimalarial Activities and Drug Distribution within Blood of a Series of 4-Aminoquinoline Drugs Using Genetic-MLR. Journal of Chemistry 2013:1–12. https://doi.org/10.1155/2013/560415.

Ono Y, Fukaya Y, Imai S and Yamakuni T (2008) Beneficial effects of Ajuga decumbens on osteoporosis and arthritis. Biol Pharm Bull 31(6):1199–1204. https://doi.org/10.1248/bpb.31.1199.

Parr MK, Botrè F, Naß A, Hengevoss J, Diel P and Wolber G (2015) Ecdysteroids: A novel class of anabolic agents? Biol Sport 32(2):169–173. https://doi.org/10.5604/20831862.1144420.

Rocha GB, Freire RO, Simas AM and Stewart JJP (2006) RM1: A reparameterization of AM1 for H, C, N, O, P, S, F, Cl, Br, and I. Journal of Computational Chemistry 27(10):1101–1111. https://doi.org/10.1002/jcc.20425.

Roy AB, Basak SC, Harriss DK and Magnuson VR (1984). NEIGHBORHOOD COMPLEXITIES AND SYMMETRY OF CHEMICAL GRAPHS AND THEIR BIOLOGICAL APPLICATIONS. Mathematical Modelling in Science and Technology, Elsevier: 745–750.

Sláma K, Abubakirov NK, Gorovits MB, Baltaev UA and Saatov Z (1993) Hormonal activity of ecdysteroids from certain asiatic plants. Insect Biochemistry and Molecular Biology 23(1):181–185. https://doi.org/10.1016/0965-1748(93)90099-e.

Sláma K, Koudela K, Tenora J and Maťhová A (1996) Insect hormones in vertebrates: Anabolic effects of 20-hydroxyecdysone in Japanese quail. Experientia 52(7):702–706. https://doi.org/10.1007/bf01925578.

Syrov VN (2000) Comparative experimental investigation of the anabolic activity of phytoecdysteroids and steranabols. Pharmaceutical Chemistry Journal 34(4):193–197. https://doi.org/10.1007/bf02524596.

Todeschini R and Gramatica P (1997) 3D-modelling and Prediction by WHIM Descriptors. Part 5. Theory Development and Chemical Meaning of WHIM Descriptors. Quantitative Structure-Activity Relationships 16(2):113–119. https://doi.org/10.1002/qsar.19970160203.

Todeschini R, Lasagni M and Marengo E (1994) New molecular descriptors for 2D and 3D structures. Theory. Journal of Chemometrics 8(4):263–272. https://doi.org/10.1002/cem.1180080405.

Turabekova MA, Rasulev BF, Dzhakhangirov FN and Salikhov SI (2008) Aconitum and Delphinium alkaloids: "Drug-likeness" descriptors related to toxic mode of action. Environmental Toxicology and Pharmacology 25(3):310–320. https://doi.org/10.1016/j.etap.2007.10.035.

Usmanov D, Rasulev B, Syrov V, Yusupova U and Ramazonov N (2020) Structure-Hepatoprotective Activity Relationship Study of Iridoids. International Journal of Quantitative Structure-Property Relationships 5(3):108–118. https://doi.org/10.4018/ijqspr.20200701.oa3.

Usmanov D, Yusupova U, Syrov V, Casanola-Martin GM and Rasulev B (2025) A Quantitative Structure–Activity Relationship Study of the Anabolic Activity of Ecdysteroids. Computation 13(8):195. https://doi.org/10.3390/computation13080195.

Usmanov DA, Ramazanov NS, Yusupova UY, Kurilov DV, Kachala VV, Krukovskaya NV, Kolotyrkina NG and Zavarzin IV (2019) Iridoids from Phlomis ostrowskiana Regel plant. Russ Chem Bull 68(12):2275–2278. https://doi.org/10.1007/s11172-019-2699-5.

Wurtz JM, Guillot B, Fagart J, Moras D, Tietjen K and Schindler M (2000) A new model for 20-hydroxyecdysone and dibenzoylhydrazine binding: a homology modeling and docking approach. Protein Sci 9(6):1073–1084. https://doi.org/10.1110/ps.9.6.1073.

Yao T-P, Forman BM, Jiang Z, Cherbas L, Chen JD, McKeown M, Cherbas P and Evans RM (1993) Functional ecdysone receptor is the product of EcR and Ultraspiracle genes. Nature 366(6454):476–479. https://doi.org/10.1038/366476a0.

Yusupova U, Sasmakov S, Usmanov D, Ramazonov N, Azimova S and Sagdullaev S (2020) Phytoecdysteroids from Silene claviformis and their antibacterial and antifungal activities. Pol. J. Natur. Sc. 35(3):313–321.

Yusupova UY, Bobakulov KM, Girault JP, Usmanov DA, Ramazonov NS and Abdullaev ND (2022) New Phytoecdysteroid from Silene popovii. Chem Nat Compd 58(6):1082–1084. https://doi.org/10.1007/s10600-022-03874-9.

Yusupova UY, Bobakulov KM, Khurramov AR, Syrov VN, Egamova FR, Karuth A, Usmanov DA, Quadir M and Rasulev B (2025) Phytochemical constituents isolated from Silene popovii Schischk. Med Chem Res 34(2):423–431. https://doi.org/10.1007/s00044-024-03345-6.

Yusupova UY, Ramazonov NS, Bobakulov KM, Egamova FR, Syrov VN and Usmanov DA (2021) Biological active components of silene tomentella and their pharmacological properties. Khim. Rastit. Syr'ya 0(1):197–202. https://doi.org/10.14258/jcprm.2021018323.

Yusupova UY, Usmanov DA and Ramazonov NS (2019) Phytoecdysteroids from the Plant Dianthus helenae. Chem Nat Compd 55(2):393–394. https://doi.org/10.1007/s10600-019-02701-y.

Zhang Y, Gong Y, Hu J, Zhang L, Benito MJ, Usmanov D, Nishanbaev SZ, Song X, Zou L and Wu Y (2024) Quercetin and kaempferol from saffron petals alleviated hydrogen peroxide-induced oxidative damage in B16 cells. Journal of the Science of Food and Agriculture 105(2):967–973. https://doi.org/10.1002/jsfa.13887.

Zhuravskyi Y, Iduoku K, Erickson ME, Karuth A, Usmanov D, Casanola-Martin G, Sayfiyev MN, Ziyaev DA, Smanova Z, Mikolajczyk A and Rasulev B (2024) Quantitative Structure-Permittivity Relationship Study of a Series of Polymers. 4(2):195–203. https://doi.org/10.1021/acsmaterialsau.3c00079.

Chapter 9
Drug Repurposing Opportunities with Phytoecdysteroids: Emerging Trends and Challenges

Durbek Abdikhoshimovich Usmanov, Ugiloy Yusufovna Yusupova, and Shamansur Sagdullaev

Abstract Phytoecdysteroids (PEs), such as 20-hydroxyecdysone (20E) and turkesterone, are natural polyhydroxylated steroids first identified as insect molting hormones but now recognized for their anabolic, cytoprotective, and metabolic effects in mammals. Unlike androgenic steroids, PEs act through non-androgenic mechanisms involving estrogen receptor-β, the ACE2/Ang-(1–7)/Mas receptor axis, and mitochondrial biogenesis, showing excellent safety across studies. Clinical data from BIO101 (pharmaceutical-grade 20E) have demonstrated improved muscle performance in older adults with sarcopenia and acceptable safety in respiratory resilience trials. From a drug repurposing perspective, PEs offer broad therapeutic potential, supporting muscle growth in frailty and cachexia, protecting the liver under oxidative stress, and enhancing cardiac and mitochondrial function. Preclinical evidence also suggests roles in pulmonary repair, neuroprotection, and wound healing. However, clinical validation remains limited, and supplement-based claims often exceed evidence. Major translational challenges include low oral bioavailability, inconsistent supplement quality, uncertain receptor dominance (ERβ vs. MasR), and evolving regulatory attention. Future development should focus on pharmaceutical-grade formulations, biomarker-guided trials, and clear pharmacokinetic–pharmacodynamic alignment. Early clinical priorities include Phase 3 studies in sarcopenia and exploratory trials in NAFLD/NASH and cachexia. Overall, PEs represent promising repurposed natural compounds capable of enhancing metabolic and functional resilience with minimal endocrine risk. With rigorous pharmacological validation and controlled clinical evaluation, they may establish a new class of safe, multitarget therapeutics for aging and chronic disease.

D. A. Usmanov (✉)
Department of Coatings and Polymeric Materials, Materials and Nanotechnology Program, North Dakota State University, Fargo, ND, USA
e-mail: durbek.usmanov@ndus.edu

U. Y. Yusupova · S. Sagdullaev
Institute of the Chemistry of Plant Substances, Academy of Science of the Republic of Uzbekistan, Tashkent, Uzbekistan

V. N. Syrov et al. (eds.), *Phytoecdysteroids II*,
https://doi.org/10.1007/978-981-95-8423-9_9

Keywords Phytoecdysteroids · Drug repurposing · Sarcopenia and cachexia · Cardiovascular protection and hepatoprotection · Pulmonary resilience · Neuroprotection · Pharmacokinetics

9.1 Why Repurpose Phytoecdysteroids Now?

PEs—notably 20E ("ecdysterone"), turkesterone, and related congeners—are polyhydroxylated steroids best known as arthropod molting hormones, but increasingly investigated in mammals for anabolic, cytoprotective, and metabolic effects (Yusupova et al. 2019; Yusupova et al. 2020; Yusupova et al. 2022; Yusupova et al. 2025; Yusupova et al. 2021). Unlike classical anabolic-androgenic steroids, PEs do not act via androgen receptors and show a markedly safer toxicological profile in preclinical models, while influencing muscle protein turnover, redox biology, mitochondria, and membrane stability. Modern repurposing interest is fueled by: (i) first signals of efficacy in humans with sarcopenia (BIO101/20E) and resilience outcomes in severe viral pneumonia; (ii) mechanistic breadth spanning ERβ-linked signaling, Mas receptor activation, and mitochondrial biogenesis pathways; (iii) an excellent tolerability profile and widespread dietary occurrence that could lower translational barriers. At the same time, hype, especially around turkesterone, has outpaced rigorous clinical data, and the regulatory environment (e.g., anti-doping monitoring) is evolving. A repurposing lens demands careful mapping of where evidence is human-ready versus still preclinical, where pharmacokinetics limit oral utility, and what formulation and regulatory innovations are needed to cross the translational "last mile." In summary, while phytoecdysteroids present a compelling convergence of mechanistic diversity, safety, and translational potential, their successful clinical repositioning will ultimately depend on bridging mechanistic insight with controlled human evidence and regulatory clarity.

9.2 Pharmacology and Mechanisms Relevant to Repurposing

The pharmacological rationale for repurposing 20E arises from convergent mechanistic hypotheses in mammalian systems, including estrogen receptor-β (ERβ) signaling, ACE2/Ang-(1–7)/Mas receptor (MasR) activation, and mitochondrial protection with redox control. Together, these mechanisms underpin the compound's anabolic, cytoprotective, and metabolic-regulatory properties across multiple tissues. The ERβ hypothesis is among the best characterized. In vitro and in vivo studies in rats and murine myotubes demonstrate that 20E enhances myotube diameter and protein synthesis, with effects dissociated from androgen receptor (AR) signaling (Isenmann et al. 2019).

Molecular docking and transcriptional analyses confirmed preferential engagement of ERβ rather than AR, suggesting that 20E acts through estrogenic rather than androgenic pathways (Parr et al. 2014). This mechanistic distinction explains the anabolic yet non-androgenic profile observed in preclinical and early human data. A complementary line of evidence centers on the MasR within the protective arm of the renin–angiotensin system (ACE2/Ang-(1–7)/Mas axis). Activation of MasR by 20E is proposed to enhance anti-inflammatory, anti-fibrotic, and mitochondrial resilience pathways (Lafont et al. 2021). The pharmaceutical-grade formulation BIO101, developed by Biophytis, specifically targets MasR to improve muscle performance and resilience in age-related weakness. Serova et al. (2024) demonstrated that 20E-induced myoblast differentiation was attenuated by the MasR antagonist A779, confirming pathway involvement. Collectively, these findings position the ACE2/Ang-(1–7)/Mas axis as a novel therapeutic interface for ecdysteroids, distinct from canonical steroid hormone receptors. In addition, mitochondrial preservation and oxidative stress control constitute a third mechanistic pillar.

Historical and modern biochemical data indicate that 20E improves oxidative phosphorylation, increases respiratory control ratio (RCR), and decreases malondialdehyde (MDA) and other lipid peroxidation products in hepatic and cardiac tissues (Todorova et al. 2024). Such effects offer mechanistic bridges to diseases characterized by energetic failure and oxidative imbalance, including sarcopenia, cachexia, non-alcoholic fatty liver disease (NAFLD/NASH), and cardiometabolic dysfunction. Across preclinical studies, ecdysteroids consistently (i) stimulate protein synthesis and favorable serum protein fractions, (ii) attenuate lipid peroxidation, (iii) preserve mitochondrial enzyme coupling, and (iv) stabilize cellular membranes against oxidative or toxic insults. Human data provide additional translational relevance. The multilayered pharmacology of 20E can be conceptually organized into three interacting mechanisms that jointly explain its anabolic, cytoprotective, and metabolic effects. Figure 9.1 summarizes this integrative framework. The first pillar, ERβ-mediated signaling, drives non-androgenic protein synthesis and muscle fiber growth. The second, Mas receptor (ACE2/Ang-(1–7)/Mas) activation, accounts for anti-inflammatory and anti-fibrotic actions, as well as enhanced mitochondrial performance. The third, mitochondrial and redox regulation, underlies improved oxidative phosphorylation, reduced ROS and lipid peroxidation, and membrane stabilization. Together, these axes converge toward systemic outcomes, muscle anabolism, hepatic protection, and cardiometabolic resilience, providing the mechanistic foundation for 20E repurposing into age- and metabolism-related disorders.

A randomized, double-blind training trial by Isenmann and others (2019) demonstrated significantly greater increases in muscle hypertrophy and strength with ecdysterone supplementation compared to placebo, prompting the World Anti-Doping Agency (WADA) to include ecdysterone in its Monitoring Program (Isenmann et al. 2019). Although these findings cannot be extrapolated directly to disease states, the physiological signal of anabolic benefit, coupled with excellent tolerability, supports the rationale for disease-focused trials in sarcopenia and cachexia (Dioh et al. 2023). Clinical development has been led by Biophytis' BIO101, a pharmaceutical-grade 20E formulation. In the SARA-INT Phase 2 study involving older adults with sarcopenia, participants receiving BIO101 exhibited

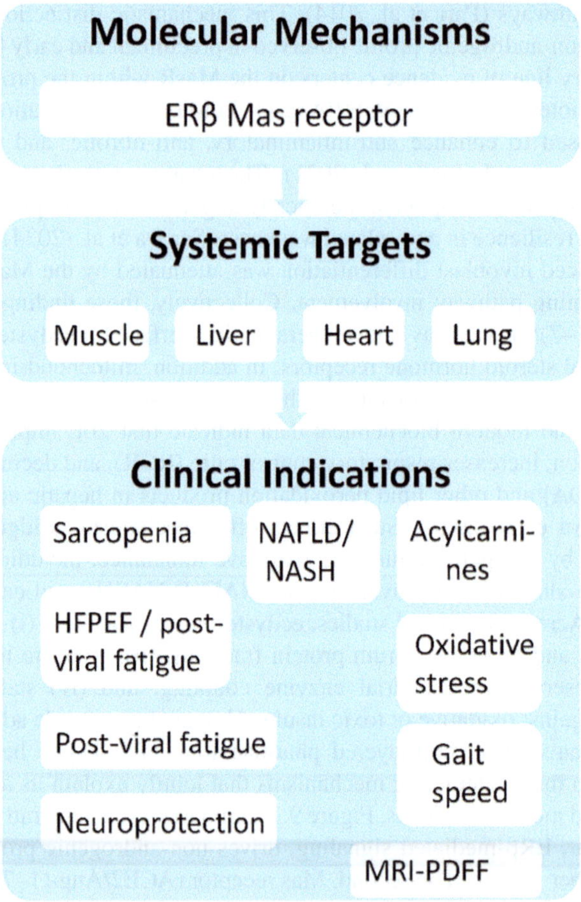

Fig. 9.1 Pharmacology and mechanisms relevant to repurposing

clinically meaningful improvements in physical performance measures, including the 400-m walk test, relative to placebo (Dioh et al. 2023). Pharmacokinetic analyses confirmed dose-dependent plasma exposure with an excellent safety profile. Complementary findings from SARA-OBS further contextualized functional benefits and mechanistic support for MasR-mediated signaling. During the COVA (COVID-19) program, a randomized Phase 2/3 trial in severe pneumonia, 20E showed acceptable safety and exploratory benefits related to inflammation control and respiratory recovery, although primary endpoints were not met.

These results broaden the safety and exposure data for acutely ill populations and inform potential repositioning into conditions involving respiratory or metabolic frailty. Taken together, evidence from molecular, preclinical, and early clinical studies highlights 20E as a unique multitarget natural steroid with distinct receptor pharmacology, redox modulation, and metabolic protection. Among current repurposing opportunities, sarcopenia and muscle-loss syndromes remain the most validated human indications, while extension to cachexia, hepatic, and cardiopulmonary disorders appears biologically plausible. Future work must clarify which mechanisms predominate in humans (ERβ, MasR, or mitochondrial) and delineate dose–response relationships through rigorous pharmacometric modeling and biomarker integration. Overall, the accumulating mechanistic, translational, and safety evidence positions 20E as a mechanistically versatile yet pharmacologically coherent candidate—one capable of linking receptor biology, mitochondrial resilience, and functional recovery into a unified therapeutic framework for age- and metabolism-related disorders.

9.3 Repurposing Opportunities by Therapeutic Domain

The pharmacological versatility of 20E opens up multiple repurposing paths across therapeutic domains characterized by metabolic stress, inflammation, and functional decline. In the musculoskeletal domain, 20E's ability to promote protein synthesis via non-androgenic signaling pathways makes it compelling for sarcopenia, frailty, and cachexia. In C_2C_{12} myotubes, 20E induces a rapid intracellular Ca^{2+} flux and sustained Akt phosphorylation, effects blocked by GPCR, PLC, and PI3K inhibitors, supporting a membrane-proximal anabolic mechanism (Gorelick-Feldman et al. 2010). In rodent models, continuous infusion of 20E increases muscle weight, indicating anabolic potential in vivo (Cheng et al. 2013). Moreover, in a model of eccentric contraction injury, oral 20E fully restores muscle function within 7 days compared to placebo, consistent with a recovery-adjunct hypothesis (Zwetsloot et al. 2021). These data support designing fully powered clinical trials for primary sarcopenia using standardized endpoints (gait speed, grip strength, SPPB). Secondary sarcopenia and cachexia (e.g., in COPD, heart failure, oncology) may benefit from additive effects on energy/redox and signaling pathways, although human data are lacking. Similarly, short-duration 20E could be tested in post-hospital deconditioning or disuse atrophy settings, using rehabilitation endpoints.

Major translational challenges include determining pharmacokinetics and dose–exposure relationships in frail older adults, ensuring pharmaceutical-grade purity (not dietary supplements), and avoiding off-target endocrine effects while delivering functional benefit. In hepatology and metabolic disease, preclinical evidence shows that 20E reduces hepatic triglyceride accumulation, modulates lipogenesis and AMPK signaling, and suppresses fat deposition in high-fat, high-fructose diet models (Buniam et al. 2023a, b). These observations align with reduced lipotoxic stress, preserved mitochondrial function, and improved oxidative metabolism. Thus, 20E could be tested as adjunct therapy in NAFLD/NASH (prefibrotic), monitoring ALT/AST, MRI-PDFF, FibroScan CAP, and redox biomarkers.

Additionally, the cytoprotective, membrane-stabilizing properties observed in rodent toxin models suggest possible utility in mitigating drug-induced liver injury. However, translating rodent results to human steatohepatitis faces hurdles, demonstrating hepatic exposure, ruling out estrogenic receptor side effects in liver, and elucidating interactions with hepatic CYPs or transporters in multidrug regimens. In cardiometabolic and vascular disease, rat and cell models suggest cardioprotective effects. In spontaneously hypertensive rats (SHR), 20E reduces systolic blood pressure and limits left ventricular hypertrophy (Phungphong et al. 2017). Such results point to mitigation of oxidative stress and preservation of mitochondrial integrity in myocardium, which is mechanistically relevant to conditions like HFpEF. Clinical translation could target functional endpoints (peak VO_2, 6MWT, quality-of-life scales) with mechanistic companion biomarker studies. However, the absence of modern RCTs in cardiovascular disease means significant investment in safety, drug–drug interaction studies, and regulatory groundwork. The pulmonary recovery and resilience domain was indirectly explored during COVID-19 trials of BIO101 (20E), which reported safety in hypoxic subjects but inconclusive efficacy. The putative link between Mas receptor signaling and anti-fibrotic/anti-inflammatory effects suggests potential in post-viral recovery, fatigue, and respiratory dysfunction. However, translation requires robust pulmonary endpoints (CPET, PROMs, and imaging biomarkers) and clear mechanistic validation. Interstitial lung disease applications should only follow after stronger preclinical evidence and interaction profiling. Finally, neuroprotection and cognitive aging remain speculative but intriguing.

The known antioxidant and mitochondrial-stabilizing actions of 20E, together with trophic signaling hypotheses, provide a theoretical basis for benefit in mild cognitive impairment or chemotherapy-induced "chemo brain." Yet, no published studies to date confirm adequate CNS penetrance, receptor engagement in human brain, or clinical benefit. Therefore, any exploratory work should first focus on PK/PD, receptor mapping, and safety before efficacy trials. Taken together, these domains represent a convergent strategy: leveraging 20E's non-androgenic anabolic signaling, redox and mitochondrial protection, and emerging receptor hypotheses to address disorders of systemic energy failure, inflammation, and functional decline. The path to clinical translation lies through rigorous pharmacokinetic/pharmacodynamic modeling, biomarker-driven designs, and careful selection of disease-specific endpoints.

The diversity of physiological actions attributed to 20E demands a structured framework to map its molecular mechanisms to organ-level functions and ultimately to clinical indications. Although most data originate from preclinical systems, converging evidence across models supports a coherent *mechanism-to-phenotype continuum*. At the molecular level, 20E appears to integrate three interlinked processes: activation of estrogen receptor-β (ERβ) and Mas receptor (ACE2/Ang-(1–7) axis) signaling, stabilization of mitochondrial energy metabolism, and attenuation of oxidative and inflammatory stress. These pathways collectively shape an anabolic-metabolic phenotype that favors cellular resilience without classical androgenic side effects. At the systemic level, this integrated signaling framework manifests as multiorgan protection, enhancing muscle anabolism, preserving hepatic and cardiac mitochondrial integrity, mitigating fibrotic remodeling in lung tissue, and potentially sustaining neuronal redox balance. Such convergence aligns with the concept of "metabolic resilience enhancers," natural agents that restore or preserve bioenergetic efficiency across diverse pathological contexts. To synthesize this multidimensional evidence into a translational roadmap, Table 9.1 summarizes the principal therapeutic domains, mechanistic rationales, evidence maturity, and next-step priorities for 20E development. It outlines where preclinical and clinical data intersect most strongly—namely in *sarcopenia*—and where additional work is required to establish exposure, mechanistic biomarkers, and pharmacodynamic endpoints (e.g., in NAFLD/NASH, HFpEF, or post-viral recovery).

Complementing the table, Fig. 9.2 visually integrates these domains into a single vertical continuum from *molecular mechanisms* through *systemic targets* to *clinical indications*. The infographic emphasizes that 20E's translational journey depends on rigorous linking of mechanistic biomarkers (such as acylcarnitines, oxidative stress indices, gait speed dynamics, and MRI-PDFF imaging) to measurable functional outcomes. By anchoring mechanistic plausibility to clinical metrics, such a framework transforms fragmented experimental observations into a coherent therapeutic development strategy. Ultimately, this mechanistic-to-clinical mapping illustrates how a traditionally "nutraceutical" compound class, such as phytoecdysteroids, can evolve into pharmacologically rational, biomarker-driven therapeutics. Properly executed, this approach could position 20E at the frontier of *resilience pharmacology*, bridging molecular bioenergetics and functional medicine in aging and chronic disease.

This integrated evidence framework demonstrates that 20E occupies a unique translational space, linking well-defined molecular targets with systemic physiological benefits. Future progress will depend on harmonizing mechanistic validation, pharmacokinetic optimization, and clinical endpoint design to transform these insights into reproducible therapeutic outcomes.

9.4 Pharmacokinetics, Formulation, and Delivery Hurdles

From a drug-development perspective, 20E presents classic small-molecule hurdles at the interface of exposure, quality, and translational pharmacology. Oral exposure is variable because 20E is highly polar and undergoes substantial first-pass handling, with human excretion studies showing rapid appearance of parent and characteristic metabolites (e.g., 14-deoxy-ecdysterone, 14-deoxy-poststerone) after single-dose administration (Tsitsimpikou et al. 2001; Kraiem et al. 2021). While clinical-grade materials (e.g., purified ≥97% 20E) manage purity and dose control, ensuring consistent systemic exposure will likely require enabling formulations (e.g., solid dispersions, lipid-based systems, and SLN/LNP platforms) that are routinely used to stabilize dissolution/absorption profiles for challenging compounds (Nguyen and Duong 2022; Rehman et al. 2024; Lüdtke et al. 2025).

In parallel, metabolic and transporter liabilities need systematic mapping: extensive UGT-mediated glucuronidation is a frequent barrier to oral bioavailability for many natural products and drugs, and comprehensive CYP/UGT/transporter panels

Table 9.1 Therapeutic domains, mechanistic rationale, and translational opportunities for 20E

Therapeutic DOMAIN	Mechanistic basis/pathways	Key preclinical and clinical evidence	Development stage	Translational priorities/next steps	Representative References
Musculoskeletal (sarcopenia, frailty, cachexia, disuse atrophy)	Non-androgenic anabolic signaling via membrane-associated GPCR/PI3K/Akt pathways ERβ/MasR crosstalk enhancing protein synthesis and mitochondrial function	In C2C12 myotubes, 20E induces Ca^{2+} flux and Akt phosphorylation, blocked by GPCR, PLC, PI3K inhibitors continuous 20E infusion increases muscle weight in mice Oral 20E restores muscle function post-injury SARA-INT trial shows mobility improvement and safety in older adults	Clinical (phase 2b completed)	Launch phase 3 RCTs in sarcopenia with gait speed/SPPB endpoints Define dose–exposure curve in frail elderly Expand to cachexia/disuse atrophy with multimodal (nutrition + exercise) arms	Gorelick-Feldman et al. (2010), Cheng et al. (2013), Zwetsloot et al. (2021), Fielding et al. (2025)
Hepatology and Metabolic Disease (NAFLD/NASH, drug-induced liver injury)	Antioxidant and membrane-stabilizing actions Reduced lipid peroxidation and improved mitochondrial respiration AMPK and lipogenesis modulation	20E reduces hepatic triglyceride accumulation and modulates AMPK signaling in diet-induced obesity models Historical protection in CCl_4 and toxin models with normalized transaminases and bile flow	Preclinical translational	Initiate proof-of-concept NAFLD/NASH trials (ALT/AST, MRI-PDFF, CAP) Evaluate DILI prevention with controlled PK/safety Characterize CYP/UGT transporter interactions	Buniam et al. (2023a, b), Guo et al. (2022)

Cardiometabolic & Vascular Protection (HFpEF, I/R injury, hypertension)	Mitochondrial preservation and antioxidant effect Anti-hypertrophic signaling and improved myocardial energetics	In SHR rats, 20E reduces BP and LV hypertrophy (Phungphong et al. 2017) Decreased necrosis and preserved mitochondrial ultrastructure in I/R models	Preclinical	Pilot phase 1b/2a CV safety and exposure studies Explore peri-ischemic conditioning add-on (CABG models) Map DDI with HF polypharmacy	Phungphong et al. (2017)
Pulmonary resilience/ post-viral recovery	MasR/ACE2 axis activation → anti-fibrotic, anti-inflammatory response Mitochondrial support in hypoxia	BIO101 (20E) tested in COVID-19 pneumonia (COVA program): Safe but mixed efficacy Preclinical anti-fibrotic effects in lung injury models	Clinical safety established; efficacy unconfirmed	Develop post-hospital rehabilitation trials (CPET, PROMs, imaging) Investigate MasR-specific biomarker expression	Dioh et al. (2023), WADA (2024)
Neuroprotection & Cognitive Aging	ERβ-linked trophic signaling Mitochondrial and antioxidant protection of neurons	20E reduces ROS/RNS and apoptosis in ischemic neuronal models No human CNS PK/PD data yet	Early preclinical	Assess BBB penetration and CNS distribution Perform neuro-PK and receptor mapping Pilot MCI or chemo brain support trials after exposure validation	Calabrese and Mattson (2024)

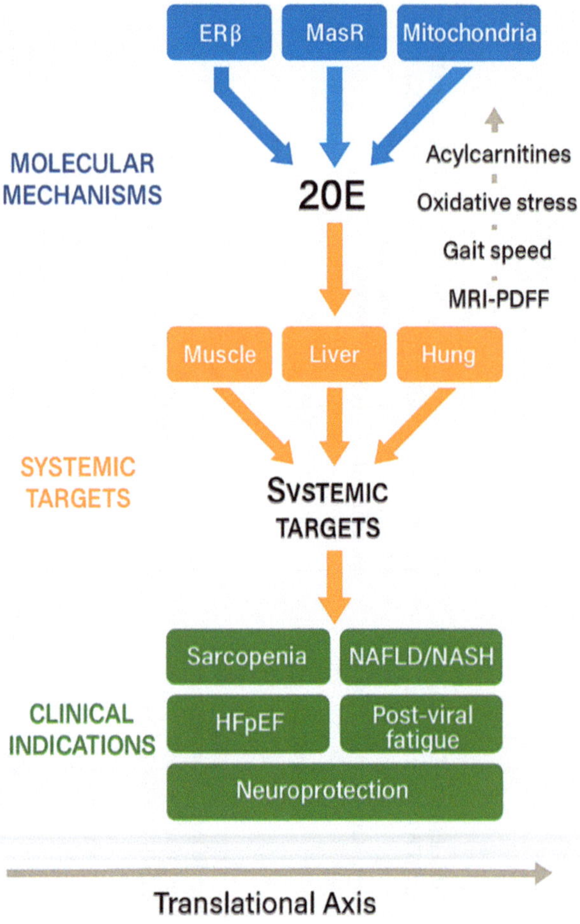

Fig. 9.2 Translational flow of 20-hydroxyecdysone (20E) from molecular mechanisms to clinical indications

early in development reduce downstream DDI surprises in polypharmacy settings typical of geriatrics and cardiometabolic care (Wu et al. 2011; Yin et al. 2024). For biomarker-anchored pharmacodynamics, repurposing programs should link exposure to clinically meaningful function plus mechanism-adjacent readouts. In musculoskeletal indications, standardized strength and mobility measures (e.g., gait speed, grip, and SPPB) pair naturally with mitochondrial-redox biomarkers such as lactate/pyruvate and acylcarnitine profiles that report on oxidative phosphorylation and β-oxidation efficiency (McCann et al. 2021; Shayota 2024).

These general mitochondrial markers are established in translational medicine and can be layered alongside pathway-specific assays once ERβ/MasR signatures are analytically validated in humans. Safety and quality considerations are twofold. First, peer-reviewed sources outside performance-athlete cohorts indicate favorable tolerability of ecdysteroid-standardized interventions, with no serious safety signals under controlled dosing (Todorova et al. 2024). Second, supplement mislabeling/underdosing is common in the commercial market, reinforcing that GMP-grade API and formal CMC controls are non-negotiable for clinical development and reproducibility (Ambrosio et al. 2020a). On the anti-doping front, ecdysterone is on WADA's Monitoring Program (not the Prohibited List) for 2025, so sponsors should anticipate analytical-detection questions for athlete subpopulations but recognize that current policy is surveillance rather than prohibition (WADA 2024, 2025; USADA 2025).

A trial-design playbook for repurposing follows naturally from this evidence base. Indication selection should begin where mechanism and early human signals converge, sarcopenia/frailty, then expand to cachexia and NAFLD/NASH with disease-specific outcomes and mitochondrial-redox panels to enhance pharmacologic interpretability (McCann et al. 2021; Shayota 2024). Dose-finding can leverage PK-guided Bayesian designs that explicitly model exposure-to-function links and incorporate candidate MasR/ERβ markers once assays are locked (Wu et al. 2011; Yin et al. 2024). Comparators and combinations should reflect standard-of-care synergies: exercise/nutrition in sarcopenia, GLP-1 s/pioglitazone frameworks in NAFLD/NASH, and pulmonary rehab in post-viral deconditioning, each supported by adherence checks via plasma 20E and quality audits to document delivered dose (Ambrosio et al. 2020b). Finally, the regulatory strategy can reasonably invoke US 505(b)(2) or EU "Hybrid" Article 10(3) pathways when prior human exposure and literature allow partial reliance on existing knowledge—while still expecting new pivotal RCTs for indication-specific claims (Prue and Mayorga 2007; Freije et al. 2020; Ramesh and Cherukuri 2023). Altogether, the path to credible repositioning hinges on exposure control (formulation + PK), clean quality systems, DDI due diligence, and biomarker-rich designs that translate 20E's non-androgenic anabolic, mitochondrial, and anti-inflammatory biology into consistent clinical benefit.

9.5 Translational Lessons and Strategic Recommendations for Repurposing 20-Hydroxyecdysone

The cumulative experience from clinical development (BIO101/SARA-INT), observational cohorts (SARA-OBS), and athlete-based anabolic studies provides a coherent translational roadmap for repositioning 20E as a therapeutic agent. Success in repurposing depends not merely on demonstrating molecular activity but on bridging pharmacology, exposure, and clinically meaningful function through rigorously controlled and biomarker-anchored trials. First, endpoint prioritization must focus on outcomes that directly translate into quality-of-life improvement. In sarcopenia and frailty, this emphasizes mobility- and strength-based measures such as gait speed, SPPB, and handgrip strength rather than purely biochemical surrogates. These tests are sensitive, validated for regulatory use, and align with EWGSOP2 and ICFSR frameworks (Cruz-Jentoft et al. 2019; Fielding et al. 2025). Second, mechanistic substudy integration is vital for linking 20E's pharmacodynamics to human physiology. Embedding mitochondrial biomarkers (e.g., lactate/pyruvate, acylcarnitines), oxidative stress panels, and pathway markers (MasR, ERβ, Akt/mTOR phosphorylation) into larger RCTs will clarify which mechanisms dominate in vivo (Gorelick-Feldman et al. 2010; McCann et al. 2021). Such multimodal biomarker integration strengthens causal interpretation and supports precision dosing in frail older adults. Third, dose–exposure modeling should use pharmacometric and Bayesian adaptive designs to connect plasma concentration with function, reducing uncertainty in dose selection and accommodating variability in absorption and metabolism (Dioh et al. 2023; Wu et al. 2011). Establishing therapeutic windows that achieve efficacy without endocrine or hepatic off-target effects remains essential before broad geriatric use. Fourth, quality assurance and product integrity are indispensable. Analyses of commercial ecdysteroid supplements reveal frequent underdosing and adulteration (Ambrosio et al. 2020a; Todorova et al. 2024). Only GMP-certified active pharmaceutical ingredients with validated LC–MS quantification and stability documentation can ensure reproducibility. All future studies must monitor plasma 20E levels to verify adherence and delivered exposure. Fifth, population stratification can improve signal detection. In SARA-INT, nominal benefits were strongest among "slow walkers" and individuals with sarcopenic obesity (Fielding et al. 2025). Targeting high-risk subgroups or employing enrichment strategies enhances trial sensitivity and cost-effectiveness. Sixth, combination-therapy frameworks should reflect real-world multimodality. Pairing 20E with exercise and nutritional interventions in sarcopenia, GLP-1 agonists in NAFLD/NASH, or pulmonary rehabilitation in post-viral deconditioning leverages mechanistic complementarity while providing pragmatic relevance (Buniam et al. 2023a, b). These designs also align with translational ethics, emphasizing additive rather than replacement therapy. Seventh, regulatory navigation requires strategic use of hybrid pathways. In the United States, the 505(b)(2) route allows partial reliance on existing safety data; in Europe, Hybrid Article 10(3) submissions under Directive 2001/83/EC serve the same purpose Freije et al. 2020; Ramesh and Cherukuri 2023). Nonetheless, new pivotal Phase 3 RCTs remain mandatory to substantiate disease-specific claims and labeling.

Finally, translational caution is essential when extrapolating from athlete or animal data to clinical populations. Physiological differences in metabolism, hormonal milieu, and oxidative stress mean that hypertrophy observed in trained adults (Isenmann et al. 2019) cannot be assumed in frail or multimorbid patients. Overgeneralization risks both inefficacy and regulatory pushback. Similarly, rodent hepatoprotection and cardioprotection models should inform, but not substitute for, human mechanistic validation and PK/PD correlation. The clinical translation of 20E requires a clear operational framework that connects molecular plausibility with reproducible human outcomes. Decades of preclinical data have established a strong mechanistic foundation, yet successful repurposing depends on how these insights are implemented in trial design, product quality, and regulatory pathways.

Table 9.2 summarizes the principal translational lessons derived from current evidence and early clinical development.

In this table, the main translational domains are synthesized into actionable guidance for repurposing 20E. Each row reflects a key operational area—from endpoint selection and biomarker validation to dose-exposure modeling and regulatory strategy, derived from existing clinical and preclinical data. The table highlights both the opportunities and limitations observed so far, emphasizing that successful translation depends not only on mechanistic insight but also on methodological rigor, product quality, and early regulatory engagement. Together, these lessons provide a concise roadmap for advancing 20E from experimental pharmacology to evidence-based clinical therapeutics.

In summary, the translational success of 20E will depend on integrating mechanistic depth with clinical precision, linking validated biomarkers, optimized dosing, and high-quality formulations to meaningful functional outcomes in patients.

Table 9.2 Translational lessons, cautions, and recommendations for 20E repurposing

Area	Key lesson	Caution/limitation	Recommended action	References
Functional endpoints	Use gait speed, SPPB, grip strength as primary outcomes	Surrogate biomarkers may not reflect patient benefit	Follow EWGSOP2/ ICFSR standards	Cruz-Jentoft et al. (2019), Fielding et al. (2025)
Mechanistic biomarkers	Include mitochondrial, oxidative, MasR/ ERβ readouts	Human validation still limited	Embed sub-studies within phase 3 RCTs	Gorelick-Feldman et al. (2010), McCann et al. (2021)
Dose–exposure	Model PK/PD links with adaptive designs	Sparse elderly PK data	Apply Bayesian or population PK tools	Dioh et al. (2023), Wu et al. (2011)
Product quality	Require GMP and verified content	Supplements often under-dosed	Audit API lots and monitor plasma levels	Ambrosio et al. (2020a); Todorova et al. (2024)
Population targeting	Enrich for high-risk groups (slow walkers)	Smaller samples limit generalizability	Stratify by frailty, obesity phenotype	Fielding et al. (2025)
Combination therapy	Combine with exercise/metabolic drugs	Additive safety unknown	Conduct pilot add-on studies	Buniam et al. (2023a, b)
Regulatory path	Use hybrid 505(b)(2)/Art. 10(3) routes	Still need pivotal efficacy trials	Engage regulators early	Freije et al. (2020), Ramesh & Cherukuri, (2023)
Data extrapolation	Athlete/animal data inform mechanism	Human efficacy cannot be assumed	Validate with disease-specific PK/PD	Isenmann et al. (2019), Begley et al. (2021)

9.6 Challenges and Risk Mitigations

In translating 20E/ecdysteroid candidates toward therapeutic repurposing, several key challenges emerge. Each challenge must be anticipated with mitigation strategies grounded in scientific precedent and regulatory prudence. First, mechanistic ambiguity is a fundamental obstacle. The competing hypotheses of ERβ versus Mas receptor (MasR) mediation complicate dose selection, biomarker prioritization, and pathway validation. Without a clear predominance, dose escalation trials might miss the optimal window for receptor engagement or misinterpret functional readouts. To mitigate this, early-phase trials must include parallel mechanistic assays, such as receptor occupancy, phosphorylation of downstream effectors, or ligand-binding biomarkers in surrogate tissues. Computational and molecular modeling studies (e.g., molecular docking, in silico dynamics) can guide which receptor hypotheses merit priority in vivo. Indeed, the molecular modeling analysis of ecdysterone–ERβ interactions supports one plausible binding mode, but also highlights uncertainties in the ligand-binding pocket conformations (Kasuya et al. 2003). Second, PK variability and low bioavailability present serious translational risk. The hydrophilic nature of 20E and its susceptibility to first-pass metabolism or transporter clearance lead to inconsistent oral exposure. Studies of ecdysteroid-containing supplements demonstrate that many products under-deliver on label claims, with wide variability in detectable levels (uPA-SCID and human volunteer excretion work), and quality control surveys show widespread mislabeling in botanical supplement categories (e.g., 89% of sports supplements inaccurately labeled). To mitigate exposure risk, development should adopt enhanced-bioavailability formulations (e.g., lipid-based carriers, cyclodextrin complexes, and nano-dispersions) and embed therapeutic drug monitoring (TDM) in Phase 2 trials to correlate plasma levels with efficacy and safety. Analytical methods (HPLC or LC-MS) for verifying 20E content should be validated, as demonstrated in recent work developing HPLC quantification of ecdysterone and turkesterone in supplements (Todorova et al. 2023). Third, quality control and supplement confusion is a reputational and regulatory risk. Dietary supplements are frequently adulterated or mislabeled with undeclared anabolic agents. Analytical surveys estimate that between 14% and 50% of "fitness/ergogenic" supplements contain undeclared anabolic steroids or other prohibited compounds. Botanical ingredient forensic reviews highlight advanced strategies of adulteration and countermeasures for detection (e.g., isotope ratio analysis and fingerprinting) (Botanical Ingredient Forensics). To mitigate, any repurposed program must insist on strict GMP sourcing, third-party, independent QC testing, and transparent CMC dossiers for every lot. Sponsors should also publicly document analytical verification (identity, purity, and impurity profiles) to maintain trust in clinical settings. Fourth, regulatory and anti-doping uncertainties pose external risks. Since ecdysteroids are on the WADA Monitoring Program, a sudden policy change to classify them as prohibited could complicate clinical development, especially in populations overlapping with athletes. Moreover, athlete-safety stakeholders may demand clear labeling and stringent contamination checks. Mitigation requires proactive dialogue with sports-medicine and anti-doping agencies, clear non-athlete indication labeling, and consistent contaminant screening protocols in clinical supply chains (WADA report). Fifth, the gap between hype and evidence—especially around analogs like turkesterone—represents a reputational hazard. Popular claims often exceed the human evidence base. For example, a four-week, 500 mg turkesterone supplementation study in active individuals showed no meaningful change in lean body mass or fat mass versus placebo. Moreover, many supplement label studies have detected the declared botanical (e.g., turkesterone) at vastly lower concentrations than promised (label vs actual). The mitigation here is to prioritize 20E (which has stronger mechanistic and human data), and insist on randomized, controlled trials before making broader anabolic or health claims. Marketing and scientific communications must emphasize qualified, evidence-based language and avoid overgeneralization.

Despite encouraging mechanistic and clinical signals, the clinical translation of 20E remains constrained by several recurrent risks. These include uncertainties in receptor dominance, whether ERβ or MasR pathways mediate the primary pharmacological response, and variable pharmacokinetics leading to inconsistent systemic exposure. Supplement-level quality issues and frequent mislabeling further compromise reproducibility, as demonstrated in independent audits of commercial ecdysteroid formulations. Regulatory ambiguity also persists, with 20E still listed on the WADA Monitoring Program rather than the Prohibited List, necessitating transparent labeling and early engagement with sports-medicine authorities. Finally, exaggerated claims surrounding related analogs such as turkesterone risk undermining scientific credibility unless supported by controlled human trials.

Table 9.3 summarizes these translational risks, their potential impact, and corresponding mitigation strategies. Each domain highlights specific actions, such as mechanistic assays, therapeutic drug monitoring, GMP-grade sourcing, and regulatory dialogue, to ensure that future repurposing of 20E proceeds on a scientifically

Table 9.3 Key risk domains and mitigation strategies in 20E repurposing

Challenge/risk	Impact	Mitigation strategy	Source
Mechanistic ambiguity (ERβ vs MasR)	Uncertain dose selection, biomarker choice	Parallel mechanistic assays, receptor occupancy, modeling guidance	Kasuya et al. (2003)
PK variability & low bioavailability	Inconsistent exposure, under-powering efficacy	Enhanced formulations, therapeutic drug monitoring	Guo et al. (2022), Erin, (2023)
Quality/supplement confusion/ adulteration	Regulatory scrutiny, nondetection of true effect	GMP sourcing, independent QC, transparent CMC	Jagim et al. (2023), Gafner et al. (2023)
Regulatory/ anti-doping changes	Clinical disruption, labeling conflicts	Early engagement with sports authorities, clean labeling, contaminant screening	WADA (2024)
Hype vs evidence (especially turkesterone)	Misleading expectations, reputational risk	Focus on 20E-based claims, demand RCT evidence	(Todorova et al. (2024)

validated and ethically robust foundation. In summary, addressing these translational risks through rigorous mechanistic validation, pharmacokinetic optimization, and quality assurance will be essential to advance 20E from experimental promise to reliable clinical application.

Collectively, proactive risk management, spanning mechanistic clarification, formulation innovation, and regulatory transparency will determine whether 20E evolves from a promising bioactive compound into a reproducible, clinically accepted therapeutic with a robust evidence base.

9.7 Outlook: From Promise to Practice

Among phytoecdysteroids, 20E currently stands as the most credible candidate for repurposed therapeutics, bridging mechanistic plausibility—via ERβ/MasR pathways and mitochondrial resilience—to emerging human signals in sarcopenia, all with a favorable safety profile (Dioh et al. 2023; Dinan et al. 2021). The path forward is pragmatic but demanding: well-powered, dose-optimized Phase 3 trials in sarcopenia must employ functional endpoints (gait speed, SPPB, grip strength) and embed mechanistic biomarker sub-studies spanning mitochondrial function, redox balance, and receptor activation. Parallel, early-phase studies in NAFLD/NASH should focus on hepatic oxidative/mitochondrial biomarkers (e.g., acylcarnitines and reactive lipid species) to validate translational relevance in metabolic disease. Carefully designed cachexia programs may combine 20E with nutritional support and exercise regimens, allowing additive or synergistic gains in muscle resilience rather than monotherapy. However, for 20E to move from associative promise to target-enabled practice, rigorous PK/PD alignment and receptor biology must converge.

The dose–response window must be tied not only to gross function but to target engagement (e.g., ERβ occupancy, MasR downstream signaling, and mitochondrial coupling). If this alignment succeeds, ecdysteroid-derived PEs could occupy a unique therapeutic niche, anabolic-metabolic "resilience enhancers" facilitating functional restoration in aging and chronic disease with minimal endocrine disruption risk. Such strategies would be conceptually consistent with models of biological resilience and hormesis, where periodic anabolic challenge and repair cycles bolster organismal robustness (Calabrese and Mattson 2024). Moreover, the broader geroscience paradigm supports targeting metabolic resilience (or "metabolaging") as a lever for age-related disease delay (Khalaf et al. 2025). Because metabolism and energy homeostasis lie at the heart of aging biology, an agent capable of strengthening mitochondrial and redox networks may serve as a foundational therapeutic adjunct across multiple age-related indications (Huffman et al. 2016; Cosarderelioglu et al. 2025). Beyond the musculoskeletal domain, mechanistic expansion explores 20E's potential in hepatic and metabolic disorders, leveraging its antioxidant and mitochondrial-preserving effects in NAFLD/NASH and other redox-linked conditions (Buniam et al. 2023a, b; Khalaf et al. 2025). Multimodal integration then envisions combination strategies with exercise, nutrition, or metabolic modulators to enhance resilience outcomes (Cosardereliogl u et al. 2025; Huffman et al. 2016). Finally, regulatory translation consolidates clinical data and manufacturing consistency into hybrid submission pathways (505(b)(2) in the U.S. and Article 10(3) in the EU), ensuring safety, GMP compliance, and exposure bridging (Ramesh and 2023; Freije et al. 2020). Table 9.4 summarizes this translational roadmap, aligning each developmental phase with its scientific focus, key objectives, biomarkers, and representative references, providing a comprehensive blueprint for advancing 20E from mechanistic promise to clinical practice.

While the translational roadmap (Table 9.4) delineates the stepwise scientific progression required to advance 20E from molecular discovery to clinical

Table 9.4 Translational roadmap for 20E from mechanistic promise to clinical practice

Development stage	Scientific focus	Key objectives	Biomarkers/tools	Representative References
Mechanistic definition	Clarify receptor engagement (ERβ vs MasR), mitochondrial and redox pathways	Validate primary molecular targets; confirm downstream signaling consistency across models	ERβ phosphorylation, MasR expression, mitochondrial membrane potential (ΔΨm), ROS/RNS quantification	Dinan et al. (2021), Kasuya et al. (2003)
PK/PD optimization	Map exposure–response link in healthy and older adults	Determine oral bioavailability, metabolic stability, and dose window	Plasma 20E and metabolites, UGT/CYP panels, LC–MS/MS	Dioh et al. (2023), Guo et al. (2022)
3. Proof-of-concept trials	Conduct phase 2b–3 sarcopenia RCTs with functional endpoints	Demonstrate clinically meaningful mobility/strength improvement and safety	Gait speed, SPPB, grip strength, serum oxidative stress markers	Fielding et al. (2025), Calabrese & Mattson, (2024)
Mechanistic expansion	Evaluate oxidative and metabolic protection in hepatic/mitochondrial disorders (NAFLD/NASH)	Confirm mitochondrial coupling and reduced lipid peroxidation	Acylcarnitine panels, ALT/AST, MRI-PDFF, FibroScan CAP	Buniam et al. (2023a, b), Khalaf et al. (2025)
Multimodal integration	Combine 20E with exercise, nutrition, or metabolic modulators	Establish synergy in muscle and metabolic resilience	Metabolomic signatures, functional composite scores	Cosardereliogl u et al. (2025), Huffman et al. (2016)
Regulatory translation	Prepare hybrid regulatory filings (505(b)(2), Art 10(3))	Leverage prior human exposure, validate safety	GMP compliance dossiers, PK bridging data	Ramesh & Cherukuri (2023), Freije et al. (2020)

validation, its ultimate significance lies in defining where these efforts translate into tangible therapeutic impact. The mechanistic and pharmacokinetic stages described above provide the scientific infrastructure upon which disease-specific programs can be built—linking receptor biology, mitochondrial energetics, and redox modulation to clinically meaningful endpoints. Building on this framework, Table 9.5 presents an *indication-specific outlook* that organizes current and emerging opportunities for phytoecdysteroid repurposing across major therapeutic domains. Each entry integrates mechanistic justification with clinical readiness, highlighting how foundational discoveries can be directed toward disease contexts marked by muscle atrophy, metabolic dysfunction, or systemic oxidative stress.

Sarcopenia and frailty remain the most mature areas, supported by human trial data demonstrating improved physical performance and safety with 20E-based interventions. Parallel preclinical evidence suggests that hepatic, cardiometabolic, and pulmonary disorders characterized by mitochondrial impairment and chronic inflammation may also benefit from similar mechanistic pathways. More speculative, but promising, are neurometabolic aging indications, where ERβ-linked

Table 9.5 Indication-specific outlook and strategic priorities for phytoecdysteroid repurposing

Therapeutic domain	Rationale/ mechanistic justification	Clinical opportunity	Key next steps	Expected impact
Sarcopenia and frailty	ERβ/MasR activation → anabolic signaling, mitochondrial protection, anti-inflammatory resilience	Phase 3 RCTs (SARA program extension)	Dose optimization; stratify slow walkers and sarcopenic obesity subgroups	Improved physical function and quality of life in elderly
NAFLD/NASH	20E mitigates lipid peroxidation, supports hepatic mitochondrial function	Early-phase proof-of-concept trials with metabolic biomarker panels	Combine with GLP-1 agonists or pioglitazone	Reduction of steatosis and oxidative injury
Cachexia (oncology/CHF)	Energy and protein turnover enhancement; anti-catabolic signaling	Multimodal trials with nutrition + exercise ±20E	PK/safety in polypharmacy patients	Preservation of lean mass, reduced fatigue
Metabolic and cardiopulmonary resilience	MasR/ACE2 axis → anti-fibrotic + antioxidant	Exploratory programs in post-viral or chronic hypoxia recovery	Pilot rehabilitation add-on trials	Better endurance, faster recovery, lower inflammation
Neuro-metabolic aging	ERβ/mitochondrial crosstalk may enhance synaptic resilience	Preclinical → early human translational studies	Brain exposure and safety profiling	Cognitive and energetic support without hormonal risk

trophic signaling and mitochondrial resilience may provide cognitive and energetic support. Together, these two frameworks, the translational roadmap (Table 9.4) and the indication-specific outlook (Table 9.5), form a coherent continuum from molecular insight to therapeutic application, illustrating how systematic, biomarker-driven development can reposition 20E from a nutraceutical compound to a clinically validated resilience enhancer. The coordinated application of these translational and indication-specific frameworks establishes a clear pathway for advancing 20E from mechanistic exploration to therapeutic validation, positioning it as a scientifically grounded candidate for promoting metabolic and functional resilience across aging and chronic disease contexts.

Overall, the successful drug repurposing of 20E will rely on integrating mechanistic understanding with rigorous pharmacological and clinical validation. Its non-androgenic anabolic and metabolic actions make it a promising candidate for improving muscle strength, metabolic stability, and physiological resilience in aging and chronic disorders. To achieve meaningful clinical translation, future research must optimize dosing and bioavailability, confirm receptor engagement through reliable biomarkers, and link these mechanisms to measurable functional outcomes. Ensuring GMP-quality formulations, standardized manufacturing, and transparent evaluation will be essential for scientific credibility. With these foundations in place, 20E could represent a new generation of repurposed natural therapeutics—bridging traditional bioactive chemistry with modern evidence-based medicine to enhance resilience across multiple disease domains.

References

Ambrosio G, Joseph JF, Wuest B, Mazzarino M, de La Torre X, Diel P, Parr MK (2020b) Detection and quantitation of ecdysterone in human serum by liquid chromatography coupled to tandem mass spectrometry. Steroids 157:108603. https://doi.org/10.1016/j.steroids.2020.108603

Ambrosio G, Wirth D, Joseph JF, Mazzarino M, de la Torre X, Botrè F, Parr MK (2020a) How reliable is dietary supplement labelling?—Experiences from the analysis of ecdysterone supplements. J of Pharm and Biomed Anal 177:112877. https://doi.org/10.1016/j.jpba.2019.112877

Begley CG, Ashton M, Baell J, Bettess M, Brown MP, Carter B, Charman WN, Davis Ch, Fisher S, Frazer I, Gautam A, Jennings MP, Kearney P, Keeffe E, Kelly D, Lopez AF, Mcguckin M, Parker MW, Craig R, Roberts B, Rush JS, Sullivan M (2021) Drug repurposing: Misconceptions, challenges, and opportunities for academic researchers. Sci Transl Med 13:5524. https://doi.org/10.1126/scitranslmed.abd5524

Buniam J, Chaiyakul S, Thongsepee N, Phungphong S (2023a) Dietary supplementation with 20-hydroxyecdysone ameliorates hepatic steatosis, dyslipidemia, and metabolic inflammation in diet-induced obese rats. Nutrients 15(14):3143. https://doi.org/10.3390/nu15143143

Buniam J, Chansela P, Weerachayaphorn J, Saengsirisuwan, V (2023b) Dietary Supplementation with 20-Hydroxyecdysone Ameliorates Hepatic Steatosis and Reduces White Adipose Tissue Mass in Ovariectomized Rats Fed a High-Fat, High-Fructose Diet. Biomedicines 11(7):2071. https://doi.org/10.3390/biomedicines11072071

Calabrese EJ, Mattson MP (2024) The catabolic-anabolic cycling hormesis model of health and resilience. Ageing research reviews 102:102588. https://doi.org/10.1016/j.arr.2024.102588

Cheng DM, Kutzler LW, Boler DD, Drnevich J, Killefer J, Lila MA (2013) Continuous Infusion of 20-Hydroxyecdysone Increased Mass of Triceps Brachii in C57BL/6 Mice. Phytotherapy Research 27(1):107–111. https://doi.org/10.1002/ptr.4679

Cosarderelioglu C, Walston JD, Abadir PM (2025) From frailty to resilience: exploring adaptive capacity and reserve in older adults. Frontiers in Aging 6:1520842. https://doi.org/10.3389/fragi.2025.1520842

Cruz-Jentoft AJ, Bahat G, Bauer J, Boirie Y, Bruyère O, Cederholm T, Cooper C, Landi F, Rolland Y, Sayer AA, Schneider SM, Sieber CC, Topinkova E, Vandewoude M, Visser M, Zamboni M (2019) Writing Group for the European Working Group on Sarcopenia in Older People 2 (EWGSOP2), and the Extended Group for EWGSOP2. Sarcopenia: revised European consensus on definition and diagnosis. Age Ageing 48(1):16–31. https://doi.org/10.1093/ageing/afy169

Dinan, L, Dioh W, Veillet S, Lafont R (2021) 20-Hydroxyecdysone, from Plant Extracts to Clinical Use: Therapeutic Potential for the Treatment of Neuromuscular, Cardio-Metabolic and Respiratory Diseases. Biomedicines 9(5):492. https://doi.org/10.3390/biomedicines9050492

Dioh W, Tourette C, Del Signore S, Daudigny L, Dupont P, Balducci Ch, Dilda PJ, Lafont R, Veillet S (2023) A Phase 1 study for safety and pharmacokinetics of BIO101 (20-hydroxyecdysone) in healthy young and older adults. Journal of Cachexia, Sarcopenia and Muscle 14 (3):1259–1273. https://doi.org/10.1002/jcsm.13195

Erin G (2023) Study Finds Most Sports Supplements Inaccurately Labeled. University of Mississippi News 7536. https://egrove.olemiss.edu/umnews/7536

Fielding RA, Dao MM, Cannon K, Desvarieux M, Miller SS, Gimness MP, Brandon DM, Villareal DT, Bruyere O, Bautmans I, Rickner K, Perry R, Kritchevsky SB, Musi N, Chehade JM, Kirstein JL, Gielen E, Pickrell P, Dilda P, Lafont R, Margalef C, Rolland Y, Signore SD, Mariani J, Agus S, Tourette C, Dioh W, van Maanen R, Veillet S (2025) BIO101 (20-Hydroxyecdysone) in sarcopenic older adults at risk of mobility disability: results from the randomized Phase 2b SARA-INT study. Journal of Cachexia, Sarcopenia and Muscle 16(2):e13750. https://doi.org/10.1002/jcsm.13501

Freije I, Lamouche S, Tanguay M (2020) Review of Drugs Approved via the 505(b)(2) Pathway: Uncovering Drug Development Trends and Regulatory Requirements. Ther Innov Regul Sci 54:128–138. https://doi.org/10.1007/s43441-019-00036-y

Gafner S, Blumenthal M, Foster S, Cardellina JH, Khan IA, Upton R (2023) Botanical ingredient forensics: Detection of attempts to deceive commonly used analytical methods for authenticating herbal dietary and food ingredients and supplements. J Nat Prod 86(2):460–472. https://doi.org/10.1021/acs.jnatprod.2c00929

Gorelick-Feldman J, Cohick W, Raskin I (2010) Ecdysteroids elicit a rapid Ca2+ flux leading to Akt activation and increased protein synthesis in skeletal muscle cells. Steroids 75(10):632–7. https://doi.org/10.1016/j.steroids.2010.03.008.

Guo C, Zhang N, Zhang X, Chi M, Liu D, Zhang J (2022) Use of liquid chromatography-tandem mass spectrometry for determination of higenamine in urine following oral administration of traditional Chinese medicine. Drug Testing and Analysis 14(8):1547–1552.

Huffman DM, Schafer MJ, LeBrasseur NK (2016) Energetic interventions for healthspan and resiliency with aging. Experimental gerontology 86:73–83. https://doi.org/10.1016/j.exger.2016.05.012

Isenmann E, Ambrosio G, Joseph JF, Mazzarino M, de la Torre X, Zimmer P, Kazlauskas R, Goebel C, Botrè F, Diel P, Parr MK (2019) Ecdysteroids as non-conventional anabolic agent: performance enhancement by ecdysterone supplementation in humans. Arch Toxicol 93(7):1807–1816. https://doi.org/10.1007/s00204-019-02490-x

Jagim AR, Harty PS, Erickson JL, Tinsley GM, Garner D, Galpin AJ (2023) Prevalence of adulteration in dietary supplements and recommendations for safe supplement practices in sport. Frontiers in Sports and Active Living 5:1239121. https://doi.org/10.3389/fspor.2023.1239121

Kasuya A, Sawada Y, Tsukamoto Y, Tanaka K, Toya T, Yanagi M (2003) Binding mode of ecdysone agonists to the receptor: comparative modeling and docking studies. Journal of Molecular Modeling 9(1):58–65. https://doi.org/10.1007/s00894-002-0113-x

Khalaf F, Barayan D, Saldanha S, Jeschke MG (2025) Metabolaging: a new geroscience perspective linking aging pathologies and metabolic dysfunction. Metabolism 166:156158. https://doi.org/10.1016/j.metabol.2025.156158

Kraiem S, Al-Jaber MY, Al-Mohammed H, Al-Menhali AS, Al-Thani N, Helaleh M, Al Maadheed M (2021) Analytical strategy for the detection of ecdysterone and its metabolites in vivo in uPA (+/+)-SCID mice with humanized liver, human urine samples, and estimation of prevalence of its use in anti-doping samples. Drug Testing and Analysis 13(7):1341–1353. https://doi.org/10.1002/dta.3032

Lafont R, Serova M, Didry-Barca B, Raynal S, Guibout L, Dinan L, Veillet S, Latil M, Dioh W, Dilda PJ (2021) 20-Hydroxyecdysone activates the protective arm of the RAAS via the MAS receptor. J Mol Endocrinol 68(2):77–87. https://doi.org/10.1530/JME-21-0033.

Lüdtke FL, Silva TJ, da Silva MG, Hashimoto JC, Ribeiro APB (2025) Lipid Nanoparticles: Formulation, Production Methods and Characterization Protocols. Foods 14(6):973. https://doi.org/10.3390/foods14060973.

McCann MR, De la Rosa GMV, Rosania GR, Stringer KA (2021) L-Carnitine and Acylcarnitines: Mitochondrial Biomarkers for Precision Medicine. Metabolites 11(1):51. https://doi.org/10.3390/metabo11010051

Nguyen TTL, Duong VA (2022) Solid Lipid Nanoparticles. Encyclopedia 2(2):952–973. https://doi.org/10.3390/encyclopedia2020063

Parr MK, Zhao P, Haupt O, Ngueu ST, Hengevoss J, Fritzemeier KH, Diel P (2014) Estrogen receptor beta is involved in skeletal muscle hypertrophy induced by the phytoecdysteroid ecdysterone. Molecular nutrition & food research 58(9):1861–1872. https://doi.org/10.1002/mnfr.201300806

Phungphong S, Kijtawornrat A, Chaiduang S, Saengsirisuwan V, Bupha-Intr T (2017) 20-Hydroxyecdysone attenuates cardiac remodeling in spontaneously hypertensive rats. Steroids 126:79–84. https://doi.org/10.1016/j.steroids.2017.08.004

Prue JC, Mayorga AJ (2007) The 505(b)(2) New Drug Application. Ther Innov Regul Sci 41:349–352. https://doi.org/10.1177/009286150704100308

Ramesh S, Cherukuri S (2023) The European Union's current regulations for the licensing of generic and hybrid medications. World Journal of Biology Pharmacy and Health Sciences 13(2):105–110.

Rehman M, Tahir N, Sohail MF, Qadri MU, Duarte SOD, Brandão P, Esteves T, Javed I, Fonte P (2024) Lipid-Based Nanoformulations for Drug Delivery: An Ongoing Perspective. Pharmaceutics 16(11):1376. https://doi.org/10.3390/pharmaceutics16111376

Serova M, Didry-Barca B, Deloux R, Foucault AS, Veillet S, Lafont R, Dilda PJ, Latil M (2024) BIO101 stimulates myoblast differentiation and improves muscle function in adult and old mice. J Cachexia Sarcopenia Muscle 15(1):55–66. https://doi.org/10.1002/jcsm.13326.

Shayota BJ (2024) Biomarkers of mitochondrial disorders. Neurotherapeutics 21(1):e00325. https://doi.org/10.1016/j.neurot.2024.e00325

Todorova V, Ivanov K, Karcheva-Bahchevanska D, Ivanova S (2023) Development and Validation of High-Performance Liquid Chromatography for Identification and Quantification of Phytoecdysteroids Ecdysterone and Turkesterone in Dietary Supplements. Processes 11(6):1786. https://doi.org/10.3390/pr11061786

Todorova V, Ivanova S, Chakarov D, Kraev K, Ivanov K (2024) Ecdysterone and Turkesterone-Compounds with Prominent Potential in Sport and Healthy Nutrition. Nutrients 16(9):1382. https://doi.org/10.3390/nu16091382.

Tsitsimpikou C, Tsamis GD, Siskos PA, Spyridaki MHE, Georgakopoulos CG (2001) Study of excretion of ecdysterone in human urine. Rapid Communications in Mass Spectrometry 15(19):1796–1801. https://doi.org/10.1002/rcm.433

USADA (2025) What athletes need to know about ecdysteroids (policy status; quality concerns). United States Anti-Doping Agency.

WADA (2024) 2025 Monitoring Program—lists ecdysterone under Anabolic Agents (In- and Out-of-Competition). World Anti-Doping Agency (effective 1 Jan 2025).

WADA (2025) 2025 List of Prohibited Substances and Methods & Monitoring Program now in force (news release). World Anti-Doping Agency.

Wu B, Kulkarni K, Basu S, Zhang S, Hu M (2011) First-pass metabolism via UDP-glucuronosyltransferase: a barrier to oral bioavailability of phenolics. J Pharm Sci 100(9):3655–81. https://doi.org/10.1002/jps.22568.

Yin H, Wang Z, Lv X, Wang Z, Wang Y, Fan W, Liu Y (2024) Inhibition of human UDP-glucuronosyltransferase enzyme by entrectinib: Implications for drug-drug interactions. Chemico-Biological Interactions 395:111023. https://doi.org/10.1016/j.cbi.2024.111023

Yusupova U, Sasmakov S, Usmanov D, Ramazonov N, Azimova S and Sagdullaev S (2020) Phytoecdysteroids from Silene claviformis and their antibacterial and antifungal activities. Pol. J. Natur. Sc. 35(3):313–321.

Yusupova UY, Bobakulov KM, Girault JP, Usmanov DA, Ramazonov NS and Abdullaev ND (2022) New Phytoecdysteroid from Silene popovii. Chem Nat Compd 58(6):1082–1084. https://doi.org/10.1007/s10600-022-03874-9.

Yusupova UY, Bobakulov KM, Khurramov AR, Syrov VN, Egamova FR, Karuth A, Usmanov DA, Quadir M and Rasulev B (2025) Phytochemical constituents isolated from Silene popovii Schischk. Med Chem Res 34(2):423–431. https://doi.org/10.1007/s00044-024-03345-6.

Yusupova UY, Ramazonov NS, Bobakulov KM, Egamova FR, Syrov VN and Usmanov DA (2021) Biological active components of silene tomentella and their pharmacological properties. Khim. Rastit. Syr′ya 1:197–202. https://doi.org/10.14258/jcprm.2021018323.

Yusupova UY, Usmanov DA and Ramazonov NS (2019) Phytoecdysteroids from the Plant Dianthus helenae. Chem Nat Compd 55(2):393–394. https://doi.org/10.1007/s10600-019-02701-y.

Zwetsloot KA, Shanely RA, Godwin JS, Hodgman CF (2021) Phytoecdysteroids accelerate recovery of skeletal muscle function following in vivo eccentric contraction-induced injury in adult and old mice. Frontiers in Rehabilitation Sciences 2:757789. https://doi.org/10.3389/fresc.2021.757789

Index

V. N. Syrov et al. (eds.), *Phytoecdysteroids II*,
https://doi.org/10.1007/978-981-95-8423-9

MIX
Papier aus verantwortungsvollen Quellen
Paper from responsible sources
FSC® C105338

If you have any concerns about our products,
you can contact us on
ProductSafety@springernature.com

In case Publisher is established outside the EU,
the EU authorized representative is:
Springer Nature Customer Service Center GmbH
Europaplatz 3, 69115 Heidelberg, Germany

Printed by Libri Plureos GmbH
in Hamburg, Germany